Nancy Knowlton

THE MOUNTAIN GORILLA

ECOLOGY AND BEHAVIOR

THE MOUNTAIN GORILLA

ECOLOGY AND BEHAVIOR

GEORGE B. SCHALLER

THE UNIVERSITY OF CHICAGO PRESS

CHICAGO AND LONDON

George B. Schaller is a research zoologist with the New York Zoological Society and adjunct associate professor at Rockefeller University. He is the author of *The Serengeti Lion, The Year of the Gorilla,* and *The Deer and the Tiger.*

TO

OLAUS J. MURIE

THE UNIVERSITY OF CHICAGO PRESS, CHICAGO 60637
THE UNIVERSITY OF CHICAGO PRESS, LTD., LONDON

PUBLISHED 1963
PRINTED IN THE UNITED STATES OF AMERICA
ISBN: 0–226–73635–0 (CLOTHBOUND) 0–226–73636–9 (PAPERBOUND)
LIBRARY OF CONGRESS CATALOG CARD NUMBER: 63–11401
86 85 84 83 82 81 80 6 7 8 9 10

CONTENTS

Preface to the Phoenix Edition xv

I. Introduction and Methods 1

INTRODUCTION / 1
ACKNOWLEDGMENTS / 3
THE COUNTRY / 6
TAXONOMY / 10
HISTORY OF THE DISCOVERY AND STUDY OF THE MOUNTAIN GORILLA / 12
PROCEDURE AND METHOD / 18

II. Distribution and Ecology 25

DISTRIBUTION / 25
ECOLOGY AND DESCRIPTION OF THE GORILLA HABITATS / 36

III. The Animals 66

AGE AND SEX CLASSES / 66
PHYSICAL ATTRIBUTES AND INDIVIDUAL DIFFERENCES / 68
INJURIES / 77
BODY POSTURES / 78
THE SENSES / 78
TEMPERAMENT / 80
ARBOREAL AND TERRESTRIAL LOCOMOTION / 81
GORILLA SPOOR / 85

IV. Population Density, Structure, and Behavior 92

POPULATION DENSITIES AND NUMERICAL ESTIMATES / 92
POPULATION DYNAMICS / 95
GROUP DYNAMICS / 102
GROUP RANGES AND MOVEMENTS / 124
THE DAILY ACTIVITY CYCLE / 137

V. Individual Activities and Behavior 149

FOOD HABITS AND FEEDING BEHAVIOR / 149
DRINKING BEHAVIOR / 169

NESTS AND NESTING BEHAVIOR / 169
TRANSPORT OF MATERIALS / 198
TOOL-USING AND HANDLING OF OBJECTS / 199
ELIMINATIVE AND SELF-CARE BEHAVIOR / 201
FACIAL EXPRESSIONS / 208
VOCALIZATIONS / 210
DISPLAYS / 221

VI. Social Behavior 237
LEADERSHIP / 237
INTERNAL ORGANIZATION OF GORILLA GROUPS / 239
INTERACTIONS OF INDIVIDUALS WITHIN THE GROUP / 240
THE DEVELOPMENT OF INFANTS FROM BIRTH TO THE AGE OF SIX MONTHS / 273
MATING BEHAVIOR / 275
AGGRESSIVE AND SUBMISSIVE BEHAVIOR / 289

VII. Responses to Environment 294
RESPONSE TO WEATHER / 294
RESPONSE TO WATER / 298
INTERACTIONS WITH OTHER ANIMALS, EXCLUDING PRIMATES / 299
INTERACTIONS WITH OTHER PRIMATES / 304

VIII. Conservation 320
CONSERVATION / 320

IX. Summary 327
SUMMARY / 327

Tables and Appendixes 337
A. ACCOUNTS BY CAPTAIN VON BERINGE CONCERNING HIS DISCOVERY OF THE MOUNTAIN GORILLA / 390
B. WEATHER DATA / 391
C. NOTES ON SKULLS OF THE MOUNTAIN GORILLA / 397
D. FOOT CANDLE READINGS AT KABARA / 404

Bibliography 405

Index 423

ILLUSTRATIONS

Following page 206

PLATE

1. A typical view of the Utu region
2. The Kayonza Forest with the Virunga Volcanoes in the distance
3. The rift mountains west of Lake Tanganyika
4. My wife watches a group of gorillas
5. The Lowa River at a point where it presents a barrier to gorilla dispersal
6. A field in lowland rain forest near Beni, Congo
7. Old secondary forest in the Utu region
8. A stand of tall bamboo
9. A typical ridge in the Kayonza Forest, showing the sparseness of the groundcover
10. A densely overgrown valley in the Kayonza Forest
11. A view of the Virunga Volcanoes
12. *Hypericum* woodland in the Virunga Volcanoes
13. The zone of giant senecios at an altitude of 12,800 feet in the Virunga Volcanoes
14. Mt. Karisimbi as seen from Kabara
15. A typical foggy day at Kabara
16. Two female gorillas and a blackbacked male in a tree
17. A female gorilla with an infant on her back
18. Ten gorillas, showing great individual differences in the facial features
19. A silverbacked male in a tree
20. A blackbacked male
21. A hole excavated by a gorilla in order to obtain a root
22. A banana stem torn apart by a gorilla to obtain the pith

PLATE
23. A typical gorilla ground nest of herbs
24. A gorilla ground nest of lobelia stems
25. A gorilla ground nest of branches
26. A gorilla nest in the crowns of bamboo
27. A gorilla nest in the crotch of a tree
28. "Symbolic feeding" by a silverbacked male preceding the chest-beating display
29. A female stands and beats her chest
30. A blackbacked male displays the strutting walk
31. A silverbacked male indicates his readiness to leave by standing motionless with legs spread
32. A juvenile rests its arms on the back of a blackbacked male
33. A silverbacked male and several other gorillas
34. Several gorillas have ascended a tree to watch me
35. An African weaving a bamboo basket in front of a bamboo hut

Photographs of gorillas are of free-living animals taken in the Kabara study area of the Virunga Volcanoes.

FIGURES

1. Regional map of the eastern Congo, western Uganda, and western Ruanda-Urundi 7
2. The present and hypothetical past distribution of the genus *Gorilla* 26
3. The geographic distribution of the mountain gorilla in 1959 . 28
4. Distribution of the mountain gorilla with respect to vegetation types 31
5. The relative abundance of bamboo shoots in the Virunga Volcanoes during 1959 and 1960 43
6. The Virunga Volcanoes of Albert National Park 49
7. A generalized vegetation map of the Uganda portion of the Virunga Volcanoes 52
8. A generalized vegetation map of the Mt. Visoke–Mt. Sabinio saddle area, Virunga Volcanoes 56
9. An aerial view of the Kabara study area, Virunga Volcanoes PLATE SECTION
10. A generalized vegetation map of the Kabara area 60
11. Shapes and individual identifying marks of noses of mountain gorillas at Kabara 73
12. Body weights of a female lowland gorilla from birth to the age of 4 years 74
13. Body weights of 4 gorillas from infancy to adulthood . . . 75
14. Foot- and knuckle prints of gorillas at Kabara 86
15. Footprints of gorillas at Kabara 87
16. Cross-sections of gorilla dung showing size and shape . . . 89
17. The joining and parting of silverbacked males in group IV at Kabara 107
18. The splitting and rejoining of group IV at Kabara 110
19. The close proximity, joining, and parting of groups VI and VIII at Kabara 113

20. Adjacent nest sites of groups VI and VIII at Kabara . . . 113
21. The routes of groups VII and XI before, during, and after meeting 115
22. The routes of groups V and VII before, during, and after meeting 117
23. The routes of groups VII and ? during a meeting 118
24. A combined nest site of groups VII and III 119
25. The extent of home ranges of 3 gorilla groups in the Kabara area 125
26. The extent of home ranges of 2 gorilla groups in the Kabara area 126
27. The extent of home ranges of 5 gorilla groups in the Kabara area 127
28. The daily route of travel of group IV from August 28 to September 14, 1959 131
29. The daily route of travel of group VII from March 15 to April 2, 1960 131
30. The daily route of travel of group VII from October 19 to November 5, 1959 133
31. The daily route of travel of group II from November 5 to 21, 1959 133
32. The routes of travel of 3 groups through the same area at approximately the same time 134
33. The daily route of travel of one group in the Kisoro area from March 29 to April 5, 1959 135
34. The daily route of travel of one group in the Kayonza Forest from June 12 to 16, 1960 135
35. A typical scattering of animals in the midday resting area . . 140
36. A typical daily route of a group, showing feeding, resting, and nesting sites 141
37. Daily feeding and resting activity patterns of gorilla groups . 142
38. Daily feeding and resting activity patterns of gorilla groups . 143
39. Body positions of resting gorillas 144
40. Body positions of resting gorillas 145
41. The daily activity cycle of gorilla groups 147
42. Two consecutive nest sites of group IV, showing overlap of nesting locations 172
43. Typical nest sites of small gorilla groups 173
44. Two Kabara nest sites, one compact, the other split 174

45. Two typical nest sites of group VII at Kabara 175
46. Two typical Kabara nest sites 176
47. Two nest sites of group VII, one compact, the other split . . 177
48. Some typical nest sites from the Kayonza Forest 178
49. Some typical gorilla nest sites 179
50. A nest site of group IV at Kabara 183
51. A split nest site of group VIII at Kabara 184
52. A nest site of group IV showing the nesting positions of the various silverbacked males 185
53. Schematic diagrams of gorilla nests on the ground in the Virunga Volcanoes 189
54. Schematic diagrams of gorilla nests on or near the ground in the Kayonza Forest 190
55. Schematic diagrams of tree nests in the Utu region 192
56. Sketches of gorillas showing body positions in the nest and during eliminative behavior 202
57. Body positions of gorillas during self- and mutual grooming . 206
58. Sound spectrograms of various gorilla vocalizations
59. Sound spectrograms of various gorilla vocalizations
60. Sound spectrograms of intensely roaring male gorillas
61. Sound spectrograms of chest beats by male gorillas
62. Sound spectrograms of chest beats by female and juvenile gorillas
} PLATE SECTION
63. Body positions of gorillas during play, and of a female carrying her wounded infant 251
64. Body positions and activities of gorillas during the midday rest period 256
65. Body positions of females and youngsters while resting near the silverbacked male 258
66. Copulatory positions of gorillas at Kabara and in the Columbus zoo 285
67. Nesting response of group VII to rain 296
68. Nesting response of group VII to rain 297
69. Cartoon 406

TABLES

1. A partial list of gorillas collected in the Virunga Volcanoes . . 339
2. Recent investigations into gorilla behavior in Uganda 340
3. Location and dates of gorilla field work 341
4. Number of visual encounters and hours of direct observation of gorilla groups 342
5. Location and dates of field work on apes other than gorillas . 343
6. The Virunga Volcanoes 344
7. Forage utilized by gorillas in mountain woodland and bamboo at Kisoro 345
8. Weights of lowland gorillas in captivity 346
9. Weights and heights of male mountain gorillas collected in the wild 347
10. The distribution of ischial callosities in six gorilla groups . . 347
11. Handedness in the chest-beating display of individual males . 348
12. Recent injuries noted in the Kabara gorilla population . . . 349
13. The frequency of tree climbing in Kabara gorillas 350
14. Relative abundance of gorilla nest sites in several areas . . . 351
15. Estimated population density of gorillas in selected areas . . 351
16. Sexual development of gorilla pairs in captivity 352
17. The occurrence of nematode eggs in gorilla dung samples from Kabara 353
18. Sex and age class composition of the Kabara population . . 354
19. The composition of gorilla groups at Kabara 354
20. The composition of mountain and lowland gorilla groups . . 355
21. Instances of inter-group contacts at Kabara 356
22. Seasonal variation in the presence or absence of bamboo shoots and gorilla sign in the bamboo zone along the trail between Rweru and Kabara 357

23. Arrival dates of gorilla groups in the study area 357
24. Extent of daily movement by gorilla groups at Kabara 358
25. Distances traveled along the route from one nest site to the following nest site by groups at hourly intervals during the day at Kabara 359
26. Gorilla group activity at various times of day 360
27. Food plants utilized by gorillas at Kabara 361
28. Food plants utilized by gorillas at Kisoro 362
29. Food plants utilized by gorillas in the Kayonza Forest . . . 363
30. Food plants utilized by gorillas at Mt. Tshiaberimu and vicinity 364
31. Food plants utilized by gorillas in the Mt. Kahuzi region west of Lake Kivu 365
32. Food plants utilized by gorillas in the Mwenga-Fizi region . . 366
33. Food plants utilized by gorillas in the Utu region 367
34. Food plants utilized by gorillas in West Africa as noted in the literature 368
35. Chemical analysis of *Peucedanum linderi* 368
36. Similarities in food habits between study areas 369
37. Presence and utilization of forage species in various study areas 370
38. Analyses of soils eaten by gorillas 371
39. Subjective evaluation of taste of some gorilla food plants . . 372
40. Height of gorilla nests above ground 373
41. Height above ground of chimpanzee and gorilla nests in the Kayonza Forest 374
42. Distances between the nearest nests in 146 nest sites at Kabara 375
43. Materials used in the construction of night nests at Kabara . . 376
44. Use of night nest construction materials in three Kabara gorilla groups 377
45. Presence and location of dung at the gorilla nest 378
46. Frequency of self-grooming and parts of the body groomed by the various age and sex classes in gorillas 379
47. Gorilla vocalizations, their probable underlying emotions, and their functions 380
48. The number of individual beats per chest-beating display in Kabara gorillas 382
49. Average number of beats per chest-beating display in gorilla males at Kabara 382

50. Occurrence of the various acts in the chest-beating sequence of the different age and sex classes 383
51. Dominance interactions between age and sex classes 383
52. Dominance interactions between the silverbacked males of group IV 384
53. Number of observations of mutual grooming in the various age and sex classes 384
54. Mutual grooming in *Papio ursinus* 385
55. The frequency of lone and social play in the various age and sex classes 385
56. Stages of behavioral development in free-living gorillas to the age of 6 months 386
57. Stages of behavioral development in the infant gorilla "Jambo" at the Basel zoo 387
58. The age of first appearance of various behavior patterns in captive gorillas, chimpanzees, and man 388
59. Estimated dates of birth of infants in the Kabara population . 389
60. Mean number of chest beats by silverbacked males in response to my presence during the first hour of contact 389
61. Precipitation at Kisoro, Uganda 391
62. Weather data, Rumangabo, Albert National Park 392
63. Weather data, Rumangabo, Albert National Park, during period of study 393
64. Kabara temperatures 394
65. Kabara cloudcover 395
66. Kabara precipitation 395
67. Weather spells at Kabara 396
68. Collecting locations of gorilla skeletal material 398
69. Measurements of gorilla crania 399
70. Measurements of newborn gorilla infants 400
71. Measurements of the deformed arm bones of a mountain gorilla 400
72. Sequence of tooth eruption in the upper jaw 401
73. Dental attrition in teeth of the upper and lower jaws . . . 402
74. Foot candle readings at Kabara 404

PREFACE TO THE PHOENIX EDITION

Fifteen years have passed since I wrote this book. Few field studies of monkeys and apes had been completed in 1961, and in fact my main sources of comparative information were the classic reports on howling monkeys and gibbons by C. R. Carpenter and the early work on Japanese macaques. But since that time hundreds of zoologists, anthropologists, and psychologists have devoted themselves to observing nonhuman primates in the field. Their studies have not only dramatically increased the amount of available information, but also generated many new and exciting ideas. So much is now known about many species that general studies are being replaced by highly detailed ones treating quantitative ecology and such topics as the relationship between social organization and habitat. The field study of Struhsaker (1975) and the symposium volume of Tuttle (1975) are good recent examples of this trend. In any expanding scientific discipline new data constantly enlarge, revise, and change the perspective of previous studies. This has fortunately happened to my work on the gorilla.

After analyzing 747 gorilla skulls, Groves (1970) concluded that there are three instead of two gorilla subspecies. The western lowland gorilla (*G. g. gorilla*) retains its classification unchanged. However, the mountain gorilla has been divided into two subspecies: *G. g. beringei* in the Virunga Volcanoes and the Mt. Kahuzi area and *G. g. graueri* in the remaining areas. The former is still referred to as "mountain gorilla" whereas the latter may best be termed "eastern lowland gorilla." A change in terminology has more than academic interest, for *G. g. beringei,* as defined in the new classification, is an extremely rare animal. Fewer than 1000 of these gorillas survive, with 375 to 400 of them located in the Virunga Volcanoes according to a recent survey (Harcourt, 1972).

My work in the Virunga Volcanoes was brief, a little over a year —only long enough to establish the broad outlines of gorilla behavior. But since 1967 Fossey (1972, 1974) has studied gorillas intensively on the Rwanda side of the volcanoes, and her work and that of her collaborators has greatly increased our understanding of the animals. Just as importantly, her presence has afforded the gorillas protection during years of political turmoil. The gorillas on Mt. Kahuzi are now protected in a national park created especially for them, and these animals are also receiving scientific attention (Goodall, 1975). Jones and Sabater Pi (1971) published an interesting report on the little-known West African gorilla, and further work on this subspecies is in progress. In the past decade, zoos have also learned how to breed gorillas successfully, thereby reducing the drain on wild populations. I reported on 6 gorilla births in captivity. Only eight years later Kirchshoffer (1970) tallied 39 births, including two sets of twins.

Although agriculturalists still encroach on gorilla habitat in the Virunga Volcanoes and pastoralists continue to be a constant threat, I find great satisfaction in knowing that a viable population of these amiable giants still rambles about its mountain home.

WORKS CITED

FOSSEY, D.

1972. Vocalizations of the mountain gorilla (*Gorilla gorilla beringei*). Anim. Behav. **20**:36–53.

1974. Observations on the home range of one group of mountain gorillas (*Gorilla gorilla beringei*). Anim. Behav. **22**:568–81.

GOODALL, A.

1975. Mountain gorillas in danger, pp. 12–19. *In* Wildlife Yearbook 1975. Edited by N. Sitwell. London: Danbury Press.

GROVES, C.

1970. Population systematics of gorilla. J. Zool. London. **161**: 287–300.

HARCOURT, A., AND GROOM, A.

1972. Gorilla census. Oryx **11**(5):355–63.

JONES, C., AND SABATER PI, J.

1971. Comparative ecology of *Gorilla gorilla* (Savage and Wyman) and *Pan troglodytes* (Blumenbach) in Rio

Muni, West Africa. Bibl. Primat. No. 13. Basel: S. Karger.

KIRCHSHOFFER, R.
1970. Gorillazucht in Zoologischen Gärten und Forschungsstationen. Zool. Gart. **38**:73–96.

STRUHSAKER, T.
1975. The red colobus monkey. Chicago: University of Chicago Press.

TUTTLE, R., ED.
1975. Socioecology and psychology of primates. The Hague: Mouton Publ.

CHAPTER 1

INTRODUCTION AND METHODS

Introduction

Probably no animal has fired the imagination of man to the same extent as has the gorilla. Over a hundred years have passed since the discovery in 1847 of this giant ape in the forests of West Africa, yet the gorilla has remained a creature of mystery. Its man-like appearance and tremendous strength, its remote habitat and reputed belligerence, have endowed the beast with a peculiar fascination and stirred popular and scientific interest.

Although numerous expeditions have traveled to both West and Central Africa to shoot, capture, and photograph the two subspecies of gorilla, encounters with the animals were usually brief and often violent. Published accounts offered little factual information other than descriptions of nests and trails; and many writers emphasized the supposedly ferocious nature of the gorilla, an aspect of the animal's behavior which was enthusiastically enlarged upon in the various mass media of communication. Early attempts to study the gorilla were undoubtedly influenced by such reports. Garner (1896) and Bingham (1932), for example, obtained only limited data, the former because he waited for gorillas to come to him, and the latter largely because he deemed it essential to track the animals with guides, gunbearers, and porters. The various difficulties encountered in studying gorillas, including the rather poor visibility in their dense forest habitat, reinforced the general consensus of opinion that a detailed study of their behavior was not feasible.

However, Mr. Harold J. Coolidge, who, as a member of the Harvard African Expedition in 1927, obtained first-hand knowledge of the mountain gorilla (*Gorilla gorilla beringei*) has over the years become convinced that a life history study of the gorilla is possible. His view was reinforced by the sporadic peaceful encounters with the apes described by Akeley (1923), Prince Wilhelm of Sweden (1923), Donisthorpe (1958) and others. Renewed interest in primate field research and recent attempts by anthropologists to compare the behavior of early hominids with that of present-day apes made it imperative to obtain some factual data on the gorilla.

The main purpose and plan of our expedition, which was under the leadership of Dr. John T. Emlen, Jr., professor of zoology at the University of Wisconsin, was:

1. A general six-month survey of mountain gorilla range in the eastern Congo, Ruanda-Urundi, and western Uganda with special emphasis on distribution, ecological diversity, and the location of the most favorable sites for intensive behavior studies.
2. Sustained and continuous observations of the life history of the mountain gorilla in one location for about one year.
3. Supplementary observations on other primates, especially other apes, whenever feasible.

The African Primate Expedition (A.P.E.) left New York on February 1, 1959, and included Dr. Emlen, myself, and our wives. After six months, at the completion of the first phase of the project, Dr. and Mrs. Emlen returned to America (Fig. 69). From August, 1959, to September, 1960, my wife and I devoted most of the time to a study of a gorilla population in the Virunga Volcanoes of Albert National Park, Congo. Work on gorillas, and also on chimpanzees (*Pan troglodytes schweinfurthi*) in other localities continued until October 25, 1960, when we moved to southeast Asia to observe orang-utans (*Pongo pygmaeus*), gray gibbons (*Hylobates moloch*), and siamangs (*Symphalangus syndactylus*). Field work was terminated on January 27, 1961.

The purpose of this report is to present a detailed account of the ecology and behavior of the mountain gorilla in its natural environment. Although in the course of the study observations were made on all ape genera and on over twenty species of monkeys, comparative data on other primate species from personal observations or from the literature are included only to illustrate, confirm, or contrast certain points.

Some aspects of the expedition's work have already been published elsewhere. Emlen (1960) discusses the project in general terms, and Emlen and Schaller (1960*a*) present detailed maps of the distribution of the mountain gorilla. Popularized accounts of the expedition's work are found in Emlen and Schaller (1960*b*), Schaller (1962), and O'Reilly (1960). Schaller and Emlen (1963) summarize certain aspects of great ape behavior and note from it various implications for hominid origins. Schaller (1961) discusses the status, distribution and behavior of the orang-utan in Sarawak.

On the following pages I have presented the facts which the expedition was able to gather. But there is unfortunately no space to reveal fully the personal pleasure which I derived from roaming through uninhabited forests and from climbing among the mist-shrouded mountains; there is no space to enlarge upon my great liking for the gorillas, which ceased to be mere subjects to be studied, but became acquaintances whose activities my wife and I discussed at the end of each day.

Acknowledgments

The support which this expedition received in the course of two years cannot be adequately acknowledged in a few words. Some help, which in itself was often small, meant the difference between success and failure in a particular endeavor. The numerous persons who aided us in the distributional aspects of the gorilla study and with the work on orang-utans have already been acknowledged elsewhere (Emlen and Schaller, 1960*a;* Schaller, 1961). To all the institutions and persons who in any way furthered the work of our expedition I wish to record my deep gratitude.

The expedition was supported by several institutions without whose help the work could not have been accomplished. Funds were primarily provided by a grant from the National Science Foundation, and supplementary funds were donated by the New York Zoological Society. The latter institution also acted as sponsor of the expedition. In Africa we were fortunate to have the patronage of the Institute of the Parks of the (Belgian) Congo and Makerere College, Uganda. In America the University of Wisconsin aided with various facilities and with equipment.

Major credit for any success that may be attributed to the expedition belongs to Dr. John T. Emlen, Jr., professor of zoology at the University of Wisconsin. The entire program, from the initial application for funds to the final completion of this report, was carried out under the direction and guidance of Dr. Emlen. In addition, he spent the first six months in the field with the expedition and was instrumental not only in establishing it in Africa but also in developing the field techniques used, in finding the most favorable study location, and in gathering a considerable amount of information on ecology and behavior which he generously permitted me to include in this report. His fine editorial pen improved the manuscript considerably.

I am especially indebted to the New York Zoological Society and its staff. With efficiency and friendliness they solved the numerous problems that beset an expedition both while still at home and in the field. Dr. Fairfield Osborn aided in many ways but most importantly with encouragement and good advice. Mr. George Merck handled correspondence and other matters, Mr. William Bridges advised on publications, and Mr. H. F. Schiemann on finances.

Mr. Harold J. Coolidge, of the National Academy of Sciences, was the driving force behind the initiation of the gorilla project. To him the expedition owes many of the original ideas, and he also helped greatly in organizing and finding support for the project at home and abroad.

To Dr. Victor van Straelen, president of the Institute of the Parks of the Congo, goes our gratitude for his unprecedented courtesy in giving the expedition a carte blanche to work in the closed sector of the Virunga Volcanoes. At Albert National Park, Commandant Marc Micha aided us in every possible way, and his hospitality and interest in our work were much appreciated. Dr. Jacques Verschuren, biologist for the Congo parks, shared freely his detailed knowledge of primates and cheerfully helped us in various ways innumerable times. He spent a week with the expedition, and I am extremely grateful for his permission to use the tape recordings of gorilla sounds which he obtained at that time.

I also wish to thank the numerous members of the staff of Makerere College, who generously aided us. Dr. Alexander Galloway, former Dean of the Medical School, and Dr. Niels Bolwig, former lecturer in zoology, helped us frequently, especially with logistic and other support. It was my privilege to spend two weeks in the field with Dr. Bolwig and I look back with pleasure on my time with this congenial companion.

Mr. Walter Baumgartel always opened the doors wide to our arrival and many pleasant hours were spent in his Travellers Rest Hotel in Kisoro, Uganda, the unofficial gorilla headquarters in Africa. His interest in these animals has resulted in his acquiring some unique information concerning them, which he always generously shared and which is included in this report.

To Mr. and Mrs. Charles Cordier, independent animal trappers formerly located near Bukavu, Congo, goes my gratitude for their hospitality and for the many ways in which they aided my work in the Utu region.

Numerous other persons helped the expedition. In the Congo I am indebted to: Dr. Urs Rahm, mammalogist at I.R.S.A.C. (Institute for Scientific Research in Central Africa), Lwiro; Mr. Guy Rousseau and Mr. P. Bouckaert, formerly on the staff of Albert National Park; Mr. André Meyer, geologist formerly in Goma, Congo; Mr. Anicet Mburanumve, the first African head of Albert National Park.

In Uganda my work was greatly facilitated by: Major Bruce Kinloch, former chief game warden of Uganda; Mr. John Blower, acting game warden; Mr. John Mills, game ranger; Mr. Allan Brooks, game biologist; Mr. Frank Poppleton, warden of Queen Elizabeth National Park.

For permission and help in observing captive gorillas I am grateful to Dr. James Vickers, Mr. Roscoe Franks, and the late Mr. Earl F. Davis, all on the staff of the Columbus zoo in 1958; Mr. M. Quinn, the keeper of the ape house at the New York (Bronx) zoo; and Dr. Ernst Lang, the director of the Basel zoo.

Further I would like to thank the various members of the first primate conference whose pertinent advice shaped much of the thinking of the project: Drs. Fairfield Osborn and John Tee-Van, New York Zoological Society; Drs. John T. Wilson, Henry S. Odbert, and Marguerite Young, National Science Foundation; Dr. C. R. Carpenter, Pennsylvania State University; Dr. William Mason, Yerkes Primate Laboratory; Mr. Harold Coolidge, National Academy of Science; Dr. David E. Davis, Pennsylvania State University; and Dr. John T. Emlen, University of Wisconsin.

For general comments, suggestions, and thought-provoking ideas I am grateful to: Dr. C. R. Carpenter, Pennsylvania State University; Dr. Raymond Dart, Witwatersrand University; Dr. L. S. B. Leakey, Coryndon Museum; Dr. A. Kortlandt, Zoologisches Institut, Amsterdam; Drs. K. Imanishi, J. Itani, M. Kawai, and H. Mizuhara, Japan Monkey Center; Miss Jane Goodall, Coryndon Museum, Kenya.

I also am indebted to the numerous Bantu and Batwa guides and porters, whom we employed at various times in Africa and whose help was essential.

Several institutions and persons identified plants and analyzed soil samples for me. I wish to thank: Dr. D. G. Troupin, I.R.S.A.C., Lwiro, Congo; Mr. Henry Osmaston, Forest Department, Uganda; Drs. E. M. Chenery and E. W. Hughes, Uganda Agricultural Research Station at Kawanda; Drs. D. Napper and B. Verdcourt, East African Herbarium, Kenya; Dr. H. Hull, Soils Department, University of Wisconsin.

For technical aid at the University of Wisconsin I am grateful to: Mr. John E. Dallman, Department of Zoology, and Mr. R. Sale, Department of Geography, who labored over many of the maps; Mr. Richard G. Smith, Department of Geography, who helped in the preparation of the photographic material; and Mr. Richard L. Penney, Department of Zoology, who handled the sound spectrograph.

I wish to thank Dr. William Laughlin for help in measuring the gorilla skeletal material collected during the expedition, and Dr. George Erikson of the Harvard Medical School, who gave me the opportunity to dissect a gorilla at his institution.

Critical comments and valuable suggestions on the manuscript were made by Drs. John T. Emlen, William Laughlin, and William Reeder, University of Wisconsin.

I also wish to thank my wife Kay, who accompanied me during the expedition. I can say without bias that her participation in the work was essential to the completion of it. She cheerfully joined into all phases of the project and contributed many valuable observations; she accepted the menial chores of handling the correspondence and collecting plants; and, importantly, she gave moral support and was a congenial companion during the long months of our life in the forest when conditions were frequently far from ideal. In addition, she typed the various drafts of this manuscript and offered valuable suggestions for its improvement. Thus, she was secretary, field hand, critic, cook, companion—a combination of jobs only willingly taken on by a wife.

And lastly, I want to express my gratitude to *Gorilla gorilla beringei,* as interesting and amiable a subject for study as an investigator can hope to find; the animal taught me many things, including the importance of knowing my own behavior more fully.

The Country

The range of the mountain gorilla encompasses one of the most varied and beautiful regions of Africa (see Figs. 1, 3, and 4).

The western portion of the range lies in the Congo basin at an altitude of about 1,500 to 3,000 feet. A seemingly endless expanse of rain forest covers the flat to undulating terrain. A view from the air presents nothing but cloud-like billows of foliage extending to the horizon. Here and there broad rivers sweep languorously past the wall of jungle. These rivers—the Lowa, Ulindi, Elila, and others—

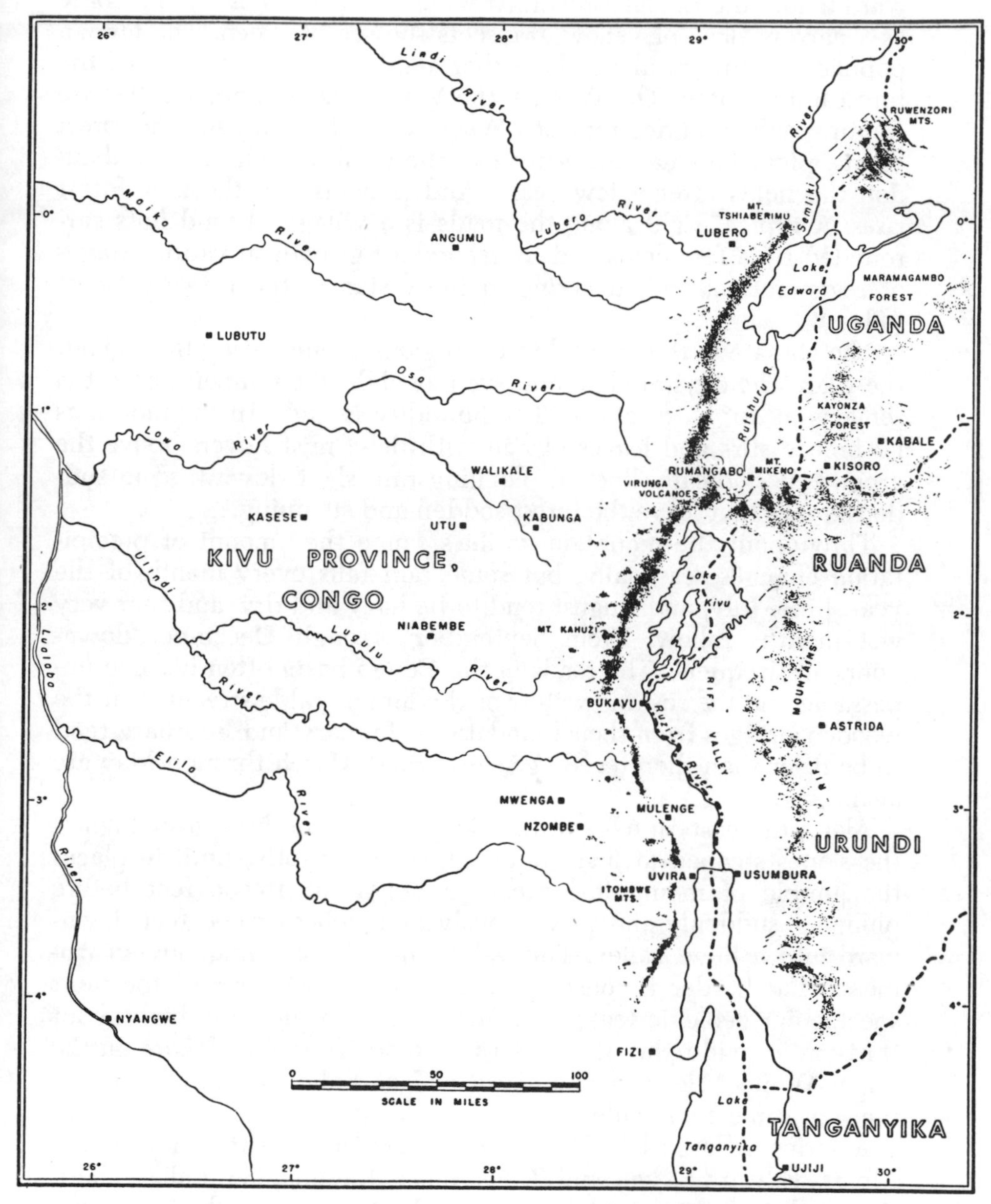

FIG. 1.–Regional map of the eastern Congo, western Uganda, and western Ruanda-Urundi, showing the major physical features in the range of the mountain gorilla.

have their sources in the mountains to the east, and they flow westward to join the Lualaba River. At wide intervals a road, visible only as a narrow slash of yellow, transects the sea of green. The human population clusters along these thoroughfares, leaving most of the forest uninhabited. The Warega, the Wanianga, and other tribes are Bantu-speaking agriculturalists. With ax and fire they fell the forest giants, plant banana and manioc in the shallow soil, only to abandon the fields after a few years. And phoenix-like the new forest rises. A typical sight along the roads is a village of mud huts surrounded by a few fields and by secondary growth in various stages of regeneration, and, towering in the distance, the primary forest (Pl. 1).

The climate in the tropical forest regions is enervating throughout the year. The days are hot, and even at night the temperature often remains at 20° C. or above. The humidity is high. In the mornings no breeze stirs and leaves glisten with dew; mist hovers above the ground. Periodically lines of plunging rain slant downward in sudden squalls that leave the forest sodden and steaming.

Throughout the mountain gorilla's range the amount of precipitation changes seasonally, but some rain falls every month of the year. June, July, and August tend to be hazy and dry, and four very wet months follow. From September through December downpours are frequent. The roads in the Congo basin often become impassable and the rushing waters of the innumerable streams tear the wooden bridges from their foundations. January and February tend to be drier than the preceding seasons and March through May are again wet.

Along the eastern edge of the Congo basin the hills grow higher, the slopes steeper, at first gradually, then abruptly, until in places the jumble of mountains reaches a height of 10,000 feet before plunging suddenly and precipitously as much as 7,000 feet downward into a huge valley. This valley and the mountainous escarpments that border it constitute the Albertine rift, one of the most spectacular geologic features in Africa. Like a gigantic ditch about thirty miles wide, the Albertine rift extends from Gondokoro on the upper White Nile to the southern end of Lake Tanganyika, a distance of some 1,000 miles.

A series of large lakes lines the bottom of the rift valley—Lakes Albert, Edward, Kivu, and Tanganyika. I remember Lake Albert, the northernmost one, chiefly for the Victoria Nile, which, just before entering the lake, surges with a roar over the Murchison Falls, and for the White Nile, which leaves the lake broad and serene on its long journey to the sea. Lake Edward evokes a picture of lake flies hovering over the shimmering waters in black clouds, and of game herds—buffalo, elephant, topi, kob—roaming the grassy plains and savannahs that border the lake on three sides. These plains are included in Queen Elizabeth and Albert National Parks. Lake Kivu is an intimate lake, nestling between the escarpments that rise like walls from the water's edge. Islands of various size break the lake's expanse and peaceful coves and inlets interrupt the shoreline. I recall mostly the burning sun while climbing through the dry scrubby forests that cover the hills patchily on the eastern shore of Lake Tanganyika, the piles of fish drying on the sandy beaches, and, at night, myriads of lights, marking the sites of fishing boats, flickering eerily far out in the lake.

Two isolated mountain massifs project from the valley floor. Between Lakes Albert and Edward and east of the Semliki River, which connects the two lakes, rise the fabled Mountains of the Moon, the Ruwenzories, to an altitude of 16,780 feet. The lower forested slopes of this upthrust peneplane are usually visible, but the bleak and rugged summits with glaciers clinging to the rocks often remain hidden in the clouds for weeks on end.

Just to the north of Lake Kivu, the chain of eight Virunga Volcanoes forms a gigantic dam across the floor of the rift valley and provides one of the divides of the Congo and Nile River systems (Pl. 2). The turbulent Ruzizi River empties Lake Kivu into Lake Tanganyika, which in turn has an outlet into the headwaters of the Congo. The Rutshuru River and several other streams meander northward into Lake Edward. Two of the Virunga Volcanoes are still active. In addition, hundreds, perhaps thousands, of small volcanic cones, some old, some new, pock the terrain around the forested mountains. The fertile volcanic soil supports one of the heaviest human populations in Africa: Bahutu, Bahunde, Bahavu, and other Bantu tribes, till the valley floor and the slopes of the escarpment, growing bananas, maize, beans, sorghum, and other crops; tall, hamitic Watutsi graze their herds of cattle; and pygmoid Batwa scour the remaining forests for game.

Uplifts and downwarps, compression, shearing and tilting, volcanic activity and erosion, all have shaped the mountains of the rift escarpment at various times and created a chaotic series of hills and valleys. The eastern margin runs southward from Lake Albert to the Ruwenzori massif. After a slight break it continues through grasslands and patches of forest to the rugged highlands of southwestern Uganda. Many of the hills are heavily cultivated, and scattered huts cling precariously to the slopes, but here and there such woodlands as the Kayonza Forest survive. In western Ruanda-Urundi bamboo still blankets the uplands in their highest parts, but as the altitude of the rift drops progressively toward the south, cultivation again takes over.

Several gaps interrupt the highlands of the western escarpment, which has a width of five to twenty miles. The most prominent of these gaps lies between Lakes Edward and Albert where a tongue of rain forest from the Congo basin reaches the base of the Ruwenzories. The top of the escarpment from Lake Edward to Lake Tanganyika presents, like the eastern side, a maze of rugged mountains, prominent among them Mt. Tshiaberimu and Mt. Kahuzi, both of which reach altitudes of 10,000 feet. The climate in the highlands is temperate, with temperatures rarely straying far above 20° C. and sometimes sinking as low as the freezing point. Many Europeans have settled around the towns of Lubero and Bukavu, and, instead of forest, coffee and tea plantations and fields of white-flowered pyrethrum cover the slopes; herds of Holstein dairy cattle graze on green meadows, which might be in Switzerland or Wisconsin. In recent years native peoples too have penetrated higher and higher, converting the forest to grassland and fields. But the mountains west of Lake Tanganyika remain wild and largely untouched. The main road winds mile after mile through gloomy valleys and fog-shrouded passes and only a few villages and some mining camps interrupt the solitude (Pl. 3).

Taxonomy

In 1847, Savage and Wyman first described the gorilla as *Troglodytes gorilla.* Geoffrey-Saint-Hilaire established the genus *Gorilla* in 1858–61, and at the generic level the gorilla has remained stable to date. However, specifically and subspecifically the taxonomic picture is in a constant state of flux, and agreement is not yet in sight.

The early part of this century saw an amazing proliferation of gorilla names, sometimes based solely on one or two skulls. Between 1903 and 1917 alone, nine new forms were recognized. Lönnberg (1917), for example, obtained seven skulls from the Virunga Volcanoes which he designated as *Gorilla g. mikenensis* to distinguish them from a single skull of *Gorilla g. beringei* (variously misspelled *berengeri, berengei, beringeri* by Elliot [1913], Galloway *et al.* [1959], and others) collected in the same mountain range less than twenty miles away.

In general, taxonomists agree that two main populations occur in the forests of Africa: one in West Africa (Nigeria, Gabon, Cameroun, Rio Muni, etc.), the other in Central Africa (eastern Congo, southwestern Uganda, western Ruanda-Urundi). The former is commonly referred to as the lowland, coast, or western gorilla, the latter as mountain or eastern gorilla. The scientific nomenclature of the two forms is confusing and three current classifications exist.

1. Coolidge (1929) measured 213 skulls comprising samples from all of the ten lowland and five mountain forms recognized at that time. His conclusion, based on the length of the palate and other minor differences, was that the eastern and western forms differed, but that these differences justified only a subspecific separation. After discarding all the old forms he recognized only:

Gorilla gorilla gorilla (Savage and Wyman)—West Africa
Gorilla gorilla beringei (Matschie) —Central Africa

Haddow and Ross (1951) re-evaluated Coolidge's data and derived the same conclusion.

2. Schouteden (1947) accepts the idea that all gorillas belong to the same species, but he divides the mountain gorilla into four subspecies, using a terminology which is also followed by Curry-Lindahl (1956).

Name	Location
Gorilla gorilla beringei (Matschie, 1903)	Virunga Volcanoes, Kayonza Forest
G. g. graueri (Matschie, 1914)	Region west and northwest of Lake Tanganyika
G. g. rex-pygmaeorum (Schwarz, 1927)	Region west of Lakes Kivu and Edward
G. g. uellensis (Schouteden, 1927)	Bondo, Uelle

G. g. uellensis is based on three skulls of doubtful origin, and the name is thus of dubious value. *G. g. graueri* and *G. g. rex-pygmaeorum* were described at a time when it was thought that the specimens were obtained from small isolated populations. In view of the limited number of skulls available for measurement and the known present-day distribution of the mountain gorilla I doubt if the distinction is justified.

3. Recently Vogel (1961) measured forty-one lower jaws of the lowland gorilla and forty-five of the mountain gorilla. He noted that the latter were larger and that the position of the foramen mentale was different. These, as well as some other minor points, led him to the conclusion that the two forms should be regarded as specifically distinct, and he suggested the names *Gorilla gorilla* and *Gorilla beringei.*

Vogel further measured seventeen mountain gorilla jaws from the Virunga Volcanoes and twenty-five from the entire region west of the rift escarpment. The ascending ramus in the two populations was consistently different, and primarily on this basis he divided the mountain gorilla into two subspecies:

Gorilla beringei beringei—Virunga Volcanoes, Kayonza Forest
Gorilla beringei graueri—all other areas

Unfortunately he measured no jaws from the Kayonza Forest or from many of the other isolated gorilla populations. He also included a jaw from "Alimbongo" with those from the Virunga Volcanoes, an error which leaves the validity of his conclusions open to question.

The current disagreements on the classification of the mountain gorilla may make the taxonomic picture confusing, but they certainly leave one a variety of choice. In this report I shall use the classification proposed by Coolidge, for it is not only the simplest of the three, but also the best documented.

History of the Discovery and Study of the Mountain Gorilla

The discovery of the lowland gorilla by Savage and Wyman (1847), the popular exploitation of this ape by Du Chaillu (1861), and other events concerning the animal, are summarized in detail by Yerkes and Yerkes (1929). Since a similar compilation is lacking for the mountain gorilla, I have attempted to outline briefly the history of discovery and study of this subspecies.

Just over a hundred years have passed since the first white explorers penetrated to the edge of mountain gorilla habitat, but only for a little over half that time has the subspecies been known to science, and only within the past few years have intensive studies been carried out. Four periods in the history of the mountain gorilla are recognizable:

1. Period of pre-discovery and final discovery	1858–1902
2. Period of hunting and collecting	1903–1925
3. Period of general scientific inquiry and collecting	1926–1955
4. Period of intensive scientific inquiry	1956–present

1. *Pre-discovery and discovery* By chance the mountain gorilla inhabited a forest area which remained unexplored until the turn of the twentieth century. Nearly every major explorer to visit Central Africa skirted along or slightly penetrated into mountain gorilla habitat without discovering the animal. When the British explorer

Speke visited Uvira, an Arab slave-trading post at the northern end of Lake Tanganyika, in 1858, he could view gorilla habitat on the rift mountains above him. In November, 1861, he and Grant traveled northward to Uganda in search of the headwaters of the Nile. Near the eastern border of Ruanda-Urundi, they observed "in the distance some bold skyscraping cones situated in the country Ruanda" and natives told them of man-like "monsters who could not converse with men" dwelling in them (Speke, 1864). Speke named the cones the Mfumbiro Mountain which today are known as the Virunga Volcanoes. Stanley (1872) and Livingstone, after their famous meeting ("Dr. Livingstone, I presume"), visited Uvira in 1871 and the former (Stanley, 1878) also glimpsed the Virunga Volcanoes from a distance in 1876. Stanley (1878, 1890) reached a bay of Lake Edward in 1876 and again in 1888. Mt. Tshiaberimu, which towered above him, harbored gorillas, but he had no inkling of their presence. The German zoologist Stuhlmann (1894) climbed in the Ruwenzori Mountains near Mt. Tshiaberimu in 1891, and in the same year, with Emin Pasha, was the first white man to reach the Kigezi District of Uganda, not far from the Kayonza Forest where gorillas occur. He also obtained in May, 1891, a better view of the Virunga Volcanoes from the floor of the rift valley south of Lake Edward than had previous explorers.

Germany, which around the turn of the century attempted to consolidate its colonial empire, disputed the exact location of the eastern Congo boundary with the Belgians. As a result German military men penetrated western Ruanda-Urundi extensively. In the fall of 1898, Captain Bethe climbed to the summit of Mt. Muhavura, the most easterly of the Virunga Volcanoes. Dr. R. Kandt surveyed the border region topographically in 1898 and 1899 (Meyer, 1955).

Several major explorers at one time traveled from Ujiji on Lake Tanganyika westward to the town of Nyangwe, established by the Arabs on the banks of the Lualaba (Congo) River as a slave collecting center in about 1860. Nyangwe saw Livingstone (Livingstone, 1874) in 1871, Cameron (Cameron, 1877) in 1874, Stanley (Stanley, 1878) in 1876, Wissmann (Wissmann, 1891) in 1882, and numerous others. This route lay just to the south of gorilla habitat. The intrepid Cameron, the second white man to traverse the African continent, crossed the Ulindi River in 1874. In 1876–77 Stanley went down the Congo River without learning that gorillas inhabited the forests less than a hundred miles to the east.

Emin Pasha, the German who became almost a sovereign over the northern Congo and southern Sudan, was assassinated by Arabs just to the west of the town of Lubutu in 1892 (Cornet, 1954) after having marched through the forest along the northern boundary of mountain gorilla range.

Probably no explorer traveled more extensively in or near gorilla habitat than von Götzen (1899), a German nobleman. On June 3, 1894, he discovered Lake Kivu and he climbed to the summit of Mt. Nyiragongo, one of the peaks in the Virunga chain, on June 11, 1894. He then headed toward the northwest, crossed the Oso River, descended the Lowa River, and finally passed south of the town of Lubutu to the Lualaba River.

Although the word "gorilla" crops up in the writings of several explorers, in most instances the animal indicated is without question the chimpanzee. Livingstone (1874) while traveling between Ujiji and Nyangwe observed natives battling with "gorillas" in 1866. Burton (1876) notes that "the Soko, which Dr. Livingstone identifies with the gorilla, extends to the Lualaba or Upper Congo in the regions immediately west of the Tanganyika Lake." Similarly, Cameron (1877) writes: "Here [west of Lake Tanganyika] I saw some gorilla (*soko*), black fellows, looking larger than men." These observations were made to the south of present-day gorilla range and refer to chimpanzees, which locally are known as *soko* (*or sokomutu*, like-man).

In October, 1876, Stanley (1878) noted that at Uhombo, which lies on the main route between Ujiji and Nyangwe, people wore "the tags of monkey skin and bits of gorilla bone" around the waist. The bones probably belonged either to chimpanzee or to man, for the nearest gorilla habitat lay nearly 100 miles to the north. In 1877, at the top of the bend of the Congo River where no gorillas exist today, Stanley described necklaces which were made of human, gorilla, and crocodile teeth. In his book of folk tales, Stanley (1893) writes about gorillas from the right bank of the Aruwimi River, which is a tributary to the Congo River, from Uganda, and from the Little Black River in the Congo "where the forest was haunted by many monstrous animals, big apes, chimpanzees, gorillas and such creatures."

In 1890 Stanley expressed to Sir Harry Johnston (Johnston in Barns, 1922) his belief that gorillas existed in northeastern Congo. Burrows (1898) supposedly photographed a dead gorilla near the Stanley Falls, but the photograph published in Coolidge (1936) is that of a chimpanzee. Johnston (Barns, 1922) crossed the Semliki River into the forests of the Congo in 1900 and was shown a photograph of a supposed gorilla taken by a Belgian official.

Grogan (Grogan and Sharp, 1902), while on his famous walk from Cape to Cairo in 1898, detoured along the foot hills of the Virunga Volcanoes to hunt big game. "During one of my elephant hunts [in the bamboo north of Mt. Mikeno] I came on the skeleton of a gigantic ape, larger than anything I have ever seen in the anthropoids, but I never saw a live specimen." Grogan was familiar with chimpanzees, and had he shipped home that skeleton instead of those of several elephants, he would undoubtedly be known today as the discoverer of the mountain gorilla.

The credit for the discovery belongs to Captain Oscar von Beringe, a German officer. During his first journey to the Virunga Volcanoes in 1899 he discovered Mt. Visoke (Meyer, 1955). In 1902 he again traveled from Usumbura, a town on the northern end of Lake Tanganyika, northward through Ruanda-Urundi. The main purpose of the trip was to impress the native chiefs and the Belgian border posts with the military might of the German empire (Appendix A). On October 17, 1902, von Beringe, accompanied by a Dr. England, attempted to climb Mt. Sabinio, one of the peaks in the Virunga Volcanoes, from the Mt. Gahinga–Mt. Sabinio saddle. Just after making camp on a narrow ridge at 3,100 meters they spotted several apes above them. The only account of this discovery, which I have never seen reprinted, is found in a German colonial newspaper (von Beringe, 1903). I have translated the pertinent paragraph freely; the exact text is reproduced in Appendix A.

> We spotted from our camp a group of black, large apes, which attempted to climb to the highest peak of the volcano. Of these apes we managed to shoot two, which fell with much noise into a canyon opening to the northeast of us. After five hours of hard work we managed to haul up one of these animals with ropes. It was a large, man-like ape, a male, about 1½ m. high and weighing over 200 pounds. The chest without hair, the hands and feet of huge size. I could unfortunately not determine the genus of the ape. He was for a chimpanzee of a previously unknown size, and the presence of gorillas in the Lake region has as yet not been determined.

2. *Hunting and collecting* Matschie (1903) described the gorilla collected by von Beringe, and his publication marks the beginning of an intensive hunting and collecting period in Central Africa. Between 1902 and 1925 at least fifty-four *Gorilla gorilla beringei* are known to have been removed from the Virunga Volcanoes alone (Table 1). Although the various expeditions provided the apes with ample publicity and furnished anatomists and taxonomists with nu-

merous specimens, accurate information concerning life history and behavior remained meager at best.

The list of visitors into gorilla habitat is a long one and I will mention only some of those who encountered the apes or noted signs of their presence. Pretorius (1947) shot a lone male on Mt. Sabinio in 1904. An expedition led by the Duke of Mecklenburg (1910) in 1907–8, which included a botanist and geologist, obtained considerable information in the Virunga Volcanoes and it was noted that "gorillas as well as tschegos [chimpanzees] haunted the margins of the upper forest" of Mt. Gahinga. Grauer spent twelve days on Mt. Sabinio in 1908 without seeing gorillas, but in 1908 and 1910 he collected sixteen of the apes in the mountains west of Lake Tanganyika (von Liburnau, 1917). In July, 1911, Meyer (1923) noted gorilla sign on the slopes of Mt. Karisimbi in the Virunga Volcanoes. Gromier (1948) observed gorillas on the same peak in 1912. Critchley-Salmonson (in Philipps, 1923) apparently heard gorillas in the Kayonza Forest in 1912. Arrhenius shot seven of the apes on Mt. Mikeno during 1913–14. Philipps (1923) saw two gorillas on the slopes of Mt. Muhavura in March, 1919. In 1920 and 1922 Barns (1922, 1923) traveled extensively in the Virunga Volcanoes and later (Barns, 1926) through the Utu region, where he shot two gorillas. Schouteden (1927) also visited the latter area. Intensive collecting by Prince Wilhelm of Sweden (1923) accompanied by Gyldenstolpe (1928) in 1921 accounted for fourteen gorillas in the Virunga Volcanoes. Akeley (1923) and Bradley (1922) also arrived in 1921 and collected five animals for the American Museum of Natural History. Burbridge (1928) captured a total of eight infants and shot a male in the Virunga Volcanoes in 1922 and 1925. Fortunately Akeley was so impressed with the beauty of the volcanoes and the supposed rarity of the mountain gorilla that he urged the Belgian government to set aside a permanent sanctuary for the animals. Albert National Park was established on April 21, 1925, but encompassed only Mts. Mikeno, Visoke, and Karisimbi. On July 9, 1929, the boundaries of the park were enlarged to include the whole volcanic chain.

3. *General scientific inquiry* The period of general scientific inquiry into the habits of the mountain gorilla begins with the 1926 Akeley expedition (Akeley, 1929; Derscheid, 1927). Although Akeley died, and lies buried in the Mt. Mikeno–Mt. Karisimbi saddle, his wife and Derscheid attempted to carry on the work. Bingham (1932), a psychologist, visited the Virunga Volcanoes in 1929 and obtained some general notes on feeding behavior and trails. Pitman

(1935, 1942) visited the Kayonza Forest in Uganda in 1934 and 1937 and published the first precise information on this population.

Collectors continued to shoot and capture gorillas in the western rift mountains and the Utu region during this period, for the Virunga Volcanoes were closed to all hunting. Coolidge (1929, 1930) shot a male as a member of the 1927 Harvard African Expedition. His summary of the available knowledge produced the first map of gorilla distribution. Other collectors in this western area included Raven (1931), Johnson (1931), and Gatti (1936). Brief accounts of gorillas were also published by Maxwell (1928), Babault (1928), Broughton (1932), de Witte (1937), Frechkop (1938), and numerous others; on the whole, these added very little to the existing knowledge.

4. *Intense scientific inquiry* The intense recent interest in the mountain gorilla was sparked by renewed emphasis on primate field behavior and the greater availability of funds, freedom of travel, and easy accessibility to mountain gorilla habitat by air and road.

In 1955, Mr. Walter Baumgartel acquired the Travellers Rest Hotel in Kisoro, Uganda, at the foot of the Virunga Volcanoes. He became fascinated by the gorillas and hoped to establish them as a tourist attraction. But he also realized the potential scientific value of the animals. It is largely through the efforts of Baumgartel that Drs. Raymond Dart and L. S. B. Leakey, of the Witwatersrand University and Coryndon Museum, respectively, provided funds for investigators to study the gorillas in the area. Intensive investigations began in 1956 and the various studies conducted in Uganda to date are summarized in Table 2. Many of the casual visitors to Kisoro were also fortunate in seeing gorillas and popular accounts include those of Blower (1956), Milton (1957), Heim (1957), Imanishi (1958), Newman (1959), Moorehead (1959), Zahl (1960), Krampe (1960), Baumgartel (1960), and Wordsworth (1961).

No intensive studies, other than the present one, have been conducted on the Congo side of the Virunga Volcanoes. Hoier (1955*a*), Mathis (1954), Imanishi (1958), Curry-Lindahl (1961), Kraft (1952), and several other persons have visited the Kabara area and some saw gorillas, but little effort was made to obtain specific data. Verschuren, the biologist for Albert National Park, spent over two months in the Virunga Volcanoes between 1957 and 1961 and collected incidental information on the animals.

A considerable amount of work on gorillas, primarily captive and dead ones, has been carried out by personnel of I.R.S.A.C., Lwiro, near Bukavu, but the information remains largely unpublished. In 1957 Mr. Charles Cordier and I.R.S.A.C. trapped several groups of gorillas in the region west of the rift and valuable observations of the animals under stress of being captured were made. Dr. Urs Rahm briefly studied free-living gorillas in the bamboo forests west of Lake Kivu during 1959 and 1960.

Procedure and Method

The data in this report are based on several sources: (1) personal observations in the field, (2) unpublished material from several reliable investigators, (3) reports from natives, (4) statements in the literature, and (5) personal and other observations on captive animals.

1. *Personal observations* The bulk of the data is based on personal observations. We visited all major mountain gorilla habitats during the first phase of the study, and I revisited several areas later for more intensive work (Table 3). In each I collected gorilla food plants and took extensive notes on trail sign and the general vegetational aspects. I encountered gorillas visually 314 times and observed them directly 466 hours primarily in the Virunga Volcanoes of Albert National Park (Table 4). Work on other ape species was limited, and the locations as well as the amount of observation are summarized in Table 5.

2. *Data from other investigators* Several investigators aided me greatly by placing their notes and information at my disposal. Emlen and I frequently worked independently in the field, and he generously permitted me to incorporate his notes into this report. Unpublished observations by Bolwig, Cordier, Baumgartel, my wife, and others furnish valuable additional data.

3. *Data from non-professional observers* Information from non-professional observers must be accepted with caution. The local human population characteristically avoids gorillas assiduously and their knowledge concerning the animals is primarily derived from trail sign and imagination. Most natives are reliable in confirming or denying the presence of gorillas (called locally *n'gagi, n'gila,* or *kinguti*) in the region known by them, and many can point out the various gorilla food plants. But I encountered not a single African who gave consistently reliable information regarding the behavior of gorillas. Rarely were the statements intended to mislead; instead, the African either gave an answer which he thought would please, or he related behavior, typical of his tribal existence, which he supposed also applied to gorillas. The writings of Garner (1896) illus-

trate well the danger of accepting statements from natives indiscriminately.

Many of our distributional records are derived from reports by natives. However, whenever we challenged an informant to show us definite sign he was always able to do so. A few food plants listed are based only on the word of Africans. I have included little detailed behavioral data without personal verification except for the changes in the composition of one group in the Kisòro area of Uganda. There the native guide Reuben Rwanzagire has seen the same few groups numerous times, and he maintains that he knows each animal individually, a claim which I do not doubt, although his interpretations of gorilla behavior must be treated with reserve.

4. *Data from the literature* The literature on free-living gorillas is embarrassingly voluminous, considering the paucity of accurate information which it contains. When descriptions limit themselves to nests and trails the information presented tends to be fairly reliable, but when they concern behavior and its interpretation the statements are often so circumstantial and based on such limited data that the serious researcher cannot rely on them. Frequently it is impossible to judge if the information presented is based on direct observation, on verbal accounts by others, on intuition, or on mere paraphrasing of the older writings, with the latter being a favorite method of padding skimpy data, thus creating a highly inbred literature giving the appearance of truth through mere repetition. Many standard texts in zoology and anthropology have drawn on such sources.

I have made no attempt to summarize the extensive literature or to list it in the bibliography, although I have consulted most of it for possible value in supplementing data collected by myself. Many important early references are found in Yerkes and Yerkes (1929) and in Ruch (1941), and the most useful papers and books containing behavioral information after 1940 are mentioned in this report. The mountain gorilla has received by far the most attention from field workers. Few concise data on the lowland gorilla are available: Merfield and Miller (1956) provide probably the most reliable general account of recent years, and authors like Petit (1920), Sharp (1929), Allen (1931), Liz Ferreira *et al.* (1945), Blancou (1951, 1955), March (1957), Sabater Pi and de Lassaletta (1958), and Sabater Pi (1960) present some useful local notes.

5. *Data from captive gorillas* To familiarize myself with certain aspects of gorilla behavior, to obtain a basis for comparing wild and captive animals, and to collect specific information not obtainable

in the wild, I visited the zoological gardens in Columbus, St. Louis, Milwaukee, New York, Antwerp, Basel, and London. Because of increased knowledge of how to raise gorillas in captivity several zoological gardens now possess adult pairs, and offspring have been produced in Columbus, Washington, and Basel. The opportunities for studying the whole breeding cycle and the development of the infants in captivity are therefore great. The detailed published notes by Lang and his associates (Documenta Geigy 1959–61; Lang, 1961*a*) of the Basel zoo should serve as a model of the type of detailed information which can only be derived from captive animals and which is essential for comparisons with observations obtained in the wild. In this report I have included information from the literature on captive animals and my own observations in zoos, and on those held in scientific institutions and private homes, only when it elucidates a certain point or presents data not readily obtainable on free-living gorillas.

Bingham (1932), Donisthorpe (1958), and other observers had failed in their attempts to observe gorillas directly for prolonged periods, and one of the main tasks which we faced was to develop a technique for watching gorillas and to find an area where this could be done. Thus, the difference between success and failure in our study hinged on the obvious points of (1) finding gorillas, (2) observing them clearly once they had been found, and, importantly, (3) keeping the animals undisturbed to prevent them from rushing away or attacking. Our techniques varied from area to area, and these were adapted to the changing circumstances of vegetation and topography, but only in one area, at Kabara, were all three points solved to satisfaction.

1. *Finding gorillas* We made a special effort to learn how to track gorillas from the excellent African guides at Kisoro, on the Uganda side of the Virunga Volcanoes, and from the Batwa in the Kayonza Forest. A knowledge of tracking the animals by such signs as a chewed piece of bark, a bent blade of grass, or a knuckle imprint is essential for any investigator, for good forest guides are rare and none are fully dependable. The ease with which gorilla spoor could be followed varied considerably from area to area. At Kabara, in the Mt. Mikeno–Mt. Karisimbi saddle area of the Virunga Volcanoes, I tracked only by myself and could without difficulty pick up a trail four to five days old and trace it to the gorillas within a few hours. In the Kayonza Forest, in the Tshiaberimu region, and at Kisoro, all areas where the trail is easily lost, we often used natives

who fanned out when the trail became obscure and searched until it was relocated.

In all study locations, except in the Utu region, we transected the forest daily by simply hiking cross-country along compass lines or contours from a centrally located base camp. In areas where the population of gorillas is fairly dense, it rarely required more than two to three days of searching to find a fresh trail. Then we tracked the animals, an activity which was frequently aided by their vocalizations and other sounds. In order to locate gorillas as soon as possible in the Kayonza Forest we sent out African scouts in teams of two in various directions. In the extensive Utu region we drove along the roads and inquired about the presence of gorillas at every village. If a group was in the vicinity we checked the sign and when possible the animals themselves. By foot I transected a fifty-mile stretch of uninhabited lowland forest between two mining roads, but the density of gorillas was too low to assure frequent contact.

2. *Observing gorillas clearly* To find gorillas is relatively easy; to observe them clearly is usually difficult, for the habitat of the animals is frequently dense and to view them in the open is often a matter of luck. In the Utu region I saw gorillas well only once when several animals climbed out of the tangled secondary growth into a tree to feed. Although Emlen and I heard gorillas at close range several times at Mt. Tshiaberimu we never caught a glimpse of them in the maze of bamboo stems. Conditions were somewhat better in the Kayonza Forest and at Kisoro, but even there prolonged unobstructed encounters were rare. If the gorillas fed on the open herbaceous clearings that cover some of the steep slopes in the Kayonza Forest, excellent observations were possible. On the other hand, I twice spent over an hour within 50 feet of gorillas which rested and fed beneath tree ferns, shrubs, and vines, and caught only occasional glimpses of a black arm or a broad back. At Kisoro our meetings with gorillas were typical of those of most other visitors in that we caught fleeting views of bolting animals or heard the roars of the male hidden behind a screen of vegetation.

At Kabara the scattered *Hagenia abyssinica* trees spread their massive branches upward, and below them lies a solid expanse of herbs only 6 to 8 feet high. Dense brush grows in isolated clumps and on some of the steeper slopes. The topography varies from gently rolling to precipitous. This limited habitat proved to be ideal for direct, prolonged daily observations of gorillas, and the possibilities offered by it were unique, for it combined the following attributes: (a) the near absence of a shrub layer made it possible to view the gorillas without difficulty, and, importantly, the gorillas could see us clearly; (b) the uneven topography greatly aided visibility, since we could watch across valleys from opposing slopes; and (c) the area lies in Albert Park where human intrusion is forbidden and the gorillas are rarely harassed. Consequently I spent nearly all my time and obtained nearly all my behavioral observations on this gorilla population. Whenever I discuss gorilla behavior without naming a precise study location I refer to this population.

3. *Observing gorillas for prolonged periods* Primates have been studied in the past by such means as baiting them in to a food supply (Imanishi, 1960) and by watching them from the obscurity of a hide or blind (Kortlandt, pers. comm.). Baiting was attempted on gorillas by Osborn (1957) and Baumgartel (1960) at Kisoro without success, for the animals refused to eat the strange foods with which they were presented. The erratic daily movement of groups over a large area make the use of permanent blinds impracticable, since gorillas may not revisit a certain location for several months. Thus the only two methods by which gorillas could be observed for prolonged periods were: (a) to approach them undetected, viewing them from the cover of dense shrubbery or a tree trunk; or (b) to advance, sit, and remain in full view of the animals, with the hope that, over a period of days and weeks, they would become habituated to the presence of the observer. I employed both methods, but the latter more extensively than the former. If conditions were such that I was able to watch the behavior of the gorillas clearly while remaining undetected by them, I chose this method, for their activity was then not affected by my proximity. However, I sometimes lost useful observations while attempting to remain hidden, and as soon as I tried for a clearer view the gorillas, with their keen eyesight, detected the movement. At such times I found that the animals were much less excited if they could see me clearly than if I deviously attempted to hide.

My technique of habituating the gorillas was simple but essential, for I could only obtain unbiased data on their behavior if they remained relatively unaffected by my presence. I usually attempted to approach the group undetected to within about 150 feet before climbing slowly and in full view of the animals onto a stump or the low branch of a tree where I settled myself as comfortably as possible without paying obvious attention to them. By choosing a prominent observation post not only was I able to see the gorillas over the screen of herbs, but, reciprocally, they could inspect me clearly, which was the most important single factor in habituating the animals (Pl. 4). Under such circumstances they usually remained in the vicinity to watch me, and even approached me to within 5 feet. I found it remarkably easy to establish rapport with the gorillas. This process was greatly facilitated by the placid temperament of the animals, and by certain conditions which I imposed on myself: (a) I carried no firearms which might imbue my actions with unconscious aggressiveness; (b) I moved slowly, and used binoculars and cameras sparingly at the beginning to eliminate gestures which could be interpreted as threat; (c) I nearly always approached them alone, leaving any companions behind and out of sight at the point where the animals were first noted; (d) I wore the same drab olive-green clothes every day; and (e) I almost never tracked the gorillas after they had voluntarily moved out of range. This last point was, I believe, of special value, for at no time were they subjected to pursuit, an action which could easily frighten them as well as increase the chance of attacks. By adhering to my conditions I not only habituated six groups to my presence quite well but also was never attacked, even though I inadvertently stumbled into the middle of a group or nearly collided with animals several times.

In spite of the unique possibilities for observing gorillas at Kabara, the vegetation and the habits of the animals made it difficult at times to see clearly and to obtain continuous observation on one group for more than a few successive hours. Even time samples of activity of any one animal revealed little, for gorillas were frequently partially hidden by a screen of herbs. Most contact periods ranged from one to three hours.

My daily routine of observation varied little. The animals were tracked from the place of encounter of the previous day until contacted, preferably between mid-morning and mid-afternoon. Frequently I observed two and sometimes three groups in one day. Heavy rains often hampered observations, and, at any rate, produced little information, for the animals merely sat motionless until it ceased. Fog frequently obscured everything.

Observations on the building of night nests required a modified technique, for even habituated animals reacted nervously to my presence at that time of day. Consequently I followed the groups cautiously at a distance of about 200 feet and attempted to watch nest building while remaining undetected. After darkness I crept closer, and on eight occasions slept beneath a tree within about 100 feet of the animals. At dawn, before the first gorillas awakened, I sought a vantage point from which to note their sleeping positions and behavior associated with rising.

I noted the actions of gorillas either with the unaided eye or with 7 × 50 Bushnell binoculars. Notes were taken in the field during or immediately after each behavioral sequence. In the evening I transcribed my observations in greater detail, and more legibly, into another notebook, which insured the presence of a duplicate set of notes. Frequently I made rough sketches of various postures of the animals. I found it unnecessary to face the problem of marking gorillas, for a large amount of physical variation made individual recognition easy. Names were assigned to most gorillas in the Kabara groups to aid in the taking of notes. Since the time difference between the Congo and Uganda is one hour (for example, 0600 in the Congo is 0700 in Uganda) all times were converted to Congo time. Because we lacked a means by which to check our watches each day, all times in this report are accurate to within about ten minutes only.

Photography is usually not compatible with behavioral observations on gorillas. I took snapshots, both in black-and-white and in Kodachrome, with 35 mm. Contax cameras, whenever it did not interfere with other work. Movie photography was not attempted during the first fifteen months of study, and thereafter only sporadically during the periods of May 25 to June 2, August 21 to 17, and September 12 to 17, 1960. Tape recordings of a few vocalizations were made between September 12 and 17, 1960.

CHAPTER 2

DISTRIBUTION AND ECOLOGY

Distribution

The genus *Gorilla* occurs in two widely separated forest habitats, one in West Africa and the other in Central Africa. Although Emlen and I studied the distribution of the gorilla in detail only in the latter area, an analysis of the data must take into account the western population and the reason for the intervening gap between the two. One must also determine, if possible, past changes in the vegetation cover, physical barriers, and other factors limiting the distribution of the gorilla. The purpose of this section is to discuss the distributional dynamics of the mountain gorilla in somewhat greater detail than presented in an earlier paper (Emlen and Schaller, 1960*a*) and to offer several suggestions to account for the discontinuity of the two subspecies.

Geographical Distribution of the Lowland Gorilla

Coolidge (1929, 1930, 1936) summarizes the available evidence for the extent of distribution of the lowland gorilla. Figure 2, which is based on a map presented by Coolidge, indicates the general geographic limits of lowland gorilla range.

In 1908, Lemarinel obtained three skulls from natives at Bondo (23°50′E), on the north bank of the Uele River some 400 miles east of the nearest present-day lowland gorilla population. Coolidge (1936) considers it probable that the skulls were obtained from lo-

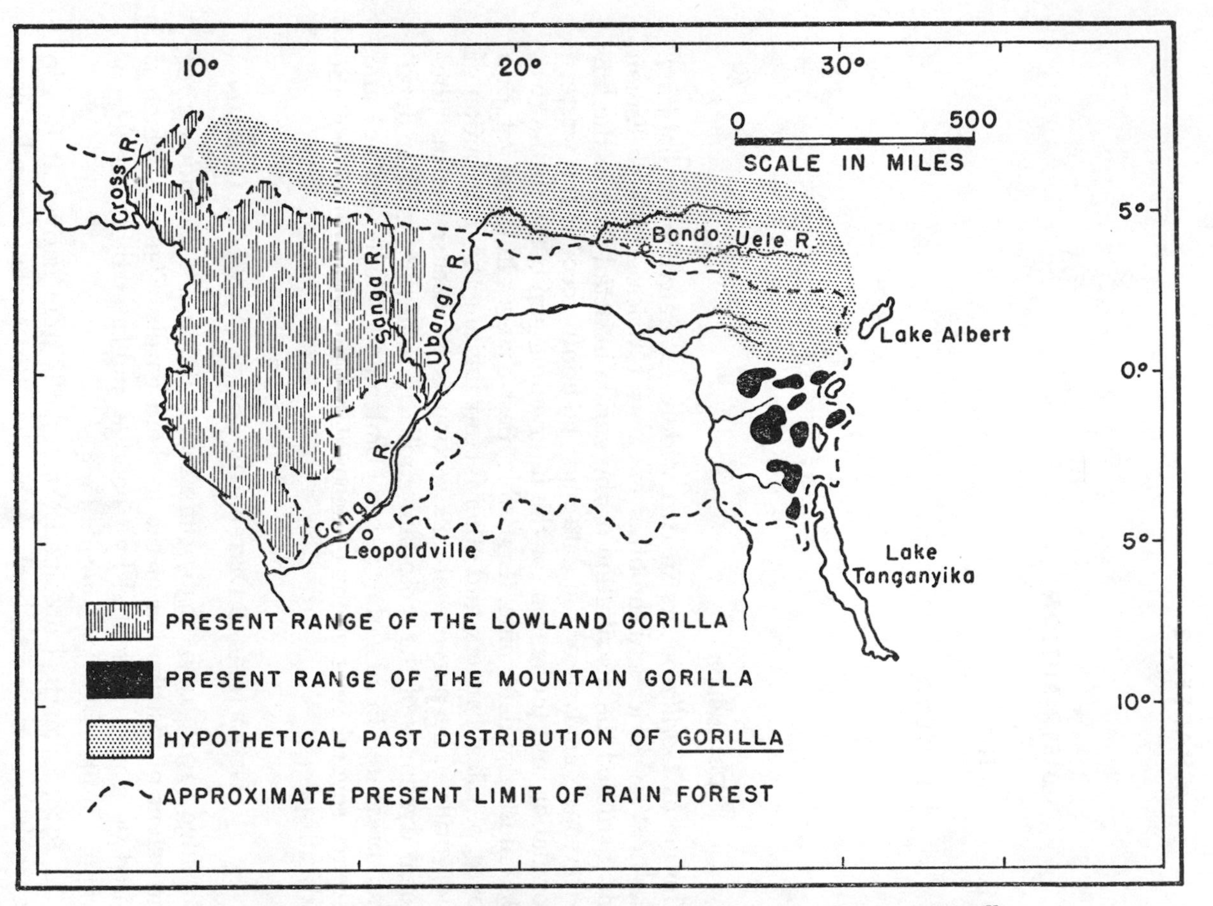

FIG. 2.—The present and hypothetical past distribution of the genus *Gorilla*

cal animals and not imported, and he states that the affinities of the skulls lie with the lowland gorilla rather than the mountain form. Allen (1939), on the other hand, believes that the affinities of the skulls are with the mountain gorilla. The population, if there actually was one, appears to be extinct.

Geographical Distribution of the Mountain Gorilla

The range of the mountain gorilla extends from the equator south to about 4°20′S latitude, a distance of a little over 300 miles, and from longitude 26°30′E to the western escarpment of the rift valley at about 29°30′E longitude. A small extension penetrates the Virunga Volcanoes and the Kayonza Forest on the floor of the rift valley and eastern rift mountains, respectively, to 29°45′E. The total east-west distance is thus about 220 miles (Fig. 3). The range is in the form of a triangle with its points near Lubutu on the northwest, Lubero on the northeast, and Fizi on the south, and it covers an area of about 35,000 square miles. The boundaries are quite distinct along the eastern and southwestern fronts. Unfortunately very little detailed information on the north boundary could be obtained, and there may well be additional pockets in the Maiko River Valley and westward toward Ponthierville.

Within the borders of this range, most of the gorillas are concentrated in about sixty more or less isolated tracts of 10 to 200 square miles each, separated from each other by distances of 2 to 30 or more miles. Groups in closely adjoining areas undoubtedly have intermittent contact, but others are isolated populations. The area actually occupied by gorillas is at least 7,500 to 8,000 square miles, or only about 21 to 23 per cent of the total area encompassed by their range. A block of about 6,500 square miles in the western region between the Lowa and Lugulu rivers apparently contains a relatively sparse but uniform population, except for concentrations of animals along river barriers and near villages. The vast stretches of forest between and around these populated areas are unoccupied except for stragglers.

Past Distribution of "Gorilla gorilla"

No fossil remains of gorillas have been found, but the minor differences between the lowland and mountain forms (Coolidge, 1929; Schultz, 1934) suggest that the divergence between the two occurred in the not too distant past. The present distribution of the two subspecies further suggests that at one time the population was

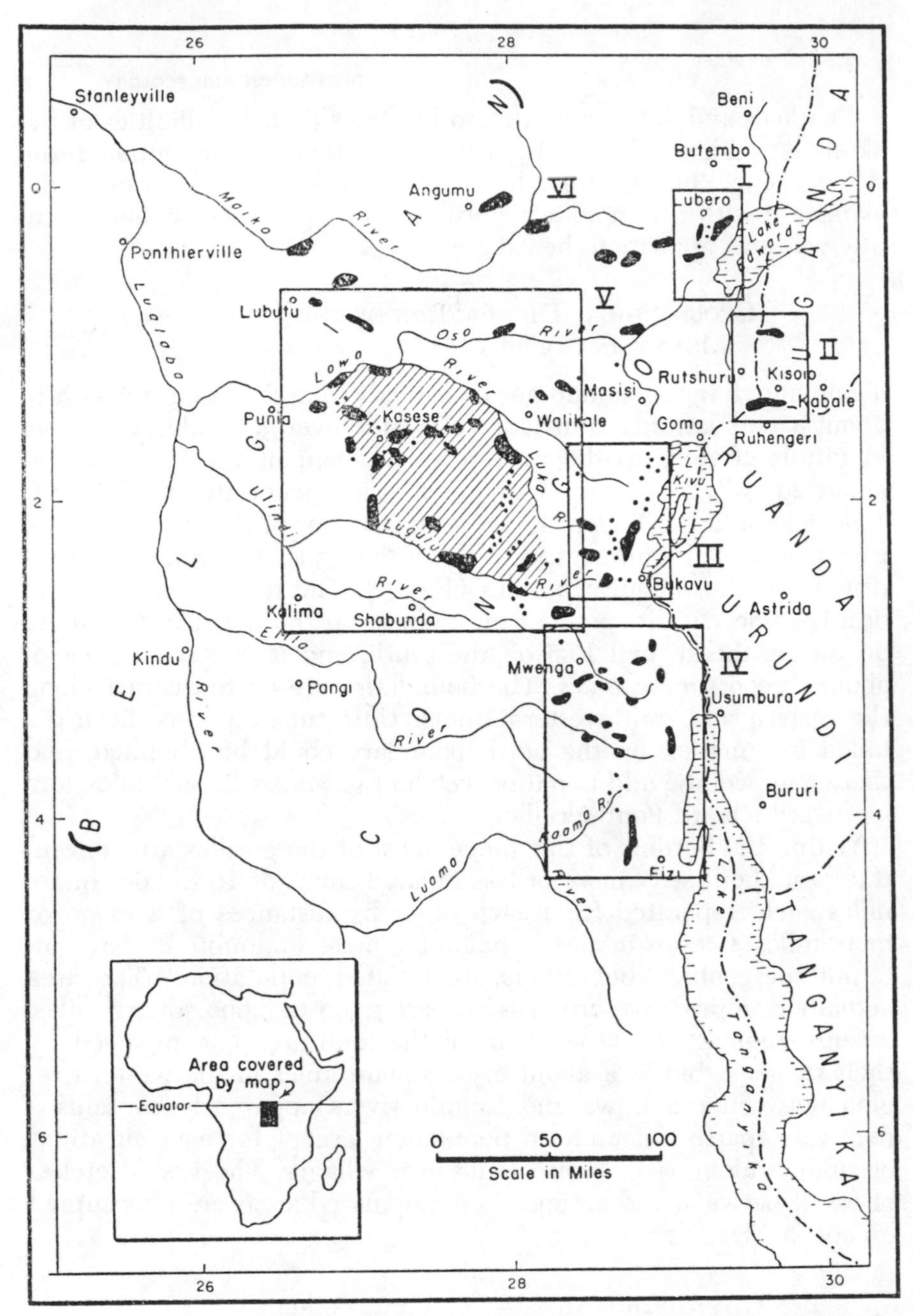

FIG. 3.—The geographic distribution of the mountain gorilla in 1959. The black areas indicate the location and approximate shape of isolated gorilla populations or of gorilla concentrations in areas of continuous distribution. The small dots represent records of gorillas outside these areas. The hatching marks a central region of continuous but sparse distribution. The numbered rectangles divide the range of the mountain gorilla into geographic regions to which the following terms are applied in this report: I, the Mt. Tshiaberimu region; II, the Virunga Volcanoes in the south, and the Kayonza Forest in the north; III, the Mt. Kahuzi region; IV, the Mwenga-Fizi region; V, the Utu region; and VI, the Angumu region. (Reproduced from Emlen and Schaller, 1960*a*.)

continuous throughout a rain forest belt from west to central Africa north of the Congo River. If these two suppositions are accepted, an explanation must be found for the gap of about 650 air miles which separates the two forms today, a gap not caused by lack of suitable habitat, for rain forest stretches without interruption from the rift mountains to the coast of West Africa (Fig. 2).

Two primary factors appear to limit the distribution of gorillas today, and supposedly also did so in the past:

1. Gorillas frequent only humid forested areas where succulent herbs and vines are readily available for forage. They do not penetrate far into extensive open woodlands and grasslands which thus present barriers for dispersal.

2. Gorillas are apparently unable to swim and they even hesitate to wade through shallow streams. Therefore, large rivers constitute barriers which can only be circumvented at the headwaters when their width becomes small enough to be bridged by a fallen tree (Pl. 5).

If the climatic conditions in the past were similar to those of today, gorillas could not have crossed the Congo and Ubangi-Uele rivers, which are wide and deep. Thus, it seems likely that at one time gorillas inhabited the northern bank of the Uele River and that penetration southward into the present home of the mountain gorilla occurred via the headwaters of that river. The records of gorilla skulls from Bondo are, therefore, of great interest, for they represent the only evidence of these apes in the gap between the two populations at a place consistent with the hypothesis expressed here.

The vegetation map of Keay (1959) shows that the Uele River lies roughly at the southern edge of a "forest-savannah mosaic" which has a width of about a hundred miles. Such habitat is not favored by gorillas today. However, slightly higher precipitation would be sufficient to extend the rain forest belt northward, and provide habitat suitable for gorillas.

Evidence from various sources indicates that at some time during the late Pleistocene the forest did indeed extend farther northward. Moureau (1954 and pers. comm.) notes that the bird fauna of the mountain forests in East Africa are, in several cases, not even subspecifically differentiated from that of Mt. Cameroon in West Africa. And on this basis he postulates that at some period in the past, presumably late in the Pleistocene, a mountain forest extended from East to West Africa along the northern rim of the Congo basin at an altitude of 2,000 to 3,000 feet. A general reduction in temperature of 5° to 7° C. would apparently suffice to create a mountain forest in the area. More direct evidence for a change in the latitude of the forest border comes from lake deposits and dunes of the Chad basin in Nigeria. During the late Pleistocene, Lake Chad was nearly five times as large as Lake Victoria is at present (Grove, 1959 and pers. comm.). To maintain such a level the lake must have received about twenty times as much water as today, from a watershed which extended eastward as far as the Sudan. The greater precipitation indicated by this observation suggests a forest-type of vegetation north of the Uele River.

Thus, it appears that at some time during the late Pleistocene, during an as yet undated pluvial period, there was a continuous stretch of rain forest suitable for gorilla habitation extending from West to Central Africa. This hypothetical population disappeared during a drier period which followed.

Distributional Dynamics of the Mountain Gorilla

An examination of Figures 3 and 4 shows that many population units are located flush against prominent physical barriers such as rivers and forest boundaries. The Lugulu River north of Shabunda, the Luka and Lowa rivers west of Walikale, and the eastern forest boundary from Lubero to Fizi are examples. But numerous populations exist as isolated units in a forest habitat which offers no obvious barriers to a more general distribution. The factors limiting dispersal at the east, south, west, and north boundaries each present special problems.

The eastern limit of the range is nicely demarcated by the rift escarpment along much of its length. The absence of gorillas in the broad tongues of grassland which extend over the rim and into the forest region north and south of Lake Kivu indicates that the actual effective barrier is not the escarpment itself but the forest-grassland boundary. Gorillas have, however, penetrated eastward to the Virunga Volcanoes and the Kayonza Forest, presumably by means of forest bridges that existed before the advent of agriculture. On the other hand, gorillas have failed to colonize the Ruwenzori Mountains even though the lowland rain forest provides an apparently suitable bridge to them at one place in the Semliki valley. It is possible that the Semliki River presents a barrier.

The forest ceases and gives way to savannah and heavy agriculture at the southern limit of gorilla range near Fizi.

There are no obvious physical barriers to prevent the gorillas from spreading westward to the banks of the Lualaba River: no ma-

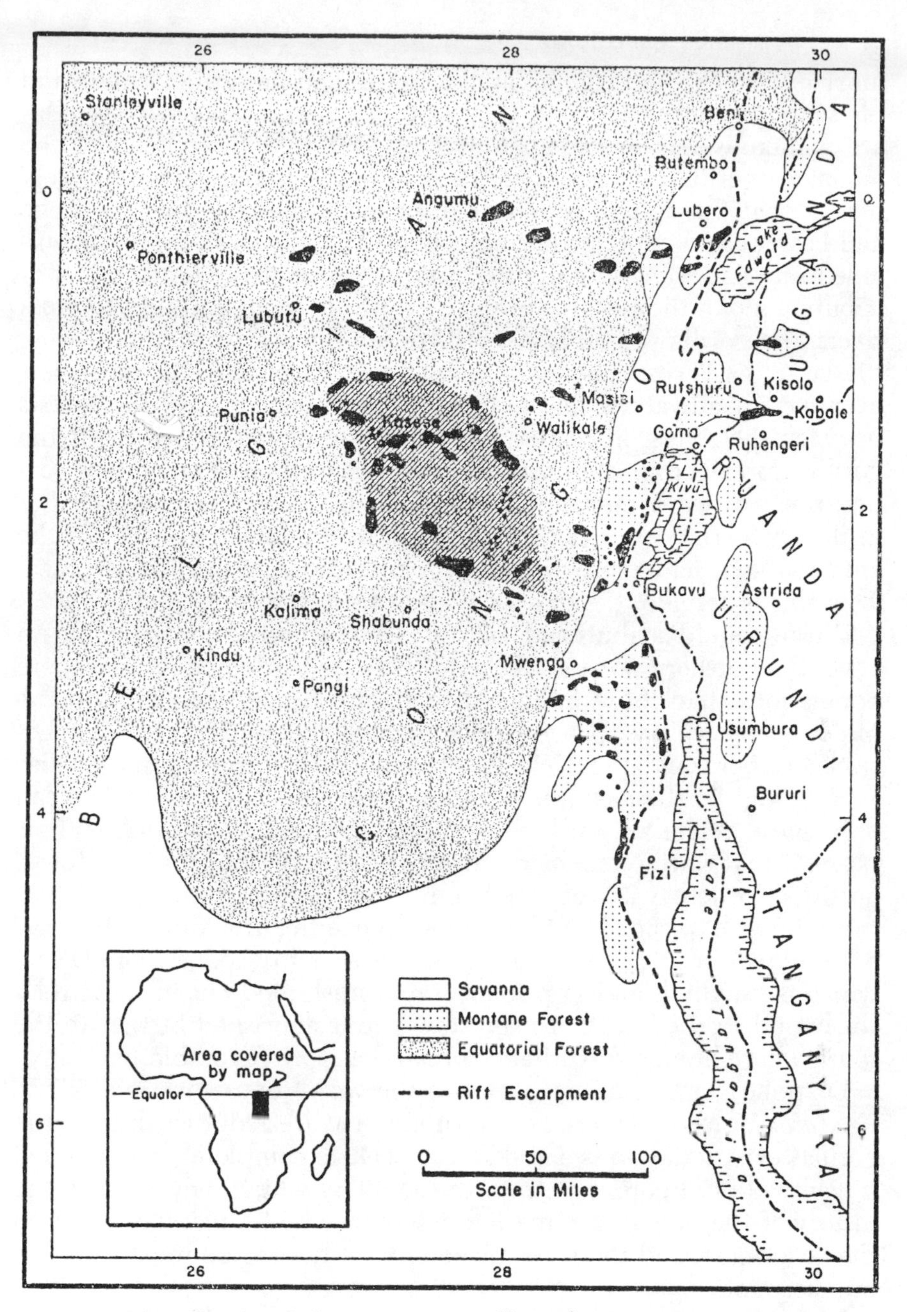

FIG. 4.–Distribution of the mountain gorilla with respect to vegetation types. The indicated boundaries of the lowland rain forest and the "montane forest" (mountain rain forest and bamboo combined) are taken in substance from Keay (1959), but refined in some details on the basis of large-scale local maps and personal observation. (Reproduced from Emlen and Schaller, 1960*a*.)

jor rivers bar the way, the habitat is continuous, and the altitude and physiography are similar to that of the Kasese area where gorillas are numerous. As an explanation I suggest that the spread of the mountain gorilla into much of the lowland forest areas is relatively recent, that the animal is or was until recently expanding its range and has not had enough time to occupy all the forested areas available to it. This view is supported by the present-day pattern of distribution along the north-south axis. With its limited ability to cross rivers, the gorilla probably penetrated southward by circumventing the large, east-west flowing Maiko, Lowa, Oso, and Lugulu rivers near their headwaters; that is to say, the colonization of the forests between the rivers proceeded from the east toward the west. If the gorillas spread from the north toward the south, as I suggested earlier, the westward movement should then have progressed farther in the northern part of their range than in the southern part, for the animals have had more time for colonization in the former area. That this is actually the case is evident from Figure 3.

During our distributional survey, Emlen and I were unable to cover the northern portion of the gorilla range adequately and more populations than those indicated in Figure 3 may well exist in the Maiko River valley. However, about 200 miles of rain forest between Lubero and the Uele River apparently supports no gorillas even though in the past a population must have been there. We can only guess at the reason for the disappearance of the gorillas. It is possible that during one of the drier climatic periods the rain forest north of Lubero was rendered uninhabitable for gorillas and that the animals have not had time to recolonize the area during the succeeding wetter stage. It is also possible that the sparse population, which is usually found in primary rain forest, died out in these relatively undisturbed northern forests while it flourished farther to the south where human disturbance created a more favorable habitat.

DISTRIBUTION OF GORILLAS IN THE VIRUNGA VOLCANOES AND IN THE KAYONZA FOREST In the two most easterly areas of distribution the gorillas exist today in isolated forest pockets completely surrounded by cultivation. In order to trace the route by which they arrived, the history of the human occupation, which has had a marked effect on the vegetation, and the volcanic activity in the region must be taken into account.

Virunga Volcanoes Gorillas occur throughout the six dormant volcanoes, but not on the two active ones to the west. This is rather surprising for the animals almost certainly penetrated the eastern sector by using the active volcanic area as a forest bridge from the slopes of the western rift escarpment. Forest has essentially disappeared from the region immediately to the west of the active volcanoes, but this appears to be a recent phenomenon brought about by heavy agriculture and repeated burning. Dr. J. Hiernaux, Université Officielle du Congo, Brussels, told me that the Bahunde, until recently the most populous tribe in the area, arrived from the east about four hundred years ago, according to their oral tradition. This suggests that a forest connection may have been still intact within the past five hundred years.

The reason for the absence of gorillas on the active volcanoes is a matter for speculation. The vegetation is, on the whole, drier and scrubbier than in the sector of the dormant volcanoes. Furthermore, the forest is irregularly broken and dissected by streams of lava rock. The amount of gorilla forage is small. Fresh lava, which gorillas might hesitate to cross, is covered by scrubby forest within fifty years of deposition.

In general, the forest around the active volcanoes is of a type not favored by gorillas, and I suggest that in the past groups may have found it adequate for penetration but unsuitable for permanent occupation.

Kayonza Forest Only 15 miles separate the eastern end of the Virunga Volcanoes from the Kayonza Forest, and gorillas probably reached this latter area via a forest bridge which has now totally disappeared. The prehistoric vegetation of the Kigezi District in southwestern Uganda appears to have been forest (Lind, 1956), but most of this has been felled by agriculturalists in the past fifty years. The Bakiga and other tribes emigrated from Ruanda into Kigezi as early as 1860. As recently as 1910 forest still covered most of the hills around the town of Kabale but today it has nearly disappeared (Purseglove, 1950).

I suggested earlier that the mountain gorilla had probably colonized various sections of lowland forest relatively recently. The little evidence available from the Virunga Volcanoes and the Kayonza Forest indicates that the invasion of these two areas also occurred not long ago, perhaps within the past several hundred years. If gorillas had penetrated the area earlier one would have expected them to colonize the extensive regions of suitable habitat available to them. Except for one very vague report by Johnston (in Philipps, 1923), there is no concrete evidence for the widespread occurrence of gorillas in southwestern Uganda, although the forests were extensive until the twentieth century. Good gorilla habitat still extends southward along the forested summits of the rift mountains east of

Lake Kivu. These forests were continuous with the Virunga Volcanoes until recently; in fact, before about 1600, when forest clearing and cattle grazing expanded with the invasion of the Watutsi into Ruanda from northern Uganda, rain forest extended almost to the town of Astrida in Ruanda-Urundi (Hiernaux, pers. comm.). Yet the complete absence of records of or references to gorillas in the many reports of explorers and settlers, past or present, strongly suggests that the animals did not occur in the area.

Local Distribution

Gorillas are not randomly distributed through the forest but are characteristically aggregated into isolated population units or concentration areas surrounded by a population of lower density. In some cases the boundaries of these areas coincide with natural barriers, such as rivers and the forest-grassland boundary, but frequently the forest continues unchanged at the boundary of an inhabited area. From a few to several dozen gorilla groups comprise such a population unit.

The distribution of an animal is a dynamic phenomenon and is undoubtedly determined by a combination of physical and environmental factors operating through time.* Population units presumably drift or are pushed about as local conditions change. Our direct observations are limited to the effects of man-made disturbances, but these are probably typical of the influences which have affected the gorillas for centuries.

The most important of these disturbances during the past hundred or more years has been the repetitive clearing of the forest in small patches to create a shifting kaleidoscopic pattern of successional stages of forest regeneration. Gorillas have responded by adjusting their local foraging activities to meet the temporarily favorable conditions created in the recently cultivated fields. There is also evidence of the expansion or movement of gorilla population units into developing agricultural or mining regions. The large concentrations of gorilla areas around Kasese and south of Mwenga correlate with a heavy influx of native families into these mining areas in recent times. A road engineer east of Kasese found gorillas moving in around the workers' villages and garden plots while preparatory construction work was still in progress, and several mining engineers told of gorillas appearing around new camp sites within a

* The following discussion is taken almost verbatim from Emlen and Schaller (1960*a*).

year or two of their establishment. Such incompletely documented anecdotes may represent instances of returns after initial displacement rather than actual invasions, but in either event they suggest a certain fluidity of movement.

The long-standing government policy of resettling natives from the deep forests on relatively narrow strips along the roadways probably accounts in part for the present concentration of gorillas along these human thoroughfares and their relative scarcity in the hinterlands. The animals have apparently moved into these disturbed areas and remained despite persistent harassment. Gorillas were abundant along sections of the Utu-Kasese road where natives had created much secondary forest, but they were distinctly scarce in that part where villages had existed for only three years or less and the primary forest had only recently been cleared from the vicinity of the road. A long-established community surrounded by fields may thus form a nucleus for a gorilla population.

Further evidence for the role of rotational forest clearing in gorilla movements is found in historical records of evacuation by gorillas following the emigration of human populations. The first consequence of an exodus of human beings may be an improved environment, freed of direct human disturbance. Thus the village of Mangombe near Kima still had a large local population of gorillas three years after it was abandoned as a mining camp and allowed to grow up to a jungle of shrubs and vines. A former resident reported that "the gorillas from the surrounding shambas moved into the village as the people moved out." Abandoned mining centers like those near Lutunguru and Alimbongo apparently had good gorilla populations during and immediately following their period of activity and have subsequently lost them.

The old Arab slave route from Angumu to Walikale resulted in vegetational disturbances which can still be detected by foresters and which apparently drew gorillas in from the surrounding forests. Reports reaching government surveyors told of gorillas persisting in this area for many decades, but current reports indicate that the animals have now largely disappeared.

The importance of rotational agricultural activity in creating secondary forest, which is the favored habitat of gorillas, cannot be overemphasized in discussing the ecology of this ape. I think it probable that one reason for the nearly continuous gorilla population between the Lowa and Lugulu rivers in the Utu region is that much of the primary forest in the area was felled in the recent past when human activity was more widespread.

Stragglers

Stragglers are somctimes found as far as twenty or more miles from the nearest gorilla population and occasionally in habitat quite atypical for the species. Most of the wandering animals are lone subadult or adult males, but small groups occur also. Natives remember encounters with such stragglers for years because of their rarity. In 1956, for example, a male remained two weeks at the edge of a dairy farm in the hills southwest of Bukavu. Another lone male injured a native on the Bukavu-Walikale road twenty miles from the nearest known gorilla population. In the bamboo of the rift mountains at the northern end of Lake Kivu gorillas are rare and the local Batwa hunters had vivid memories of a small group and four lone males which they had encountered over a period of ten years. An isolated subadult male penetrated into the fields south of Mt. Tshiaberimu and remained there until killed in a wattle tree grove on May 9, 1959. Some other records of stragglers are indicated in Figure 3, and the supplement to our distribution paper (Emlen and Schaller, 1960*a*) lists our numerous records of them.

The erratic wandering of these scattered individuals and small groups is probably the principal method by which new territory is colonized. If a group settles sufficiently far from neighboring populations, it may form the nucleus of a new population unit more or less completely isolated spatially from other such units.

Ecology and Description of the Gorilla Habitats

Although the range of the mountain gorilla is small, it possesses considerable ecological diversity. The animals inhabit several major types of forest which are stratified altitudinally from the Congo basin eastward to the summits of the mountains bordering the Albertine rift. The purpose of this section is (1) to present a general description of the major forest types frequented by gorillas and (2) to provide a detailed account of each area in which gorillas were studied intensively.

General Description of the Major Forest Types

Keay (1959) recognizes two major forest types in the range of the mountain gorilla (Fig. 4): the montane forest and the moist evergreen forest at low and medium altitudes. I find it convenient to use the following three major types, which include all habitats permanently occupied by gorillas.

Forest type	Approximate altitude (in feet)
Lowland rain forest (or moist evergreen forest)	to 5,000
Mountain rain forest (or moist montane forest)	5,000 to 11,500
Bamboo forest	8,000 to 10,000

About 75 per cent of the mountain gorilla's range lies in the lowland rain forest of the Congo basin. The upper limit of this forest is irregular, but, following Chapin (1932), I place it at about 5,000 feet to separate it from the physiognomically similar mountain rain forest. Lebrun (1935) distinguishes a transition zone between the two types, roughly between 4,000 and 5,000 feet.

The mountain rain forest and the bamboo forest grow primarily in the rift mountains and in the Virunga Volcanoes. The transition from one type to the other may be abrupt or gradual, and the altitude at which it occurs varies from area to area. Sometimes stands of forest grow scattered throughout the bamboo, and in the Virunga Volcanoes and at Mt. Tshiaberimu subtypes of mountain forest are also found above the bamboo zone.

Even though gorillas frequent floristically diverse types of forest, the habitats utilized by them are similar in being lush and damp with an abundance of forage near ground level throughout the year. The animals have shown considerable ecological adaptability within the subtypes and seral stages of the various habitats, but they have remained entirely within the moist forest zones.

LOWLAND RAIN FOREST Lowland forest covers the central Congo basin from the west coast of Africa eastward for some 1,500 miles to the base of the rift mountains. It is a vast area with seemingly endless forests, broken by an irregular and intricate patchwork of small, ephemeral native fields.

I made most of my observations on the vegetation along a hundred miles of road between the towns of Kabunga (28°10′E, 1°40′S) and Kasese (27°7′E, 1°38′S), and in an expanse of uninhabited forest some fifty miles in width between the mining camps of Utu (27°55′E, 1°37′S) and Niabembe (27°44′E, 2°15′S). These two areas appear to be typical of the lowland forest region as a whole and include numerous samples of the seral stages of regeneration from cultivated field to mature forest. Methods of vegetation analysis were limited to qualitative description. Specimens of the dominant plants were collected and later identified either by the I.R.S.A.C. station at Lwiro or the East African Herbarium in Nairobi.

The clearing of land practiced in the lowland forest region is of the slash and burn type, which results in an irregular mosaic of forests in various stages of regeneration. Large trees are felled eight to fifteen feet above ground, leaving the buttresses standing (Pl. 6). After the smaller trees, shrubs, and herbs have been cut and burned, mountain rice (*Oryza*), banana, and manioc (*Manihot*) are planted in the shallow-turned soil between the fallen trees. Since humus decomposes rapidly on direct exposure to the sun and the mineral contents are quickly leached, the plot is usually abandoned after three or four years. A luxuriant secondary growth of "weeds" springs up, and develops rapidly through a series of stages until after eighty years the vegetation again resembles primary forest (Chapin, 1932). Where population pressure is high, however, the regenerated forest on old fields is often recut every fifteen years and sometimes every twelve years.

Emlen and I obtained descriptive notes on former fields abandoned for a known number of years. Our data supplement the general discussions on secondary succession found in such publications as Chapin (1932), Robyns (1948*b*), Lebrun and Gilbert (1954), and Ross (1954). In general, succession follows about six main stages:

I Cultivation stage: Bananas form a fairly irregular canopy about 15 feet in height below which shrubby manioc usually grows in profusion.

II 0–1 year after abandonment: Numerous herbs, grasses, and vines invade the understory.

III 2–3 years: Saplings and shrubs become prominent among the herbs and vines, all forming an extremely dense tangle which gradually crowds out the bananas.

IV 4–20 years: Trees are conspicuous, especially the umbrella tree (*Musanga cecropioides*), which dominates until about 20 years of age when it is 50 to 60 feet tall (Ross, 1954).

V 20–50 years: The forest trees grow taller, but the shrubs, vines, and herbs become progressively more sparse under the high canopy.

VI 50+ years: The forest takes on more and more the character of primary forest with an upper canopy about 120 feet above ground, several indistinct lower tree strata, and a sparse ground cover.

Little precise information on rate of growth, species composition, and changes in composition of secondary forest on a year-to-year basis is available. The general information presented below is based largely on my field notes.

Although mountain rice is frequently planted for one year in a newly cleared field, bananas and manioc constitute the major food plants of the Bantu agriculturalists in the area. Bananas require about eighteen months from the time the shoot is planted until the harvest of the fruit (Troupin, pers. comm.). Usually the tree is then cut and a new shoot set out. In order to utilize the clearing to the full, natives frequently grow manioc beneath the banana trees, thus creating a maze of intertwined branches to a height of 6 to 7 feet.

If the field is to be abandoned within a year, usually no attempt is made to halt the invasion of grasses and herbs, with the result that secondary forest succession is in progress while the harvest continues. Native women still dig up manioc tubers in fields densely overgrown.

At two to four years bananas have almost been crowded into obscurity by the rapidly growing saplings of *Terminalia superba, Musanga cecropioides,* and *Caloncoba welwitschii,* which form a very irregular canopy some 10 to 25 feet in height. The lower stratum consists of an almost impenetrable tangle of *Aframomum, Costus, Megaphynium,* and *Marantochloa,* all herbs characteristic of disturbed situations. *Mikania,* a climbing composite, and *Olyra latifolia,* a climbing grass, entwine the herbs as well as the shrubby manioc.

At five years *Musanga cecropioides* has emerged as the dominant tree species, reaching a height of about 40 feet (Lebrun and Gilbert, 1954). Other such trees as *Caloncoba welwitschii, Terminalia superba, Croton mubango, Macaranga spinosa, Maesopsis eminii,* and *Myrianthus arboreus* form an irregular second canopy at a height of 10 to 20 feet, below which herbs, vines, and shrubs grow in such profusion that strata are not distinguishable. Bananas have largely been choked out, but some manioc survives.

At seven to ten years *Musanga cecropioides* and *Terminalia superba* remain as the dominant tree species, the former reaching a height of sixty feet, the latter forty feet. Both species produce a rather thin canopy over a lower discontinuous tree stratum which is about twenty to twenty-five feet high. Included in this stratum are not only such secondary forest genera as *Trema, Macaranga, Chlorophora, Harungana, Croton, Polyscias,* and *Ficus,* but also saplings of those species which ultimately will achieve dominance in the forest. The lower strata below a height of ten feet present a seemingly inextricable tangle. Tall slender stems of *Vernonia conferta,* with tufts of enormous leaves at the apex, are conspicuous. The ubiquitous herbs *Aframomum, Costus,* and *Megaphynium,* the shrubby fern *Marattia,* various shrubs, and numerous vines all grow in profusion. The ground cover of herbs and grasses is often sparse with the exception of a few ferns and the large-leafed *Palisota.*

At twenty to thirty years the dominant *Musanga* dies out (Ross, 1954), being replaced by other trees which produce a more or less continuous canopy about sixty feet high. Although many of the secondary forest trees persist, the dominant species of the primary forest have also become prominent. Numerous saplings form a poorly defined lower stratum.

As the height of the trees increases and the canopy becomes more continuous, less and less direct sunlight penetrates to the ground, with the result that many of the vines, shrubs, and herbs cannot survive in the gloom. Consequently the dense mass of vegetation near the ground thins out considerably, except for *Megaphynium macrostachyum* and various such ferns as *Marattia.* Only in those places where the canopy is broken does the luxuriant vegetation remain.

At sixty or more years it is difficult for an untrained observer to distinguish secondary stands from primary forest. It seemed to me that primary forest in the areas which I visited was limited to a few isolated stands and that old secondary forest covered most of the terrain. The following description includes both types. The trees are 120 to 180 feet high and the canopy almost continuous. According to Lebrun and Gilbert (1954) the trees are nearly always mixed as to species, the characteristic ones being *Brachystegia laurentii, Diogoa zenkeri, Diospyros* sp., *Heisteria parvifolia, Isolona thonneri, Maba* sp., *Pavetta tetramera, Polyalthia suaveolens, Psychotria brevipaniculata, Schaudtia stipitata,* and others. The trees in the sub-canopy are rather sparsely foliaged and extend from a height of 60 to 70 feet almost to the ground. Lianas entwine the branches of the trees and epiphytes are common.

The forest near groundlevel is relatively open. Although sunshine filters to the ground in places, the amount is usually insufficient to support a herbaceous layer. Various saplings such as *Pycnanthus angolensis* and occasional shrubs grow scattered in the shade. Much of the ground is either bare or covered with a carpet of dead leaves. The most prominent survivors from the young secondary forest appear to be the cabbage-like *Palisota,* the fern *Marattia,* and a broad-leafed *Graminae.*

Sometimes, however, the canopy is broken, perhaps by the death of a forest giant (Pl. 7). Many herbs immediately invade the open, sunny glade. *Aframomum, Costus,* numerous vines, and many other species quickly establish themselves and a new succession is initiated as a small island in the old forest.

MOUNTAIN RAIN FOREST Mountain rain forest differs from the lowland rain forest primarily in the somewhat smaller size of the trees, fewer lianas, and the presence of gymnosperms. We visited extensive stands of mountain forest west of Lake Tanganyika between the towns of Mwenga and Fizi and in the Kayonza Forest of Uganda. Several subtypes of mountain rain forest of special importance in this study occur in the Virunga Volcanoes and on Mt. Tshiaberimu. These will be discussed fully in conjunction with descriptions of these study areas.

Cultivation has considerably altered the distribution of the mountain rain forest. In areas where the human pressure is low the cycle of rotational forest clearing and the subsequent stages of regeneration are essentially similar to those described for the lowland rain forest, making further discussion unnecessary. However, in areas with a high human population, the forest has frequently been completely eliminated over large areas. The rich soil is suitable for long-term cultivation in many parts with the result that villages are fairly permanent and the same plot continues to be tilled year after year. Abandoned fields often revert to grassland instead of forest, the secondary succession being deflected either by repeated grazing or by fire.

Mature stands of mountain rain forest have an upper canopy 80 to 140 feet high, which tends to be continuous on the more gentle slopes and ridges but incomplete on bluffs and in narrow valleys. Tree species are numerous, including *Newtonia hildebrandtii, Chrysophyllum* sp. and *Carapa grandiflora.* Several such trees as *Polyscias fulva, Neoboutonia macrocalyx,* and *Hagenia abyssinica* grow commonly along roads, indicating disturbed situations (Lebrun and Gilbert, 1954). *Galiniera coffeoides,* and *Alchornea hirtella* are prominent among the trees that form the indistinctly defined lower stratum. Where the tree canopy is continuous, herbs and shrubs are sparse, although saplings frequently create dense stands. As in the lowland rain forest, dead leaves cover the ground together with scattered ferns and such herbs as *Palisota.* The more open valleys and other locations with a discontinuous canopy exhibit a dense tangle of vegetation, with the shrub *Mimulopsis arborescens,* the vine *Urera hypselendron,* and the herb *Aframomum* sp. being the most characteristic species. Tree ferns (*Cyathea*) are also common on the lower slopes and along streambeds.

BAMBOO FOREST Bamboo (*Arundinaria alpina*) forms a definite vegetation zone in the mountains from an altitude of about 7,500 or 8,000 feet upward to about 10,000 feet. Unfortunately no detailed ecological studies have been made of bamboo in Central Africa, and it is not known what position this grass occupies in the vegetational succession. In some localities isolated patches of mountain forest break the expanse of bamboo; in others single tall trees of *Neoboutonia macrocalyx* and *Polyscias fulva,* both characteristic of secondary forest, are prominent.

The physiognomy of bamboo varies considerably from area to area but two main types can be recognized: (a) tall stands reaching a height of 25 to 35 feet; and (b) stunted, scrubby stands, heavily overgrown with vines, and rarely exceeding 20 feet in height.

a) Tall stands of bamboo grow primarily on gently sloping to flat terrain. The shallow and rather translucent though continuous canopy varies from fairly open to quite dense. In the more open stands the vertical stems may reach a diameter of 3 inches and grow far enough apart so that progress by foot is not appreciably hindered. The groundcover under these conditions consists primarily of grasses and an array of horizontal and sub-horizontal bamboo stems (Pl. 8). The canopy of a dense stand permits only suffused light to filter through, with the result that much of the soil is merely covered by a layer of dead leaves and occasional patches of dark-green *Selaginella.*

b) On steep slopes the irregularly clustered vertical and sub-vertical bamboo stems are often spindly, bent, and interwoven. Such vines as *Urera hypselendron, Mikania,* and *Clematis* cling to and even cover the clumps of bamboo, creating a quite inextricable tangle. The groundcover is sparse or absent when the canopy is continuous, but scrubby bamboo stands are characteristically broken by small clearings and trails which support various grasses and herbs.

Bamboo usually reproduces by means of vegetative shoots. The abundance of bamboo shoots varies seasonally, a fact which is indicated in Figure 5 for the Virunga Volcanoes. During 1959–60, shoots were generally most abundant during the season of heavy rains from October through December, and absent during the height of the dry season in July and August.

Descriptions of the Main Study Areas

Our intensive studies were concentrated in a few selected localities which encompassed samples of all the major forest types inhabited by the apes. The length of time which we spent in each study area varied from one week to over ten months, with the result that the following descriptions are either somewhat cursory or quite detailed depending on the intensity and duration of our work. For each area I present a brief account of the terrain, climate, vegetation, and utilization by gorillas.

UTU REGION The Utu region lies in the vast expanse of lowland rain forest between the Lowa and Lugulu rivers in the Congo basin. In the eastern portion, near the rift mountains, the terrain is steeply rolling, but toward the west the hills grow lower and lower until they disappear almost entirely in the vicinity of the town of Kasese.

During both my visits to the Utu region, in June–July, 1959, and August–September, 1960, it was humid and hot. The mornings were usually clear, but clouds gathered by noon, followed often by a

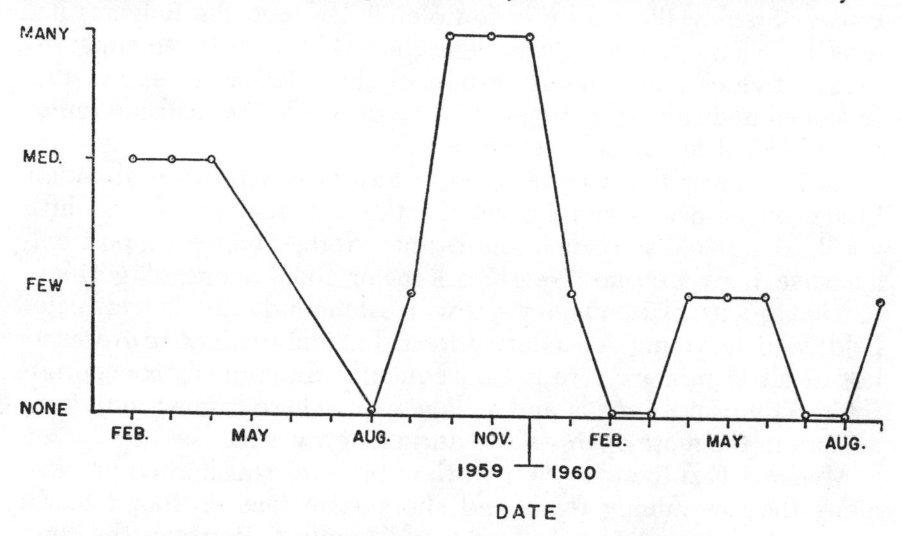

FIG. 5.—The relative abundance of bamboo shoots in the Virunga Volcanoes (Kabara and Kisoro combined) during 1959 and 1960.

heavy downpour. My wife and I took temperatures intermittently. Our lowest reading was 20°C. (68°F.) at 0600 and the highest was 30°C. (86°F.) at 1600 near Utu, which lies at an altitude of about 2,400 feet. The I.R.S.A.C. station at Irangi (roughly 1°40′S, 28°20′E), situated at an altitude of about 2,600 feet, recorded in 1959 a daily mean temperature of 21.8°C. The amount of precipitation at Irangi in 1959 was 2,320 mm. (58.9 inches); Yangambi (0°45′N, 24°34′E), which lies near Stanleyville at an altitude of 1,380 feet, recorded an average of 64.7 inches over an eleven-year period (Chapin, 1932).

The vegetation in the Utu region reflects the historic changes in the distribution of human occupation. Today nearly all villages cluster along or within a few miles of the sparse network of roads that transects the forests. Between the towns of Utu and Kasese, for example, villages composed of mud huts are encountered along the road every two to three miles. A patchwork of cultivated fields and secondary forest in various stages of regeneration surrounds each village (Pl. 1). However, large areas which today are completely uninhabited also support much secondary forest. During a hike that took me over fifty miles southward from Utu, I passed not a single village, but palm trees and relatively open valleys indicated that cultivation had been practiced throughout the region in the not too distant past. Apparently villages were sparsely scattered throughout the forests until just before the first world war when the Bukavu-Stanleyville road was completed. By 1920 the Belgians had resettled many of the natives near the roads in order to supervise their activities more closely. Much of the interior forest has thus remained undisturbed for over forty years, while the roadside forests are subjected to frequent recutting.

Gorillas favor those parts of the forest which provide abundant forage on or near ground level. In the Utu region I found little gorilla sign in old secondary and primary forests which support only a sparse herb stratum. Nearly all major food plants of gorillas—banana, *Ficus, Aframomum*—grow most abundantly in cultivated fields and in young secondary forest, but only thinly scattered or not at all in primary forest. Consequently the animals concentrate their activity near roads and villages, or, where human activity is absent, in the more open valleys and along river courses.

MWENGA-FIZI REGION Our work in this area was limited to travel along the few mining roads and short excursions on foot into the surrounding mountain rain forest and bamboo. Between the town of Fizi and the Elila River, a distance of about sixty air miles, the topography is rolling. Extensive grasslands break the continuity of the forest, creating numerous forest islands. Along the road the percentage of grassland to forest changes from mile to mile, but over all it is about fifty-fifty. From the Elila River northward the mountains become rugged and the grassland ceases. Except for some mining camps and a few villages, the area is uninhabited until just south of Mwenga where the altitude drops and cultivation becomes extensive.

Mountain rain forest, the physiognomy of which has already been described, covers the more rugged portions of the terrain. The canopy is continuous except where the steep slopes are broken by rock bluffs. Above 8,000 feet the mountain forest gives way to bamboo, which caps the high ridges. Secondary succession of the forest along disturbed roadsides and on former fields is similar to that described for the lowland rain forest.

We found gorilla sign most abundantly in secondary forest and in the open river valleys which were heavily overgrown with tree ferns, vines, and herbs. Bamboo was also utilized.

MT. TSHIABERIMU Near the northern end of Lake Edward the rift mountains rise to an altitude of 10,000 feet before dropping abruptly in elevation. This terminal mountain massif is part of Albert National Park, which dips briefly away from the shores of Lake Edward to bring a section of the rolling uplands under its protection.

The weather during our visit in May, 1959, was cool and wet. Seven temperature readings at 0600 by our campsite, which lay at an altitude of 8,600 feet, gave a mean of 10.8°C. (9° to 12°). Our highest recorded temperature was 19.5°C. Rain fell on twelve of the sixteen days, and thirteen days were cloudy.

The lower slopes of the escarpment facing Lake Edward are rather dry and barren, being covered with bracken fern, grass, and open woodland. Cultivation is heavy in parts. The fertile uplands on top of the rift have been converted into fields so extensively that cultivation borders the park boundary on three sides. The only large tract of undisturbed forest extends southward from Mt. Tshiaberimu for about twenty miles.

The dominant vegetation on Mt. Tshiaberimu and along the steeply undulating hills to the south consists of bamboo, which covers the land with even soft-green stands 20 to 30 feet high. Only screened light penetrates to the ground. An understory is almost absent except for a few scattered shrubs of *Mimulopsis arborescens* and the herbs *Impatiens* and *Laportea.* During rains the soil is unbelievably slippery and the numerous streams are full, flowing beneath the solemn canopy of bamboo.

Where the canopy thins out or gives way to little clearings, numerous vines, herbs, and shrubs spring up. *Smilax kraussiana, Physedra bequaertii, Rumex ruwenzoriensis* and *R. bequaertii, Philippia trimera, Urera hypselendron, Sonchus schweinfurthii, Gunnera perpensa, Canthium venosum, Erlangea globosa,* and other species occur abundantly in the open glades. Small marshy sedge-grass meadows, surrounded by *Usnea*-laden *Hypericum* trees, nestle along the northern base of Mt. Tshiaberimu.

Lone *Podocarpus milanjianus* trees tower above the bamboo to a height of 70 feet throughout the area. Isolated patches of mountain rain forest huddle in protected valleys, along some river beds, and on a few slopes. At the northern and western sides of the Tshiaberimu massif, and along the upper edge of the escarpment facing Lake Edward, a band of forest grows below the zone of bamboo. Much of this forest is somewhat scrubby, only 40 to 60 feet high, although a few tall stands exist. Such species as *Pygeum africanum, Podocarpus milanjianus, Hypericum lanceolatum, Grumilia megistosticta, Macaranga kilimandscharica, Galliniera coffeoides,* and *Xymalos monospora* are prominent. Tree ferns (*Cyathea* sp.) are not abundant. *Hypericum lanceolatum* grows also near the summit of Mt. Tshiaberimu together with grasses and sedges, some wild celery (*Peucedanum*), and the shrub *Pycnostachys goetzenii.* On the peak an almost pure stand of tree heath (*Erica*) reaches a height of 10 feet.

Gorillas were feeding primarily on bamboo shoots during our short visit. I found no gorilla sign on the steep escarpment facing Lake Edward. During the dry season, when bamboo shoots are absent, the animals probably concentrate their activity in the clearings and in the patches of mountain forest.

KAYONZA FOREST At the Congo-Uganda border between Lakes Edward and Kivu the eastern escarpment of the rift rises abruptly from the valley floor, forming a rugged mountain area covered with forest. The terrain is steeply undulating, with a maze of ridges and valleys aligned in every direction, so that a person unfamiliar with the country and lacking a compass is easily lost (Pl. 2). The precipitous narrow valleys, often harboring small streams, lie at an altitude of about 5,500 feet; but the ridges, whose crests are usually narrow, rise steeply to 6,500 and 7,000 feet and occasionally as high as 8,000 feet.

During our visits to the forest the days were warm and the nights cool. Between April 17 and 27, 1959, the days were partially cloudy with intermittent heavy showers (totaling about fifteen hours) on seven days. Between June 9 and 19, 1960, scattered clouds usually appeared by mid-morning but no rain fell during the period except for one brief sprinkle. Temperatures near our camp, at an altitude of about 6,700 feet, were:

Time	No. of Readings	Range	Mean
0600..........	10	12.0–15.5	14.1
1200..........	6	20.0–26.5	23.8
1600..........	9	18.0–23.0	21.2
1800..........	9	15.0–18.5	16.7
2000..........	8	15.0–16.0	15.2

Our camp was located on a knoll from which the tall trees had been removed. Temperatures in the valleys and under the canopy are undoubtedly somewhat lower.

The town of Kabale, which lies at an altitude of 6,140 feet and about 25 miles southeast of our Kayonza Forest camp, recorded the following temperatures and precipitation over a forty-one-year period (Saben's Handbook of Uganda, 1959):

Mean annual maximum....................	23.3° C.
(average of highest temperature each day)	(74° F.)
Mean annual minimum....................	10.0° C.
(average of lowest temperature each day).	(50° F.)
Average precipitation....................	41.0 inches
Wettest month—April....................	5.4 inches
Driest month—July....................	0.9 inch

The main block of the Kayonza comprises a forest reserve with an area of 96 square miles. I visited the eastern, western, and central parts of the reserve, and the following description of the mountain rain forest is based on my field notes.

In the eastern portion the valleys are relatively broad and the hills much lower than in the other parts of the forest. Large trees are few in number, most being 80 feet or less high. Some of the open valleys seem to have been disturbed in the past by logging. *Chrysophyllum, Olea, Olinia, Pygeum,* and *Polyscias* are some of the conspicuous tree genera. *Mimulopsis arborescens* is the dominant shrub in the valleys and it, together with numerous vines, other shrubs, and herbs, creates an extremely dense stratum to a height of 8 to 10 feet. A few small patches of bamboo occur there.

In the western portion many of the ridges and some of the slopes near the Congo-Uganda border have been cleared for cultivation in the past and these are now covered with lone trees of *Polyscias, Pygeum, Olea,* and *Strombosia. Mimulopsis arborescens, Aframomum,* numerous vines, and grasses cover the ground.

In the central portion a few slopes were cleared in the past and now support little more than bracken fern and some scattered *Myrianthus arboreus, Xymalos monospora,* and *Myrica kandtina* trees. However, much undisturbed forest remains. On the ridges, trees reach a height of 120 to 140 feet and only intermittent sunlight penetrates the canopy of such species as *Entandrophragma* sp., *Newtonia hildebrandtii, Chrysophyllum* sp., *Podocarpus* sp., and *Carapa grandiflora.* Ill-defined lower tree stories extend from a height of about 80 feet almost to the ground. Prominent among the saplings and trees of the lower strata are *Galiniera coffeoides, Coffea spathicalyx, Derplatzia lutea,* and *Alchornea hirtella.* Although the branches of the shrubs and saplings entwine at times so intricately that one can walk over them several feet above ground, the upper slopes and ridges are usually relatively open and hiking there is an easy matter (Pl. 9). Groundcover is sparse—a few scattered herbs like *Palisota* and ferns like *Dryopteris* and *Conchitis.* A thick layer of dead leaves covers the ground.

On the lower slopes and in the valleys, large trees are scattered or absent. Tree ferns (*Cyathea deckenii*) are common, reaching a height of 20 feet. They sometimes form such dense stands that their canopy excludes nearly all direct light except for occasional shafts of sun that penetrate the gloom to reveal a chaotic jumble of dead stems. Away from these dense stands, shrubs, herbs, and vines grow in such profusion as to nearly justify the area's popular name of "Impenetrable Forest" (Pl. 10). *Mimulopsis arborescens* and *Brillantaisia nyanzorum* are mixed with tree ferns, *Vernonia,* and *Rubus pinnatus.* The shrubs, young trees, fallen logs, and stumps are overgrown with masses of such vines as *Urera hypselendron, Momordica foetida, Fleurya ovalifolia, Mikania cordata, Basella alba, Piper capense, Culcasia scandens* and *Senecio syringifolius.* Among the herbs are numerous yellow-flowered composites, a few lobelias, *Rumex usambarensis, Impatiens,* and *Aframomum* as well as several broad-leafed grasses and lianas which trail from the lower branches. Thus the ground is covered completely by one highly irregular stratum varying from 5 to 15 or 20 feet in height.

Gorillas utilize the whole forest although they appear to be scarce in the western and southern parts. In the central portion they follow a fairly set though not invariable pattern of activity, feeding on the lower slopes, where herbs and vines are abundant, and bedding down for the night on the upper half of the slopes where the numerous saplings furnish excellent materials for nest construction. They cross ridges to move from one valley to another, but, because of the scarcity of food, rarely tarry on them.

THE VIRUNGA VOLCANOES The Virunga Volcanoes (also called Kivu, Mfumbiro, Birunga, or Kirunga volcanoes) constitute the classical home of the mountain gorilla, for the subspecies was discovered on their slopes and the animal has been more frequently visited and thoroughly studied there than in any other habitat. The volcanic range comprises eight peaks which span the bottom of the Albertine rift transversely from east to west just to the north of Lake Kivu. The total extent of the range is about 48 miles (29°2′E to 29°42′E) and forms in part the international boundary between the Congo, Uganda, and Ruanda-Urundi. All of the peaks are included in Albert National Park with the exception of parts of the three eastern ones which belong to Uganda.

The Virunga Volcanoes are divided geographically into the eastern, central, and western sectors (Table 6). Numerous names and spellings exist for the various peaks, and the given altitudes vary from source to source. Names, spellings, and altitudes are listed as given by Meyer (1955), except for the altitudes of the volcanoes which lie partially in Uganda. For their height I have consulted the 1:50,000 Uganda Lands and Surveys map of 1957. Gorillas are limited to the six eastern and central volcanoes covering an east-west

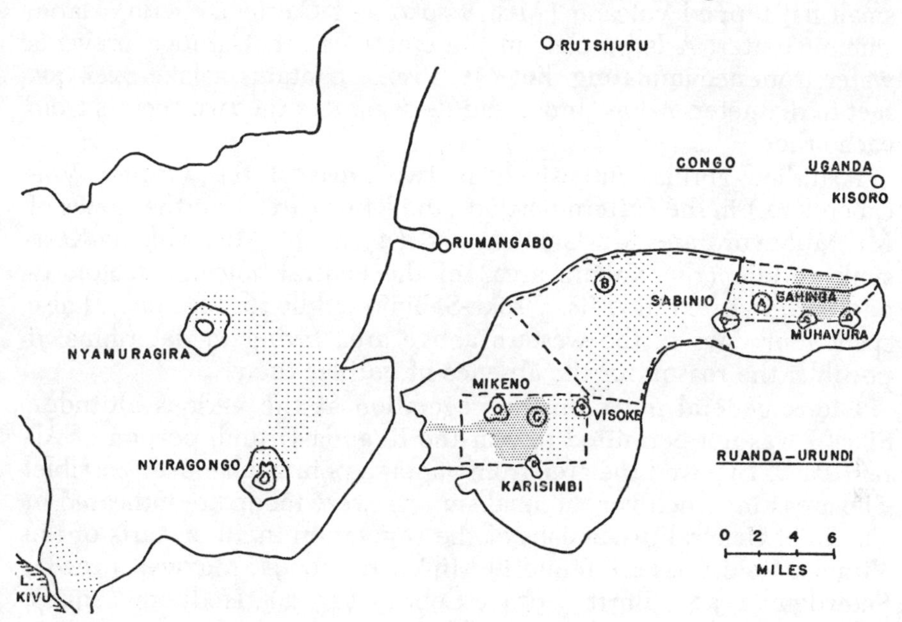

FIG. 6.—The southern sector of Albert National Park, Congo, which comprises the Virunga Volcanoes.

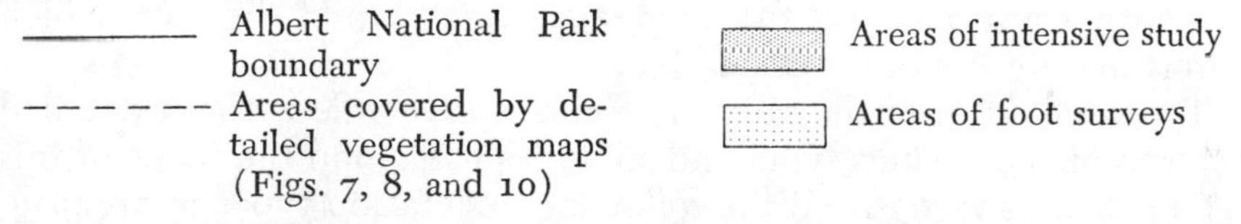

distance of about 25 miles (29°20′E to 29°42′E) and a north-south distance of 4 to 12 miles (roughly 1°20′S to 1°36′S). Forest is continuous along the whole chain of mountains although it narrows to a mile-wide strip at the site of the Goma-Rutshuru road in the saddle between the western sector where no gorillas occur, and the central sector (Fig. 6).

The two eastern sectors, comprising three volcanoes each (Pl. 11), show a striking similarity to each other. Each exhibits one old volcano (Mts. Mikeno and Sabinio) from which the outer walls have eroded away, leaving only the central precipitous rock plug carved with deep canyons. Each has one tall perfect volcanic cone (Mts. Karisimbi and Muhavura) on which the summit is dimpled with a small crater. The one on Mt. Muhavura is filled with water and is about 60 to 80 feet in width. Mt. Karisimbi has also the large subsidiary Branca Crater which lies at a lower altitude to the east of the main cone in Ruanda-Urundi. And finally each group has a small flat-topped volcano (Mts. Visoke and Gahinga) with a large summit crater. A large cleft in the crater of Mt. Gahinga prevents water from accumulating, but Mt. Visoke contains a lake over 400 feet in diameter. A low, broad saddle separates the two sectors from each other.

I studied gorillas intensively in two areas of the Virunga Volcanoes: (1) in the eastern portion (the Kisoro area) on the slopes of Mt. Muhavura and Mt. Gahinga and (2) in the Mts. Mikeno-Karisimbi saddle (the Kabara area) of the central volcanic sector. In addition, I visited the Mts. Visoke-Sabinio saddle for four days. I also spent four days in the western active area trying to determine, if possible, the reason for the absence of gorillas there.

I took general notes on the vegetation at the various altitudes. Since I was not permitted to visit the Ruanda-Urundi portion of Albert Park, I viewed the area from vantage points whenever possible. The areas in which vegetational surveys were made are indicated in Figure 6. General discussions of the vegetation in all or parts of the Virunga Volcanoes are found in Milbraed (1909), Maxwell (1928), Snowden (1933), Burtt (1934), Lebrun (1942), Hedberg (1951), Blower (1956), Donisthorpe (1958), and several others. An inventory of species for Albert Park is given by Robyns (1947–55), and this author also presents the only vegetation map of the area, which unfortunately is quite inaccurate.

The vegetation of the volcanic chain is diversified, and a detailed analysis of the various types and subtypes is beyond the scope of this report. I have recognized the following vegetation types in my principal study areas:

Mountain Rain Forest
- Mountain woodland
- Dry colonizing woodland
- *Hypericum* woodland
- *Hagenia* woodland

Meadows
- Long grass meadow
- Short grass meadow
- Herb meadow

Lava
Bamboo
Tree Heath
Giant Senecio
Alpine

The Kisoro study area The precipitous northern slopes of Mts. Sabinio, Gahinga, and Muhavura and the saddles between the peaks, an area of 9 square miles, have been set aside by the Uganda government as a forest and game reserve for the protection of the gorillas. The rocks are primarily of leucite, basanite, and lavas, covered in the Mts. Gahinga-Sabinio saddle with Pleistocene boulder deposits. Soils are of the potassic ash and lava type (Watt, 1956).

From March 26 to April 7, 1959, we stayed in the Mts. Muhavura-Gahinga saddle at an altitude of about 9,700 feet. The mornings and evenings were usually clear but around midday clouds tended to obscure the peaks. The lowest temperature recorded was 4.5°C., the highest 17°C. Heavy but usually short showers fell on seven of the thirteen days. The average amount of precipitation over a fifteen-year period at Kisoro, which lies at an altitude of 6,200 feet about 7 miles from the base of the mountains, was 66.06 inches (Appendix B, Table 61).

Four types of forest grow on the slopes of the mountains in the Kisoro area. Figure 7 presents a rough sketch of the distribution of the vegetation zones.

Mountain woodland: Most of this vegetation type lies below an altitude of 9,000 feet and outside the forest reserve. It is being extensively cut at present and will undoubtedly disappear within the next few years. A patch occurs also above the bamboo zone on the slopes of Mt. Muhavura between about 10,200 and 11,000 feet. The forest is quite dense. *Hagenia abyssinica, Hypericum lanceolatum, Peddiaea* sp., *Pygeum africanum, Ilex mitis, Nuxia congesta, Agauria salicifolia, Pittosporum spathicalyx, Bersama abyssinica, Rhamnus prinoides, Xymalos monospora,* and others produce a fairly continuous canopy of evergreen foliage at a height of 40 to 60 feet. A poorly defined tree layer below this canopy consists of saplings of the above and such additional species as *Maesa lanceolata* and *Cluytia* sp. Shrubs like *Erlangea globosa, Pycnostachys goetzenii,* and *Vernonia adolfi-frederici* are common and reach a height of 20 feet. Patches of bamboo (*Arundinaria alpina*) occur throughout. Logs, shrubs, stumps, and the lower branches of trees are overgrown with various vines such as *Clematis, Mikania cordata, Urera hypselendron, Galium simense, Stephania abyssinica,* and *Crassocephalum bojeri. Impatiens, Senecio, Peucedanum, Rumex, Laportea, Cynoglossum,* and numerous other herb genera cover the ground densely though irregularly to a height of 4 to 5 feet. In general the forest growth is quite luxuriant and so profuse that progress by walking is slow and visibility poor.

Some of the woodland trees like *Pygeum*, *Pittosporum*, *Hagenia*, and *Hypericum* create tall stands up to about 70 feet in height along the rims of the various canyons on Mt. Sabinio. These groves are

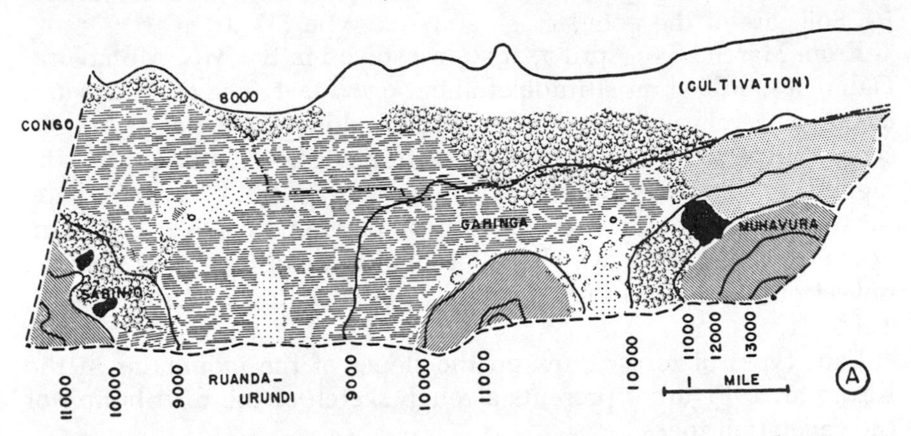

FIG. 7.—A generalized vegetation map of the Uganda portion of the Virunga Volcanoes.

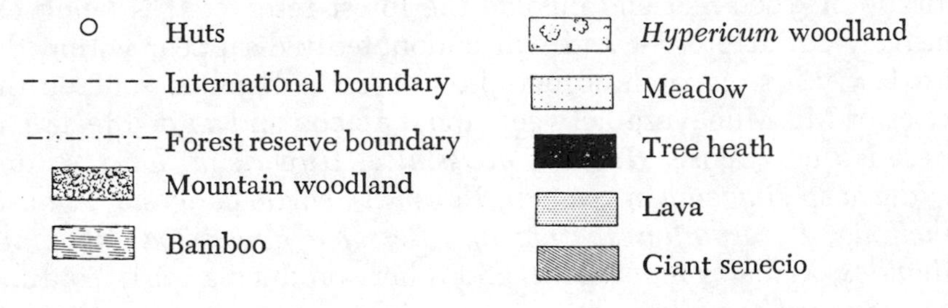

frequently quite open beneath the canopy and the ground is covered primarily with grass, although shrubs are prominent in some places.

Bamboo: Bamboo covers the slopes of Mts. Gahinga and Sabinio between about 8,500 and 10,000 feet, but it barely reaches Mt. Muhavura. Over much of the area bamboo grows in dense, almost monotypic stands to a height of 15 to 20 feet, broken here and there by lone trees or small clumps of *Neoboutonia macrocalyx*, *Pygeum*, or other trees found in the forests at lower altitudes. Groundcover is frequently sparse, except for *Selaginella* which appears to thrive in the shade. Where the canopy is disturbed numerous herbs such as

Cynoglossum, Alchemilla kivuensis, Viola abyssinica, and various grasses make their immediate appearance. Much of the bamboo is densely overgrown with vines, especially *Clematis, Urera, Mikania,* and *Gynura scandens.* Well-beaten buffalo and elephant trails, as well as small grass-sedge meadows, provide open situations which eventually may become colonized by herbs characteristic of both lower and higher altitudes, and by the shrubs *Senecio maraguensis, Volkensia ruwenzoriensis, Tephrosia* sp. and *Polyala elliotii.*

Hypericum woodland: The forest habitat at an altitude of about 10,000 feet in the Mts. Muhavura-Gahinga saddle surrounds a large, marshy sedge-grass meadow. Most of the species in this zone also occur in the mountain woodland. However, stunted *Hypericum lanceolatum* trees, reaching a height of 20 to 30 feet, are dominant (Pl. 12). *Vernonia adolfi-frederici* and *Pycnostachys goetzenii* are the most abundant shrubs. The canopy is discontinuous, for the dense stands are interrupted by small meadows, buffalo wallows, patches of scraggly bamboo, and small clearings of grasses and herbs. The latter reach a height of 6 to 8 feet and include the following species: *Senecio chiovendeanus, Cynoglossum geometricum, Senecio maranguensis, Impatiens* sp., *Stachys aculeolata, Helichrysum nandense, Cineraria grandiflora, Conyza newii, Pycreus nigrans, Carduus kikuyorum, Laportea alatipes, Senecio trichopterygius, Droquetia iners, Rumex nepalensis,* and *Andropogon amethystinus. Carex petitiana* and *Cyperus* sp. grow in clumps here and there.

Giant senecio: At an altitude of 11,000 to 11,500 feet trees and shrubs cease rather abruptly and the slopes become very open. The only tall plants are the arborescent senecios (*Senecio erici-rosenii, S. alticola*) and giant lobelias (*Lobelia wollastonii* and *Lobelia* sp.). These grow either scattered on the ridges or in dense stands in the shelter of ravines, where they reach a height of about 10 to 15 feet (Pl. 13). The groundcover consists of a dense mat of *Alchemilla cinerea* and *Alchemilla cryptantha. Peucedanum kerstenii* and *Helichrysum* sp. are prominent herbs. Several grasses (*Festuca, Anthoxanthum, Deschampsia*), sedges, mosses, and lichens occur throughout.

Other vegetation types: Several vegetation types are limited in extent and of little importance to gorillas.

1. Long grass meadow: Three large tussocky grass-sedge meadows, which tend to be quite swampy during the height of the rains, occur in the saddles between the mountains.

2. Tree heath: *Erica arborea* and *Philippia johnstonii,* reaching a

height of 20 to 30 feet and heavily laden with *Usnea* lichens, grow patchily in monotypic stands along the rocky ravines of Mt. Sabinio above the mountain woodland. The groundcover consists mostly of moss. On the northern slope of Mt. Muhavura, tree heath less than 10 feet high is quite abundant in a few spots at an altitude of about 11,000 feet. Sphagnum-like moss covers the ground; grasses too are common in the area.

3. Lava: The steep eastern, southern, and much of the northern slopes of Mt. Muhavura are essentially dry and barren between an altitude of 9,000 and 11,000 feet. The dominant vegetation consists of tufts of grass which grow between outcrops of lava. Heath, 2 to 3 feet high, once covered much of the slope but has been destroyed by fire.

Utilization by gorillas: Gorillas utilize in varying degrees all of the vegetation zones from the mountain woodland to the giant senecios. Tracking and contact with the animals revealed that the same groups move back and forth between these vegetation types during the course of their foraging activities. The primary criterion for the extensive use of an area appears to be the presence of herbaceous forage, and only secondarily the extent or type of canopy.

Mountain woodland furnishes a large variety and an abundance of food plants, and if the time spent by gorilla groups in this vegetation zone is indicative, it represents their preferred habitat. Bamboo appears to be of secondary importance, for in the absence of bamboo shoots which are seasonal (see Fig. 5) little forage is available, except for some vines and occasional *Cynoglossum*. This is best illustrated by Table 7, which shows the kinds of food eaten in the bamboo zone as compared to the mountain woodland. The table indicates that in the mountain woodland a proportionally greater number of vines and *Peucedanum* were obtained than in the bamboo. Both Donisthorpe (1958) and Kawai and Mizuhara (1959*b*) maintained that little or no celery is eaten when bamboo shoots are available. The evidence in Table 7 contradicts this contention. Scarcity of food probably explains the infrequent occurrence of gorillas in the extensive stands of bamboo in the Mts. Gahinga-Sabinio saddle area. During the dry season gorillas frequently travel from the mountain woodland upward to the *Hypericum* woodland without stopping in the bamboo.

The giant senecio zone is only sporadically visited by gorillas. However, I saw some sign on or near the summits of all three peaks. The number of food plants in this vegetation zone is small (about five) and the supply is short. Gorillas frequently break down giant senecios and lobelias to extract the tender pith of the stems. Continuous utilization of these slow-growing plants by gorillas would rapidly deplete them, for De Heinzelin de Braucourt (1953) estimates that giant senecios require about 200 years to become fully grown.

The other vegetation types are rarely used. Mr. Baumgartel told me that he has two records of gorillas crossing the lava field on Mt. Muhavura. I saw gorilla sign in the tree heath on a few occasions. The animals were never noted on the large meadows, although they frequently skirted the edges where the grassland was broken by clumps of trees. Gorillas find little or no forage in these three minor habitats.

In summary, it appears that in the Kisoro area good gorilla habitat exists only in the mountain and *Hypericum* woodlands. These comprise perhaps 2 to 3 square miles at present and the area will be even smaller when the forest outside the reserve has been destroyed.

The Mts. Visoke-Sabinio saddle The relatively flat saddle between Mts. Visoke and Sabinio stretches for some 8 miles at an altitude of 8,000 to 8,500 feet. Several small hills rise abruptly from the level terrain, and a ridge, highest near Mt. Sabinio and lowest near Mt. Visoke, runs along the Congo-Ruanda divide. Several small volcanic craters lie near the foot of Mt. Visoke. I visited the area at the height of the dry season in July, 1959, when the vegetation was extremely dry and water was found only in several permanent water holes. The distribution of the vegetation types is shown in Figure 8.

Dry colonizing woodland: This rather dry forest on lava is found only at low altitudes along the park boundary. Scattered *Neoboutonia macrocalyx* and other trees grow to a height of 60 to 70 feet. Below these, and forming a continuous sub-canopy about 15 feet above ground, is a dense tangle of *Mimulopsis arborescens, Vernonia* sp. and *Acanthus pubescens,* a prickly shrub with purple blossoms. Grasses and a few herbs are found in the lowest story.

Bamboo: *Arundinaria alpina* covers the major portion of the saddle. In the flat areas the stands are about 30 to 35 feet tall and form a fairly continuous but relatively open canopy below which only grass grows in abundance. The bamboo at somewhat higher altitude along the border hills is scraggly and heavily overgrown with vines. Numerous cattle trails and small openings covered with grass and herbs break the continuity of the bamboo. Similar vegetation occurs on the Ruanda-Urundi side of the park.

Long grass meadows: A large meadow lies in the central portion of the saddle, interrupted by numerous islands of bamboo and patches of *Hypericum*. Fingers of this meadow reach to the border hills and to the park boundary.

Herb meadow: A large open area covered with herbs 4 to 6 feet high, broken here and there by clumps of the shrubs *Vernonia* and *Pycnostachys*, stretches near the foot of Mt. Visoke. In general this vegetation type resembles the *Hypericum* woodland of Kisoro in

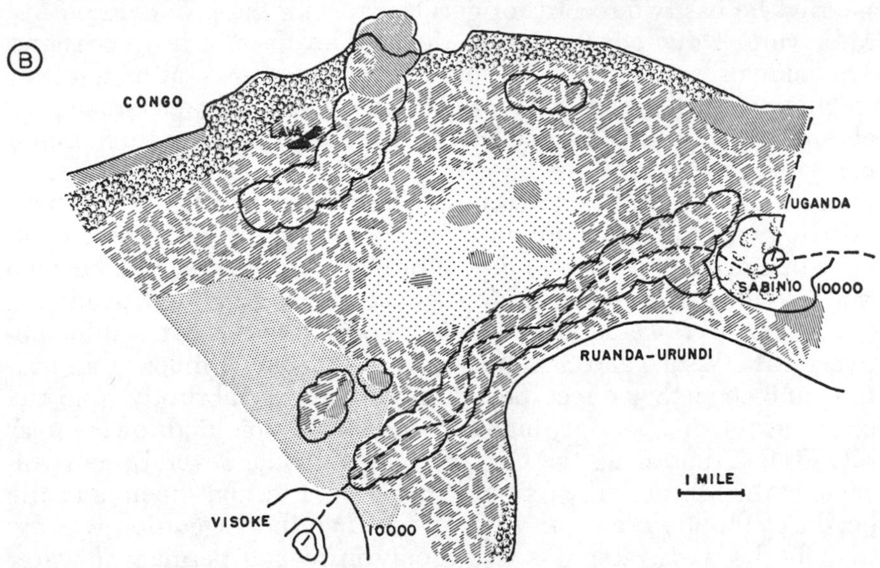

FIG. 8.—A generalized vegetation map of the Mt. Visoke–Mt. Sabinio saddle area, Virunga Volcanoes.

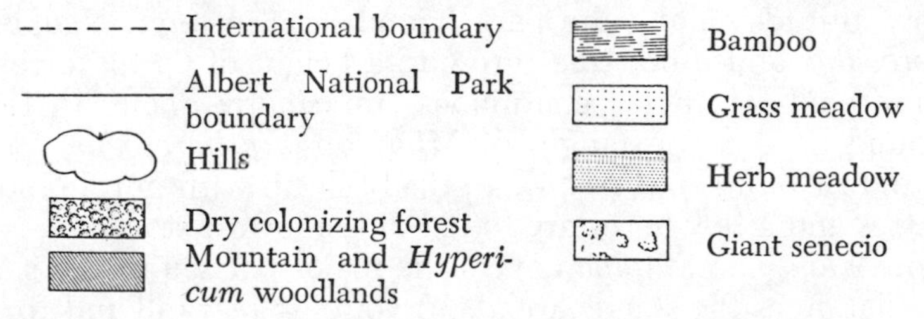

physiognomy and species composition except that the tree stratum is absent.

Lava: The only volcanic eruption in the dormant volcano area during historic times occurred in 1957 when Volcan Mugogo suddenly appeared as a lava plug in the bamboo forest and spewed forth a flow three-quarters of a mile long (Verhaeghe, 1958). Only lichens had colonized the lava to any extent in 1959.

Hagenia, mountain, and *Hypericum* woodlands: These woodlands occur only in several negligible patches, often heavily mixed with bamboo.

Utilization by gorillas: The main activity of gorillas centered in the herb-bamboo complex of the border hills and in the herbaceous meadows near the base of Mt. Visoke. I found sign of one group in the dry colonizing forest; another group traversed the tall bamboo. In general, the saddle area with its extensive bamboo and grassland is poor gorilla habitat, since forage, especially during the dry season, is scanty.

The Kabara study area The usual route of access to our camp at Kabara was from the village of Kibumba at the southeastern foot of Mt. Mikeno. The gradient of the trail is low at the beginning, following the edge of a canyon in an easterly direction, then it steepens suddenly and after a long climb culminates in a grassy knoll surrounded by trees. Rweru, as this spot is named, is the approximate half-way point to Kabara and a traditional resting place for the porters and their loads. The course of the canyon swings to the north at this point and is interrupted by two large waterfalls. A large side canyon from the slopes of Mt. Mikeno joins the main one, and after a mile of steady climbing the trail leaves the edge of the gorge, following the base of the Mt. Mikeno slope upward into the saddle.

The Kabara meadow, some 600 feet long and 80 feet wide, broken by small outcrops of lava and clumps of *Hypericum,* lies in the saddle. A shallow lake, 60–70 feet wide, nestles in the northern end of the meadow; a wooden cabin stands in the middle. About 100 feet from the cabin is the grave of Carl Akeley, who became ill and died here in 1926 just as he was starting a study of the gorillas. Toward the southeast and Mt. Karisimbi, the forest sweeps upward about 1,200 feet in a little over a mile (Pl. 14) before ceasing abruptly at the large Rukumi meadow, which spreads, broken by stands of tree heath and *Hypericum,* along the northwestern slopes of the mountain. About 2 miles north of Kabara the Bishitsi bluff drops about 500 feet to the forested plains below. The saddle area is not precipitous. The altitude of the rolling terrain drops only slightly toward the north, but it falls somewhat more steeply toward the southeast. The aerial photograph (Fig. 9) illustrates the carved slopes of Mt. Mikeno, the Rukumi meadow, as well as Bishitsi and other localities and their altitudes.

The Kabara study area is thus bounded by several prominent features: the Bishitsi bluff on the north; the Rukumi meadow on the southeast, the upper end of the bamboo zone on the southwest, and the upper slopes of Mt. Mikeno on the northeast.

The weather during our ten months at Kabara was generally cool and wet. My wife and I took temperatures at 0600, 1200, 1600, 1800, and 2000 each day, and the high, low, and mean for each month is listed in Appendix B (Table 62). Temperatures changed little throughout the year, but the daily fluctuation usually ranged between 4°C. (39.3°F.) and 15°C. (59°F.). Temperature data from Rumangabo, the Albert National Park headquarters, which lies 12 miles to the northwest at an altitude of 5,637 feet, are presented in Appendix B (Tables 62 and 63).

A considerable difference in nighttime temperatures exists between the point 5 feet above the surface, where all temperatures in Table 64 were taken, and the ground, on which gorillas usually lie. Verschuren recorded the following night temperatures with maximum-minimum thermometers:

Date	Minimum 5 Feet above Ground	Minimum on Ground
September 11–12, 1960....	7.5°C.	5.0°C.
12–13.........	7.5	5.0
13–14.........	6.0	1.0
14–15.........	7.5	5.0
15–16.........	5.0	2.5
16–17.........	3.0	−2.5

At higher altitudes the temperature probably falls to the freezing point, or near it, every night. Snow falls on the summits of Mts. Karisimbi and Mikeno occasionally and remains until mid-morning. Donisthorpe (1958) noted ice around the edges of the lake on the summit of Mt. Muhavura on July 19, 1957. In mid-June, 1960, Verschuren recorded a minimum temperature of −4°C. near the summit of Mt. Muhavura. On March 13, 1960, the moss on the upper slopes of Mt. Mikeno was still frozen at 0900.

The mornings are often beautifully clear. But the mountains apparently draw moisture from the surrounding plains, for by 0900 clouds usually drift up the valleys and slopes, and soon fog engulfs the forest (Pl. 15). By noon the sky is all or partially cloudy but tends to clear in the late evening and often remains so throughout the night. Not a single day (0600–1800) was entirely clear, and 72.6 per cent of the time there were heavy clouds for most of the day (Appendix B, Table 65). June and July are reported to have many sunny days, but no work was done at Kabara during those months.

Precipitation fell frequently at Kabara. Out of a total of 267 days listed in Appendix B (Table 66), rain for more than half an hour in duration was recorded on 176 (65.9 per cent) days. During the time of the study we averaged nearly two hours of rain per day. The average amount of precipitation at Rumangabo is about 1,750 mm. per year (Appendix B, Tables 62 and 63); at the 3,000 meter level it is about 1,500 mm. per year; on the summit of Mt. Karisimbi about 943 mm. per year (Robyns, 1948*a*). Much of the rain was in the form of drizzles, but occasionally steady downpours occurred for hours. Most rain fell during the day; only in April, 1960, was it common at night. Hail was noted at Kabara on the average of once every three to four weeks. Snow fell on Mts. Mikeno and Karisimbi in conspicuous amounts also about once every three weeks on the average. It never fell at Kabara; in fact I never saw it below an altitude of about 13,500 feet.

Winds blew usually from the north-east. Most were mere breezes, but occasionally strong blasts funneled through the saddle and rattled the tin roof of our cabin. Gusts are often violent higher on the mountains where the slopes are more open and the force of the wind is not broken by trees.

The sunny and rainy weather usually came in spells during which one type predominated (Appendix B, Table 67).

I took detailed notes on the vegetation and collected numerous specimens around Kabara; Figure 10 summarizes the results. My discussion of the lower altitudes is based largely upon samples studied along the trail from the park boundary to Kabara. A transect to the summit of Mt. Karisimbi or Mt. Mikeno includes the following vegetation types and their approximate altitudes:

Vegetation type	Altitude
Dry colonizing woodland	6,800– 7,300
Bamboo	7,300– 9,200
Mountain woodland (along canyons only)	7,400–11,500
Hagenia woodland	9,200–11,000
Hypericum woodland	11,000–11,400
Tree heath	10,000–11,500
Short grass meadow	about 10,000
Long grass meadow	11,400
Giant senecio	11,300–14,000
Alpine	14,000+

Dry colonizing woodland: From the park boundary toward Kabara the trail winds through about half a mile of rather dense shrubby habitat. *Neoboutonia macrocalyx*, the dominant tree species, reaches a height of 60 to 70 feet and covers the lower strata with a very broken and intermittent canopy. *Myrica salicifolia, Bersama ugandensis*, and *Agauria salicifolia* trees occur also, especially in ravines. The ubiquitous *Mimulopsis arborescens* is the dominant shrub, providing a rather solid canopy at a height of 10 to 20 feet. Small grassy clearings, herb-covered glades, and clumps of bamboo are scattered throughout this vegetation zone. The prickly *Acanthus pubescens* grows primarily along trails and in other disturbed locations.

Bamboo: The bamboo is scrubby, only about 10 to 15 feet high, and heavily cluttered with such vines as *Urera hypselendron, Clematis wightiana, Stephania abyssinica,* and *Mikania cordata.* Buffalo and elephant trails, small marshy clearings, and herbaceous

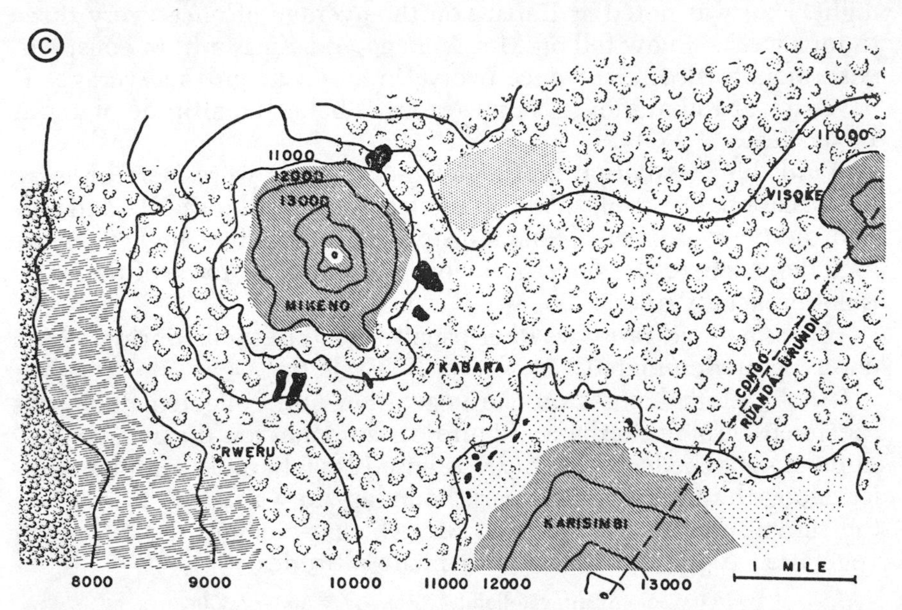

FIG. 10.—A generalized vegetation map of the Kabara area, Virunga Volcanoes

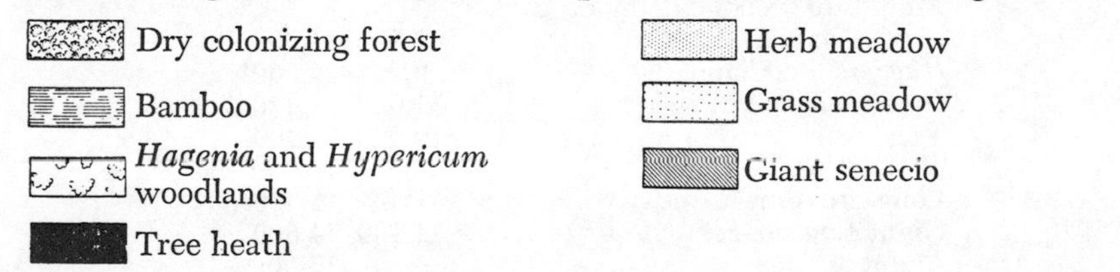

meadows overgrown with *Peucedanum, Cynoglossum, Senecio,* and other plants dissect the bamboo throughout. Lone trees of *Neoboutonia* rise above the canopy of bamboo up to the 8,200 foot contour. Above that altitude the first *Hagenia abyssinica* as well as *Hypericum lanceolatum, Pygeum africanum,* and *Clerodendron* sp. appear. In the more dense patches of bamboo the sparse groundcover is strongly dominated by *Selaginella kraussiana.* At the upper end of its altitudinal range bamboo becomes more and more intermixed with the vegetation of the *Hagenia* woodland, and finally ceases about half a mile above Rweru, and at a somewhat lower altitude on the southeastern slopes of Mt. Mikeno.

Hagenia woodland: *Hagenia abyssinica* is characteristically a tree of secondary succession in mountain forest. But between an altitude of 9,200 and 11,000 feet on the slopes of Mikeno, Karisimbi, and Visoke this species grows in unique and almost monotypic stands. Robyns (1948*a*) classifies the *Hagenia* woodland as a "forest-prairie," for the vegetation consists primarily of two stories: an upper one of trees and a lower one of herbs.

The trunks of *Hagenia* trees in these stands are massive, some 3 to 8 feet in diameter, but the height of the canopy rarely exceeds 60 to 70 feet. The main limbs flare semi-horizontally from the trunk near the ground before veering skyward. Clusters of long pinnate leaves and grape-like pendants of flowers hang primarily from the apex of the branches. The foliage is thin and the crown quite open. The huge semi-horizontal limbs support soft cushions of moss (*Dicranum affine, Antitrichia kilimandscharica*), ferns (*Polypodium*) and other epiphytes. Tattered beards of *Usnea* lichens hang from the branches.

The only other tree species of prominence in the *Hagenia* woodland is *Hypericum lanceolatum*, which grows to a height of 40 to 60 feet throughout this vegetation zone, especially near such clearings as the Kabara meadow. At a higher altitude *Hypericum* replaces *Hagenia* as the dominant tree species.

Hagenia woodland is open, resembling parkland. The canopy covers at best only about 50 per cent of the area, a fact clearly revealed in Figure 9.

The herb layer is characteristically 5 to 8 feet high and very dense. The most prominent herb and vine species are: *Senecio trichopterygius, Impatiens eminii, Stachys aculeolata, Lobelia giberroa, Carduus afromontanus, Galium simense, Peucedanum linderi, Plectranthus laxiflorus, Rumex ruwenzoriensis, Laportea alatipes,* and *Cineraria grandiflora*. Most of these species are succulent and their stands are so dense that any large mammal that walks through them leaves behind a trail of crushed vegetation which remains conspicuous for many days. Generally the yellow-flowered *Senecio trichopterygius* is the dominant herb, but in certain situations others become almost monotypic. In shady, moist glades *Rumex ruwenzoriensis* is often prominent; on open, dry slopes *Carduus afromontanus* is conspicuous; and just north of Kabara and at the western base of Mt. Visoke *Laportea alatipes* grows to a height of 8 feet and covers several acres so uniformly that little other vegetation survives. My hikes through these stinging nettle fields are among my most unforgettable experiences. Stands of *Lobelia giberroa* are especially prevalent northeast of Kabara. *Galium simense* creeps over fallen logs, covers stumps, and trails from the lower branches of trees, sometimes in dense mats.

The shrubs *Vernonia adolfi-frederici* and *Pycnostachys goetzenii* are abundant, especially in open disturbed locations as along steep ravines and on the precipitous slopes of Mt. Mikeno.

Hypericum woodland: At about an altitude of 11,000 feet, or a little higher on the slopes of Mt. Mikeno, *Hagenia* trees become progressively less abundant until they cease entirely and *Hypericum* becomes the dominant species. At this altitude the first tall *Senecio erici-rosenii* appear, and the shrub *Rubus runssorensis* becomes conspicuous. Otherwise the general appearance of the vegetation is similar to that of the *Hagenia* woodland. *Hypericum* is also common in the forest islands on the Rukumi meadow.

Giant senecio: Although one specimen of the arborescent *Senecio erici-rosenii* grows at Kabara, the zone designated by that name commences at about 11,400 feet, where *Senecio alticola, S. erici-rosenii, S. refractisquamatus, Lobelia wollastonii,* and *L. lanuriensis* compose the tallest stratum, usually to a height of 10 to 15 feet but occasionally as high as 20 feet. These arborescent herbs grow scattered over the slopes, but in the ravines they form dense groves with entwined moss-laden branches and a maze of dead limbs. Slippery *Sphagnum planifolium* covers the ground. Silver-gray *Alchemilla cinerea* carpets the open slopes to a depth of 2 feet. Grasses are also common.

Other vegetation types: Several types of vegetation have a limited distribution and are of little importance to gorillas.

1. Mountain woodland: The edges of deep ravines in the *Hagenia* woodland support mixed stands of *Pygeum africanum, Pittosporum lanatum, Rapanea pulchra,* and *Cornus volkensii* which resemble those found on the slopes of Mt. Sabinio. Grass is frequently the dominant groundcover under the dense canopy, but sometimes various shrubs are also found.

2. Tree heath: *Erica arborea* and *Philippia johnstonii* occur, (a) along steep, rocky ridges and on rock bluffs; and (b) in patches on and surrounding the Rukumi meadow. On rocky ridges heath often forms a dense canopy about 15 feet high below which little except moss and some grass survives. On the Rukumi meadow, tree heath reaches a height of 30 to 35 feet. The trees grow individually or in small clumps often associated with the arborescent senecios and *Hypericum*. The ground tends to be covered with grass or remains bare; *Rubus* occasionally grows in profusion.

3. Long grass meadow: The Rukumi meadow, which resembles the grassland of the Kisoro area, comprises the only extensive meadow near Kabara. Its expanse is broken by numerous forest islands. Grasses and sedges form a continuous cover 1 to 2 feet high. Much of the meadow is damp to swampy and in the wet depressions large tussocks of *Carex runssorensis* flourish. Several small grass meadows also lie at the southwestern base of Mt. Visoke.

4. Short grass meadow: Small openings covered with closely-cropped green grass are scattered sparsely throughout the *Hagenia* woodland (Pl. 14). Of these the Kabara meadow is the largest. *Agrostis mildbraedii*, *Poa annua*, and *Poa leptoclada* are some of the common grasses. The meadow is dotted with numerous low-growing flowers: *Trifolium rueppillianum*, *Oxalis corniculata*, *Helichrysum* sp., *Alchemilla cryptantha*, *Ranunculus* sp., and a white daisy. Lava outcrops are covered with yellow-flowered *Sedum ruwenzoriense* and *Galium spurium*. Photographs published by Chapin (1932), Bingham (1932), and others in the past show that the Kabara meadow was then more overgrown with shrubs than at present.

5. Alpine: Above about 14,000 feet giant senecios are absent or survive only in stunted form. *Alchemilla* remains as the most abundant plant of the groundcover, but even it ceases at about 14,500 feet. *Festuca*, *Anthoxanthum*, *Carex*, a few herbs, and various lichens and mosses are all that persist on the exposed lavas and soils near and on the summit of Mts. Mikeno and Karisimbi.

Utilization by gorillas: I never saw sign of gorillas in the dry colonizing forest. Natives remembered that several years before one group visited the fields at the park boundary, but they agree that this was an unusual incident. In fact, I never noted gorilla sign below an altitude of 8,300 feet, which suggests that gorillas rarely visit the lower altitudes southeast of Mt. Mikeno. The occupation of the bamboo zone by gorillas appears to be sporadic and to some extent seasonal, being correlated with the abundance of bamboo shoots (see Table 22).

The *Hagenia* woodland with its abundant and continuous supply of forage is without question the most favored gorilla habitat in the Virunga Volcanoes. Most groups in the central volcano area confined their activity to this zone, even though they left it occasionally for other zones above and below. This observation contrasts sharply with Barns (1922), Burbridge (1928), Schäfer (1960), and several others who have maintained either that gorillas seldom visit *Hagenia* woodland or that the area furnishes poor gorilla habitat.

Gorillas visit the giant senecio zone only sporadically and for a few days at a time. I tracked one group to an altitude of 13,500 feet on Mt. Mikeno, and I noted old sign on the summit of Mt. Visoke. On Mt. Karisimbi, however, the Rukumi meadow seems to act as a barrier, for I never saw sign above 11,400 feet on that mountain.

The other vegetation types do not support gorillas for long periods. The mountain woodland along the canyons supplies little forage. Similarly the number of food plants in tree heath is small. Grass meadows are usually circumvented. I followed one group across an arm of the Rukumi meadow which was at that point about 300 feet in width. Another time a lone silverbacked male moved across the grassy expanse of the Rukumi meadow and fed on the *Rubus* berries that grew in the forest islands. No groups crossed the Kabara meadow except at the extreme ends. Since gorillas do not seem to ascend the mountains higher than to an altitude of about 13,500 feet, they fail to penetrate the alpine zone.

The active volcanoes The terrain surrounding Mts. Nyiragongo and Nyamuragira is transected by numerous broad lava flows of recent age, augmented every few years by eruptions from the slopes of the peaks and from streams of lava that well up from the forest floor. The two mountains are in constant activity—a lava lake bubbles in the depths of Nyiragongo, and steaming fumaroles in the huge crater of Nyamuragira attest to the internal turmoil. Robyns (1948*a*) classifies the vegetation on and surrounding the volcanoes as shady mountain forest and dry forest and brush, with the latter predominating especially on the lava flows. Lebrun (1942) presents a general summary of the vegetation on Mt. Nyiragongo. I spent four days in the area and climbed both peaks via the regular trails up the northeastern slopes (Fig. 6).

Along the park border *Neoboutonia macrocalyx, Dombeya goetzenii,* and *Croton macrostachys* are the dominant tree species, reaching a height of 60 to 80 feet. Under their broken canopy, such shrubs as *Acanthus pubescens, Mimulopsis arborescens, Rhus incana,* and *Solanum indicum* reach a height of 20 feet and these are heavily overgrown with numerous vines (*Clematis simensis, Piper capense, Thunbergia stuhlmaniana*). Herbs of the genera *Impatiens, Thalictrum,* and *Geranium,* as well as several grasses, constitute the lowest story.

All forest in the vicinity of the volcanoes grows on old lava. Our trail led across several of the recent flows. In 1954 a wide, fiery river of molten rock cut a swath through the forest, bypassing several low hills which remained as forested islands in a sea of gray lava. During this century eruptions in the vicinity of the two active volcanoes occurred in 1904, 1912, 1938, 1948, 1950, 1951, 1954, 1957, and 1958 (Bourlière and Verschuren, 1960). However, new vegetation rapidly finds a foothold on the dead rock. A hoary coat of lichens covered the lava of the 1954 flow, and ferns grew from some of the fissures. A dense scrubby forest over 40 feet high was found on the 1904 flow. On old lava flows at the base of Mt. Nyamuragira, trees up to 80 feet high occasionally form a continuous canopy with a groundcover primarily of grass. *Myrica salicifolia, Faurea saligna, Agauria salicifolia, Maesa rufescens* and *Clausena anisata* are some of the prominent tree species.

On Mt. Nyamuragira, between roughly 8,200 and 9,000 feet, *Hypericum* is the dominant tree. It is scrubby, only some 30 feet tall, and the groundcover beneath the broken canopy consists mostly of grasses, *Senecio maranguensis,* and herbs like *Helichrysum* and *Kniphofia.* Above this zone and extending to the summit, the slopes are covered with lava on which grasses and tree heath grow sparsely, the latter reaching only 3 to 4 feet in height.

A mountain forest, consisting of such tree species as *Alangium chinense, Neoboutonia macrocalyx, Podocarpus* sp., *Cornus volkensii, Polyscias fulva, Pygeum africanum,* and *Dombeya goetzenii,* covers the slopes of Mt. Nyiragongo between about 8,500 and 9,500 feet. The trees provide an almost continuous canopy about 75 feet high over the shrub layer, which consists of *Dracaena afro-montana, Maesa rufescens, Trichilia* sp., *Canthium* sp., and *Galiniera coffeoides.* Vines are conspicuous (*Urera, Clerodendron, Schefflera*), and *Carex* sp., *Mimulopsis violacea, Laportea alatipes,* and *Coleus silvaticus* are abundant. Between roughly 9,500 and 10,600 feet a zone of *Hagenia* woodland, similar to that at Kabara, is found. Tall tree heath grows on the final summit slope, but ceases at about 10,800 feet. The last stretch is almost bare lava except for a few stunted heaths and grasses. A small patch of bamboo exists on the southern slopes of Mt. Nyiragongo.

Much of the terrain around the active volcanoes supports lava fields sparsely covered with lichens, grasses, and ferns, dry scrubby forests, and tall forests with a groundcover of grass—all vegetation types which provide little or no gorilla forage. Only the upper slopes of Mt. Nyiragongo are clothed with forest of a type generally favored by gorillas.

In conclusion, in the 6 dormant Virunga Volcanoes only about one-third of the area provides abundant forage throughout the year. About one-third of the region is covered with bamboo, which has only moderate value as gorilla habitat. The final third comprises vegetation types that are visited by gorillas only intermittently or not at all.

CHAPTER 3

THE ANIMALS

Age and Sex Classes

Haddow (1952) divided red-tailed monkeys (*Cercopithecus ascanius*) into infants, juveniles, subadults, adults (males and females); Carpenter (1934, 1940) recognized infants, juveniles, and adults (males and females) in howler monkeys (*Alouatta palliata*) and gibbons (*Hylobates lar*). In default of precise standards and because gorillas present special problems, I have used the following categories:

Age and Sex Class	Approximate Age (in years)	Definition and Characteristics
Infant	0–3	Any animal which is carried by or rides on a female for prolonged periods; any animal which remains closely attached to a female and weighs less than about 60 pounds
Juvenile	3–6	Any animal which is larger than an infant and smaller than an adult; any small gorilla which lacks prolonged close contact with a female; any animal weighing roughly 60 to 120 pounds
Subadult and adult	6+	Any animal advanced beyond the juvenile stage
Female	6+	Any animal which persistently transports an infant; any large gorilla with the following physical attributes: round body and smooth curves; sagging breasts with nipples up to an inch long
Blackbacked male	6–10	Any female-sized animal with angular and muscular body; any black-coated gorilla lacking prominent breasts and nipples, but showing a conspicuous *pectoralis major* muscle over which the skin is usually taut
Silverbacked male	10+	Any gorilla of very large size (often over 300 pounds) with a prominent sagittal crest, muscular and angular build, and with the pelage of the saddle and sometimes legs, neck, and sides gray to silver in color

Infants

I traced the development of one infant from birth to the age of one and a half years in the wild and that of numerous others of variable ages for ten to twelve consecutive months. This gave me a means of comparing the sizes and behavior of the infants, and, I believe, a fairly accurate means of estimating age. For further comparison, I obtained weight data on captive animals from several zoos (Figs. 12 and 13), which show a fairly steady increase from birth to the age of about five years. By comparing the size of captive gorillas of known age with those of wild gorillas I derived another crude scale by which to age them. The transition from infant to juvenile is slow and occurs between about two and a half and three and a half years. As a median value, and for the sake of convenience, I have chosen three years as the dividing line between the two in this report. The sex of infants is usually not determinable in the wild.

Juveniles

Records from zoos and observations in the wild indicate that females reach adult size at or soon after the age of six years and that they may conceive soon after they have reached this stage in their development. Thus in females the period between three and six years of age encompasses their juvenile and most of the subadult stage. Males do not attain their conspicuously large size until the age of about nine to ten years, an age at which they probably also become sexually mature in the wild. In view of the disparity in development, I have placed both males and females between the ages of three and six years into the category of juveniles, and I have not recognized a subadult category as such.

The transition from juvenile to the older age class is gradual, occurring between the ages of five and a half to six and a half years. I have arbitrarily used the age of six years to divide juveniles from the older animals. Juveniles can frequently be sexed by their physical appearance and behavior. Although I often felt reasonably certain of their sex, I was unable to confirm my observations, and consequently deleted most references to sex in juveniles.

Subadults and Adults

Female Any female aged six years or older was placed into the subadult-adult category. I was sometimes able to classify these females as young, middle-aged, and old, distinctions which will be used later in discussing certain aspects of behavior (Pl. 17).

Blackbacked Male Most males between the ages of six and nine or ten years of age are probably sexually subadult animals. The young males resemble females in size and shape until the age of about nine years when quite suddenly their secondary sexual characteristics become prominent. The slow replacement of black with gray hair in the saddle of the back is one of the most conspicuous of these changes. The pelage change provides a useful morphological character by which to divide the males into two classes. Hence, any male with a predominantly black saddle is referred to as a black-backed male (Pl. 16).

SILVERBACKED MALE Males ten years and older have a prominent gray or silver saddle. Because of their large size and appearance, these males are unmistakable (Pl. 19).

Physical Attributes and Individual Differences

The gorilla is the most robust and massive of the primates, and its gross physical features are readily apparent from the photographs in this report. Rather than present a systematic morphological description of the type found in many anthropological texts, I concern myself in this chapter primarily with the individual differences that I observed within groups or populations, and with the features which aid in distinguishing the various age and sex classes.

Individual recognition of animals is essential in a study of social behavior. Fortunately individuals in a population of gorillas vary conspicuously in their physical attributes. In learning to recognize individuals, I found it convenient to divide the group into the respective age and sex classes and then to mark down the outstanding features of each animal, followed by a detailed sketch of the nose. After a series of encounters, individual gorillas become so familiar that recognition need not be based on a single feature, such as a prominent scar or color deviation, but rather on a whole series of physical and behavioral traits which in their general configuration are distinctive of one animal alone.

HANDS AND FEET

The hands are extremely broad. The thumb, although relatively shorter than in man, is well developed and readily opposable. Gorillas use the thumb extensively in conjunction with the other fingers when plucking forage, when climbing, and when picking up small objects. I have never seen a gorilla pick material up between index and middle finger in the wild, although I observed it several times in zoos. The fingers are webbed nearly to the first joint.

The foot of gorillas exhibits a moderately well-developed heel and a great toe which is widely separated from the others and opposable.

The other toes are webbed to the first joint. Schultz (1934) and Straus (1930) discuss the structure of the foot in detail. On the whole, the foot is not well adapted to arboreal prehension. Although the great toe is opposable and sometimes used to steady the foot on a branch, its grasping ability in adults appears limited. However, infants use the feet to clasp the hair of their mothers.

Skin

At birth the entire skin of gorillas is pinkish-gray, or, as described by Thomas (1958), light to medium brown. Pigment accumulates rapidly so that by the age of one and a half to two months the skin is dark brown to black, although the palms of one infant were still gray at the age of three months. Large infants, juveniles, and adults are usually black on such unhaired or sparsely haired portions of their body as the face, ears, fingers, chest, and soles of hands and feet. The chest is either gray or black, with the latter more common in males. The skin beneath the densely haired portions is gray. Some adults have dark brown or dark gray fingers and soles.

Three out of fifty-four infants had large, irregular, bright pink spots on their palms and soles. One two-month-old infant exhibited such spots on all four extremities. One three-month-old infant had two pink spots on the sole of its left foot, and another showed them on the palm of one hand. An infant, obtained by Cordier in the Utu region, showed similar spots on palms and soles at the age of nearly one year. The marks seem to fade or disappear with advancing age. I saw one adult female with faint pinkish spots on the sole of one foot.

The tongue and the skin on the inside of the mouth of gorillas is pink, black, or mottled.

Pelage

The length, color, and distribution of the hair on the gorilla's body shows conspicuous variation.

Altitudinal Variation My observations on museum specimens and on captive and free-living mountain gorillas revealed that the gorillas living in the Utu region and other habitats at low altitudes have shorter hair than those living at high altitudes in the Virunga Volcanoes. The longer hair at high altitudes is probably an adaptation to cold temperatures.

Color and Length Infants are sparsely covered on much of the body with soft, black hair. Thomas (1958) found that the hair on the crown became lighter at the age of one and a half months in one zoo specimen of lowland gorilla. By two to two and a half months infants at Kabara had brown hair on their crowns. The length of hair on the head of a newborn infant which I found dead was 3 cm. At about four months of age a new growth of shorter black hair begins to replace the brown hair on the crown. In five infants I noted the beginning of this conspicuous molt between the ages of fifteen and seventeen weeks. The new growth replaces the old one from the center of the crown outward with the result that infants between about seventeen and thirty weeks of age appear to be wearing a black cap surrounded by a wild fringe of brown hair. The crown is completely black by the age of eight to nine months.

A white tuft of hair is present on the rump of infants at birth and persists at least to the age of four years.

The hair of gorillas is characteristically black. In many animals the black is of a rather subdued color and often exhibits brown and reddish overtones. One female had a brown crown, another a brown rump, a third had brown hair between the breasts. Some individuals have a shiny blue-black pelage, which is especially intense and striking in silverbacked males. Adult animals turn gray with age. Males eight to nine years old sometimes have a light touch of gray below the ear, and soon thereafter white hairs appear in the saddle. By the age of about twelve years the saddle of males is completely gray or silver. With advancing age the flanks, sides, neck, head, and the abdomen lighten, until old animals are almost entirely gray, except for the arms which remain black. Females too turn gray, usually lightly just below the ear, and on the neck, brow, and crown; in old individuals the shoulders are also peppered with white. Some adults have a small wedge of light colored hair in the center of the brow, the shape and size of which is sometimes useful in individual recognition.

The length of the hair varies on different parts of the body and individual variation is great. That of the back is generally short; that of the legs, rump, and abdomen is relatively long; and that of the arms appears longest, reaching a length of 20 cm. in males. Some gorillas are woolly-looking; others seem shorn. A few animals possess conspicuous Franz-Josef side whiskers. The hair on the crown of some is long and unkempt but others have a stubbly growth like a crew-cut. Some gorillas have scraggly goatees.

Distribution Newborn infants are covered with hair on the head (except the face), back of neck, back, and the outside of the arms and legs. The rest of the body is essentially bare. By the age of two months a considerable increase in hair is noticeable and by three months the abdomen and chest are covered. The inside of the forearms and the inside of the legs often remain sparsely haired until six to seven months of age.

Gorillas grow no hair on their ears, lips, nose, palms, and soles. The section between upper lip and nose is sparsely haired as is the armpit, and the groin where the thigh joins the abdomen. The chest, from the sides of the neck to just below the rib cage, is often quite bare in adults and in the older juveniles.

Features of Head and Face

The head of gorillas is massive. The forehead is low, and bulging supraorbital ridges overhang the eyes like a cornice. The jaw projects snout-like forward and downward. A saggital crest, especially prominent in adult males, furnishes an attachment for the large temporal muscles which move the lower jaw.

A few gorillas have flat faces and others are extremely prognathous; some have large round faces, some oval ones, others long ones. The size of the sagittal crest is highly variable among silverbacked males, two of those we observed being so large that they resembled hairy miters. Large females occasionally have low sagittal crests.

The ears are small, black or brown, and resemble those of man in shape. They frequently are cut and torn, even though they are set close to the head and almost hidden in the long hair.

The whole visible part of the eye is dark brown, and only if the animal rolls its orbs far to one side does some white become noticeable in one corner.

The eruption of teeth is discussed in Schultz (1930), Thomas (1958), and Lang (1959, 1960*a*). None of the animals which I was able to observe closely showed injuries to the large canines, missing teeth, or obvious abnormalities. I was unable to note caries, for the teeth of adults are usually covered with a black, brown, or reddish tartar. Infants have white teeth as do many juveniles, but those of most older animals are at least in part discolored. One silverbacked male, however, retained sparkling white teeth.

During our first encounters with gorillas, Emlen noted that their noses appeared distinctive, and later I found that the nose was the best single character for recognizing individuals. In general, the bridge of the nose is flat, flaring out toward the nostrils. The nostrils open directly forward and are surrounded by cartilagenous rings which merge with the upper lip. The nasal wings are thick and separated by a thin septum.

The nasal wings have three basic shapes: (1) closed all around and pointed at the top (Fig. 11A); (2) closed all around and rounded at the top (Fig. 11B); and rarely (3) rounded or pointed at the top but open at the bottom so that no distinct dividing line exists between nostril and upper lip. (Fig. 11C). The edge of the nasal wings may be smooth, irregular, or roughly serrated. The shape of the nostrils varies from oval to square.

The nasal region between brow and nostril is highly distinctive. Most gorillas have a divided nose bridge separated by a shallow trough. Occasionally only one ridge is present and sometimes there are three (Fig. 11, No. 9). The trough between the ridges is usually furrowed and wrinkled. Figure 11 (Nos. 1–13) illustrates roughly the most prominent wrinkles of the nose in all juvenile and adult members of group VII. Several of the photographs (Pls. 17, 18, and 34) likewise show the variable characteristics in the structure of the nose.

Additional marks on or near the nose aid further in distinguishing gorillas. One juvenile lacked a visible nasal septum. One juvenile and one silverbacked male had torn the edges of their nostrils, leaving conspicuous scars. One silverbacked male had a small black, wart-like growth in his nostril, and another showed a similar defect between the nose and eyelid.

Weights and Measurements

An analysis and listing of the various body measurements of the gorilla is beyond the scope of this report and the publications on the subject are listed by Ruch (1941). More recent information is found in Schultz (1942), Randall (1943–44), Haddow and Ross (1951), Straus (1950), Gregory (1950), Didier (1951), Ashton and Zuckerman (1952), Ashton (1954), Steiner (1954), Galloway *et al.* (1959), Mills (1960), Le Gros Clark (1960), Chagula (1961), Hall-Craggs (1961*a, b*), and others. Measurements were obtained from the skeletons of two gorillas which I found dead, and from parts of 17 other gorillas collected by Emlen and myself in native villages. The data for these specimens are summarized in Appendix C.

Weights Few precise weights exist of mountain gorillas collected in the wild. Attention of hunters is usually focused on shooting the largest males with the result that nearly all data fall into this age and sex class. Most of my information on females and subadults is based on zoo animals.

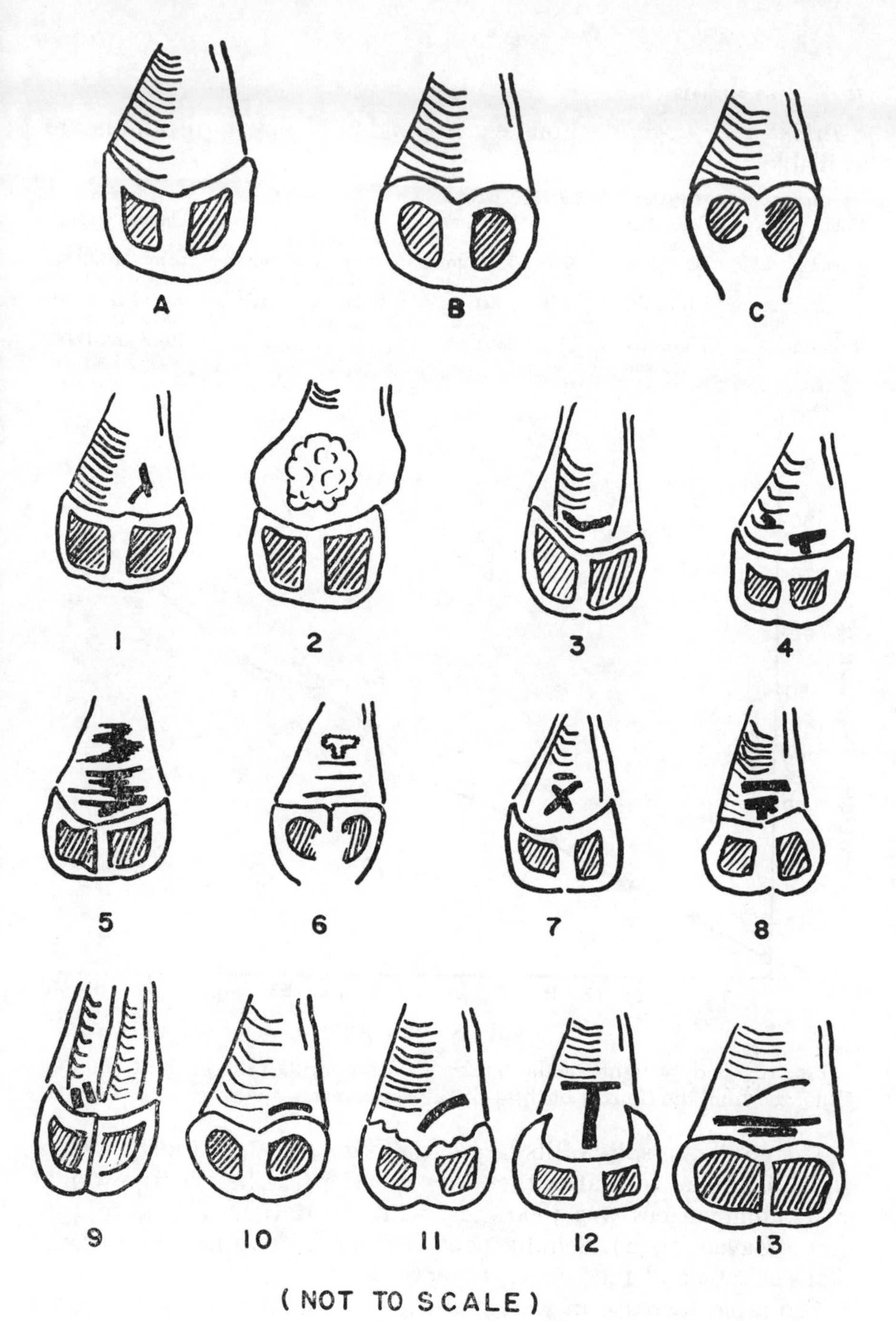

FIG. 11.—Shapes and individual identifying marks of noses of mountain gorillas at Kabara. *A*, *B*, and *C* represent the 3 basic shapes of the nasal wings as noted in the Kabara gorilla population. Nos. 1 to 13 show the nose variations of the members of group VII. No. 3 is the silverbacked male; 2 and 10, blackbacked males; 1, 5, 6, 7, 8, and 9, females; and 4, 11, 12, and 13, juveniles. The heavy black marks indicate prominent wrinkles.

Infant and juvenile Four exact weights of newborn infants are available:

Name	Location	Sex	Weight	Comments	Source
Colo.........	Columbus zoo	♀	1743 gms.	Weighed 24 hours after birth	Thomas (1958)
Goma.......	Basel zoo	♀	1820 gms.	Weighed 36 hours after birth	Lang (1959)
....	I.R.S.A.C., Lwiro	?	1965 gms.	Killed by female at birth	van den Berghe (1959)
Tomoka.....	Washington zoo	♂	2403 gms.	....	Reed and Gallagher (in press)

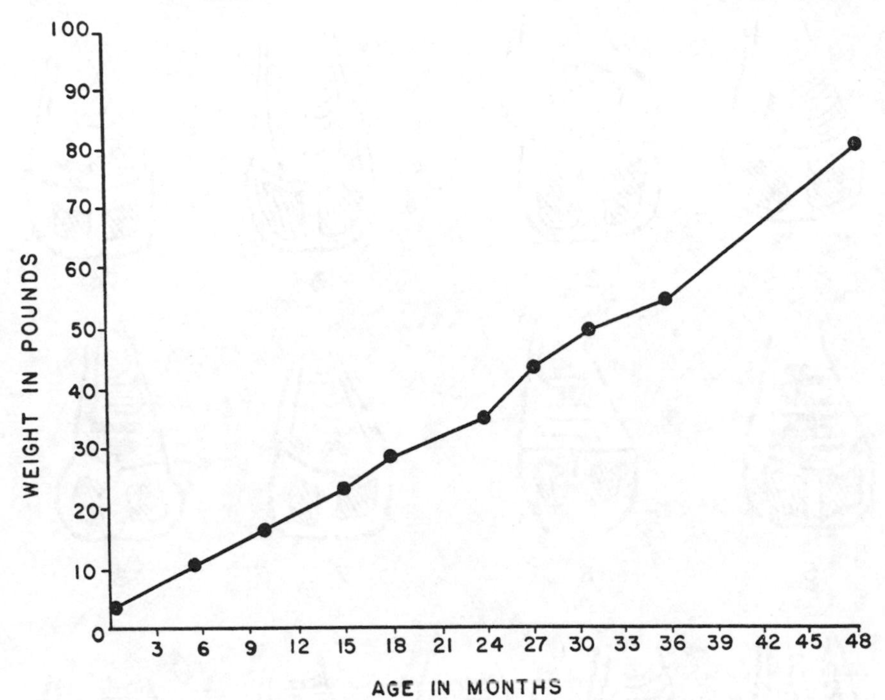

FIG. 12.—Body weights of the female lowland gorilla Colo at the Columbus, Ohio, zoo from the time of birth to the age of 4 years.

The gorilla, in spite of its large adult size, has thus a birth weight which is similar to that of the other great apes. The average weight of 42 chimpanzees was 1800 gms. with a variation of 1130 to 2370 gms. (Gavan, 1952). Schultz (1938) recorded two newborn orangutans at 1265 and 1600 gms., respectively.

The rapid increase in weight during an infant's life is illustrated in Figure 12, which is based on measurements of the female Colo at the Columbus zoo. Other zoo weights of infants show a similar increase, although the ages of the animals are less precisely known.

The weight increase of four animals (two males, two females) in the New York (Bronx) zoo is shown in Figure 13. On the basis of zoo weights I suggest the following very approximate age-weight correlation, assuming, of course, that the animals are in full physical and mental health, and that they are neither obese nor undernourished.

Year	Approx. Average Weight (Lbs.)
1	15–20
2	35
3	60
4	80
5	120

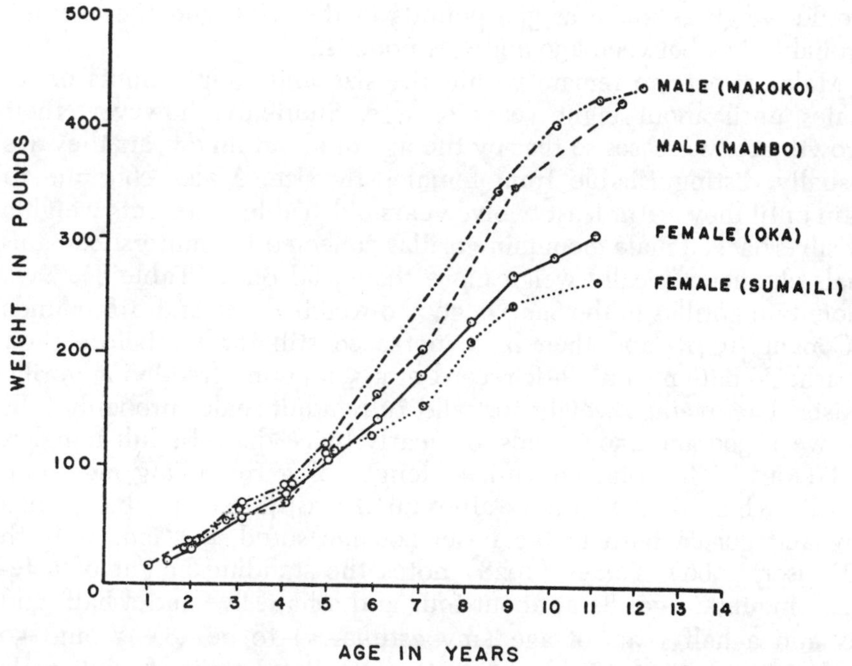

FIG. 13.—Body weights of four gorillas at the New York (Bronx) zoo from the ages of about 1 to 11 years. The female Sumaili is a mountain gorilla; all others are lowland gorillas.

By the sixth year there is greater disparity in weight, probably due to individual differences, but males and females in the wild remain of similar size.

Adults Females probably attain their adult size between six and seven years of age, although they may continue to gain weight slowly for several years. One adult female shot in the Virunga Volcanoes weighed 214.5 pounds (Gyldenstolpe, 1928); another, a female of the lowland race, weighed 159.5 pounds (Grzimek, 1957*a*). Yerkes (1928) weighed a female mountain gorilla, which was captured as a juvenile in the wild, and obtained the following successive yearly weights: 65, 128, 160 pounds. According to Yerkes' estimate, the gorilla was five to six years old on arrival. However, after perusing the published photos of the animal I suspect that its age was overestimated, being at most only about four and a half years old on arrival. The weights of zoo animals (Table 8) are not necessarily comparable to free-living ones, for captives tend to be excessively obese. I think that only an exceptionally large mountain gorilla female would weigh as much as 300 pounds in the wild, and the average probably lies between 150 and 250 pounds.

Males appear to remain within the size and weight limits of females until about eight years of age. Suddenly, however, their growth rate increases so that by the age of about nine years they are visually distinguishable from females by size. Males continue to gain until they are at least twelve years old. Table 9 presents weights of silverbacked male mountain gorillas collected by hunters. Zoo animals characteristically weigh more than wild ones (Table 8). Two mountain gorillas in the San Diego zoo weighed 585 and 618 pounds (Conant, 1941) and their bulk increased still further before their death. To date no authentic record of a 500-pound free-living gorilla exists. The average weight for wild fully adult males probably falls between 300 and 450 pounds, or nearly twice that of adult females.

HEIGHT The total crown-heel length of a free-living mountain gorilla which died two days after birth was 48 cm. (Table 70); a lowland gorilla born in the Basel zoo measured 47.5 cm. at birth (Hauser, 1960). Yerkes (1928) notes the standing height of a female mountain gorilla at about four and a half, five and a half, and six and a half years of age (my estimates) to be 38, 47, and 50 inches, respectively. Table 9 lists the standing height of adult male mountain gorillas. Most males fall between 140 and 185 cm. (54 to 73 inches). Eight lowland gorilla males varied from 142 to 182.9 cm. in height (Coolidge, 1936; Grzimek, 1957*a*). One female lowland gorilla was 142 cm. high (Grzimek, 1957*a*).

ISCHIAL CALLOSITIES

Some gorillas possess small ischial callosities or sitting pads. These were noted only in adult animals even though the infants and juveniles in six groups were scrutinized carefully for their possible presence. Table 10 presents the distribution of this character in the groups.

A total of 6.8 per cent of the animals checked had ischial callosities. The pads were hairless, brownish-gray, and roughly rectangular, and the skin on them appeared dry and sometimes cracked. Estimated size of the pads in females was about 10 × 5 cm. and in one male about 8 × 8 cm.

Handedness

Yerkes (1928) tested a female mountain gorilla and found that she used her right hand two-thirds of the time when handling objects. Oka and Makoko, two lowland gorillas in the New York (Bronx) zoo, used the right hand more frequently than the left (Riess *et al.*, 1949). Haas (1958) found that the three juvenile lowland gorillas tested by him were primarily right-handed, but that one of these animals was left-handed in certain activities.

At first I attempted to note if feeding gorillas showed a preference in the use of hands. However, my evidence was inconclusive, for the animals reached with whatever hand was nearest the source of food; and if the food was directly in front of them, they were ambidextrous—observations which are similar to those made by Grzimek (1949) on captive chimpanzees and orang-utans. Next I observed which hand was used first to strike the chest during the chest-beating display. The limited data in Table 11 suggest that several males are primarily right-handed in this respect. The right hand was used first to strike the chest in 90 out of 110 observations recorded in the various age and sex classes.

Injuries

Schultz (1939) has shown that gibbons and orang-utans fracture their bones quite frequently. Gorillas too suffer occasionally from serious injuries. One adult male lowland gorilla had lost the right arm above the elbow but remained as the leader of his group (Merfield and Miller, 1956). The same authors also noted a lone male with a deformed leg and a female with a deformed foot and withered arm. Rousseau (pers. comm.) encountered a male with a broken arm and three broken ribs in the Virunga Volcanoes. Wallis (1934) describes an adult male gorilla with the upper anterior part of the lower jaw missing. Schultz (1937) studied the clavicles and limb bones of 127 gorillas and found a total of 9 fractures; Randall

(1943–44) noted that 2 per cent of 123 females and 6 per cent of 213 males had healed fractures of the skull. I found the skeleton of one adult male gorilla at Kabara in which one forearm was severely deformed (Appendix C, Table 71).

Most of the fourteen injuries listed in Table 12 were minor, probably caused by the animal's bumbling into a branch, stepping on a sharp stick, or falling down a slope. Some of the serious injuries (Nos. 11 and 12) are perhaps also attributable to accidents. However, two of the wounds (Nos. 5 and 14) appeared to be bites.

Body Postures

Gorillas commonly assume body postures which closely resemble those of man. I sketched gorillas in the wild in their typical attitudes, and Figures 39, 40, 56, 57, and 63–66 illustrate crudely some of the common positions employed. The photographs of gorillas scattered throughout the report show the quadrupedal and bipedal stances as well as some other postures not included in the sketches. Detailed descriptions of the behavioral context in which various positions are employed are found in several chapters.

The Senses

My acquaintance with gorillas supplied some knowledge of their senses which, though more anecdotal than precise, nevertheless offered some information on their responses to the environment. I encountered the animals under many different conditions, and I found that, in general, their acuteness of sight, hearing, and smell was roughly comparable to my own. It must be emphasized, however, that gorillas are so well attuned to their environment that to the untrained observer their senses seem superior to those of man. When I had established rapport not only with gorillas but also with the sights, sounds, and smells of the forest around me, the animals never reacted visibly to stimuli from the environment which failed to affect me similarly.

Sight

Sight is the most important sense in gorillas, and the animals appear to rely almost entirely on it for final discrimination except in situations obviously needing no visual confirmation. Their vision is acute and they are quick in spotting slight movement. However, their capacity to do so is not superior to that of a man who is reasonably conscious of the surrounding forest.

Apparently gorillas possess greater power of eye accommodation than man. When grooming, they usually watch the activity of their fingers from a distance of 6 inches or less. Similarly, when they examine small objects, such as bits of matter picked from nose or teeth, they hold them so close to the eyes that the animals appear short sighted.

Hearing

Gorillas possess a well-developed sense of hearing, not a poor one as stated by Donisthorpe (1958). However, the animals usually respond only to sounds which are either strange or out of context with their own activity. Thus, when the group feeds noisily, it is not necessary to approach silently, for gorillas do not differentiate a branch breaking under the weight of a shoe from one snapped by a gorilla. If, on the other hand, the group rests, they become alert to unusual noises in their vicinity. A strange sound is immediately pinpointed and checked visually. Once I approached a sleeping group quietly and in full view to within 50 feet. When I sat, the clasp on my rucksack jingled. The male and at least three other animals jumped to their feet. On another occasion a large male failed to hear my camera as he walked by me at 30 feet, but at about 40 feet he jumped forward and ran when it clicked again. The animals heard the whir of my movie camera as far as 100 feet, and its sound startled groups before they saw me. When our campboy unknowingly began to chop wood within 200 feet of a resting group, the gorillas looked in the direction of the sound and craned their necks in trying to spot the cause of it; finally they silently filed away.

The only sound to which I have seen gorillas react with immediate flight is the human voice. Emlen, concerned over my prolonged absence one day, called my name loudly. All members of the gorilla group which I was observing rose silently and fled. On another occasion, an African talked loudly to me as we walked through the forest. A group near our trail again fled silently and continued to move at a fast walk for over a quarter of a mile. Similarly, Akeley and Akeley (1932) noted that ". . . at the sound of my voice all gorillas bolted."

Smell

Although I several times approached resting gorilla groups downwind to within 50 feet or less, the animals failed to detect my presence. Twice, however, they appeared to smell me, and both times

I had perspired profusely. In one instance the group rested in a shallow ravine and the wind eddied unpredictably. I stood motionless behind a tree, with the nearest gorilla at 60 feet and slightly uphill. Suddenly the male grew restless; he displayed four times and the group rose. The animals faced downhill but did not focus on the tree behind which I stood. They milled about, obviously uneasy, until they spotted me. My general impression was that the sense of smell in gorillas is relatively weak, but that it is sufficiently acute to detect strong odors in the environment. Unidentified stimuli are then confirmed visually.

In captivity, gorillas regularly bring objects to their nose as if to smell them. Yerkes (1927) writes about a captive female mountain gorilla: "Visual inspection seldom convinced Congo that an unfamiliar object might not be eaten. If acceptable, it was promptly carried to the nose and mouth, usually both." Lang and Schenkel (1961*a*) noted that the infant Goma at the Basel zoo began to smell and taste unknown objects at the age of four months. Similar behavior can be noticed in man in response to strange food items. In the wild, gorillas sometimes brought an object to the face, but I was unable to determine if the animal merely looked at it, smelled it, or both. However, I saw no instance in which a gorilla raised an object to the nose and definitely seemed to smell it.

Temperament

Any investigator accustomed to the volatile behavior of chimpanzees, so fully described by Köhler (1957) and Yerkes (1943), is apt to find the gorilla an enigma at first. The popular literature ascribes two different temperamental attributes to the gorilla: the animal is said to be excitable, unpredictable, and vile-tempered, or, as expressed by Barns (1922), "of a silent, morose, and even phlegmatic disposition." Their actual temperament will, I hope, become clearly evident from the numerous direct and indirect references to it throughout this report, especially in the section on response to man.

Free-living gorillas exhibit great individual variation in their affective behavior, and the spectrum of emotions exhibited by the members of the same group to my presence varied from excitement to seeming disinterest; in other words, there were nervous individuals and calm ones, aggressive individuals and shy ones. However, in spite of the individual variation, several basic temperamental traits characteristic of the subspecies as a whole soon become readily apparent.

Most gorillas possess an outwardly placid nature which is not easily aroused to excitement. Their usual facial and body expressions, even in disturbing situations, are ones of repose. Gorillas, unlike chimpanzees, never had diarrhea, often a good indicator of internal emotions, in response to my approach. Only once, when I persistently tracked and drove a lone blackbacked male for nearly one mile, was his third defecation atypically soft.

Gorillas, according to Yerkes (1927), have "a markedly shut-in or introverted personality" and this temperamental trait is particularly evident in the paucity of overt social interactions between the adult members of a group. In spite of their close association, they give the impression of independence and of self-dependence; the terms reserved, stoic, and aloof describe them best.

The description of one captive female mountain gorilla given by Yerkes (1927) conforms well with my observations and impressions of gorillas in the wild:

> Whether lacking in affectivity or merely inexpressive of her emotions, she appeared to be strangely calm, placid, even-tempered, and self-dependent. It seemed at times as though she were repressing or inhibiting acts. Especially when confronted with trying situations, such as insoluble problems, disappointment, or disagreeable stimuli, she exhibited often a degree of self-control which was suggestive of stolidity. Her aloofness and air of independence suggested also superiority. She seldom acted impulsively, and a fit of temper such as young chimpanzees and orang-utans frequently exhibit, was never observed.

Arboreal and Terrestrial Locomotion

Gorillas are primarily quadrupedal and terrestrial. Although they climb readily into trees to feed, nest, rest, or just to sit and look around, they do so cautiously. When danger threatens they descend and flee on the ground, a fact also noted by Merfield and Miller (1956) in West Africa.

In discussing the gorilla's arboreal and terrestrial adaptations and the evolutionary importance of the animal's locomotion, Schultz (1927) writes "that the gorilla has changed comparatively recently from an arboreal to a largely terrestrial mode of life. It has been shown that the hand and the foot of gorillas possess in late fetal stages most of the features typical of arboreal, though not extremely arboreal, primates. . . . This all rather strongly suggests that, while attaining its unique colossal size, which rendered tree life impracticable, the gorilla swerved in its evolutionary trend toward that [mode of life] which was most likely followed by the human precursors."

Quadrupedal Locomotion on the Ground

Gorillas spend about 80 to 90 per cent of their waking hours on the ground, either lying, sitting, squatting, or walking around. In the typical stance, and the one used almost exclusively during progression, both in walking and running, the soles of the feet are placed flat on the ground while the anterior part of the body is supported along the whole length of the middle phalanges of the fingers. The quadrupedal gait is similar to that of dogs and human infants, i.e., the animal raises the two diagonally opposite feet and advances them while the other pair supports and propels the body forward. When walking through dense vegetation gorillas frequently push herbs aside with their hands without breaking their stride. When running, they commonly progress slightly sideways with one shoulder in front of the other and the head turned to one side, thus protecting the face from the vegetation. Occasionally gorillas advance the two forelimbs in or almost in unison, immediately followed by the two hindlimbs. This curious gallop is most commonly seen at the beginning and at the end of a run, and in playing young. Females carrying infants with one arm progress like three-legged dogs. For brief periods, especially when holding an object, gorillas sometimes place the entire back of the hand on the ground.

Bipedal Locomotion on the Ground

Gorillas spend roughly one-third of the day in a vertical position on their hindlimbs with their hands freed for such tasks as feeding and grooming. The hindlimbs provide the principal support when the animal squats, either with or without further support by the rump, when it rises to beat its chest, or when it stands to observe something. Playing youngsters advance at each other bipedally for two to five steps. Large infants and juveniles hold on to the rump of females with their hands and progress on their legs alone, sometimes for 300 and more feet. Gorillas surprised by my sudden appearance sometimes wheeled about and ran several steps bipedally, thus gaining momentum before continuing on all fours. One female walked on her legs 15 to 20 feet into the vegetation from the rest area with both hands clenched against her brow—as if she had a headache.

Prolonged bipedal locomotion is rare. Interestingly, on both occasions when I observed an animal to walk 20 or more feet in an upright position it was raining and the animals seemingly hesitated to wet their hands or chest on the sodden herbs. One silverbacked male walked hunched over, arms folded, and with a shuffling gait, for about 20 to 25 feet. A female, holding an infant to her chest, hiked 60 feet slightly uphill and then sat beneath the bole of a tree, thus remaining dry.

Bipedal locomotion confers no obvious selective advantage in the dense vegetation and on the precipitous slopes of the mountains. In fact, such a gait would appear to be undesirable in these situations, and man too is often reduced to the gorilla's quadrupedal mode of progression.

Arboreal Locomotion

Gorillas ascend trees with ease, though rather slowly and deliberately. They grasp a branch and find a foothold for one leg while advancing with the other arm and leg. On vertical trunks without branches they simply walk up, behavior greatly aided by their long arms and short legs. In difficult situations gorillas sometimes advance both arms and then both legs at a time. Vines, branches, and irregularities in the trunk are all used for support. Sometimes a support of dubious quality is seemingly tested by jerking it before the animal trusts it with the whole weight of its body.

In trees gorillas remain quadrupedal. On the whole, their climbing is extremely cautious, in sharp contrast to the chimpanzee. The bulk of adult gorillas is correlated with a reduced agility; at least they rarely jump from branch to branch or run around. They place their hands carefully and firmly. However, their feet are incapable of holding onto branches securely, and I have several times seen gorillas misstep or slip and only save themselves from falling with a firm handhold. Similarly, they appear to be poor judges of what branches can support their weight, and small or dead limbs commonly break beneath them.

Infants and juveniles sometimes hang by one arm alone, slide down vines, turn upside down, and swing back and forth in trees, but adults rarely attempt such feats. On one occasion a silverbacked male suddenly swung free and hung by his arms alone. As he reached for another branch, the supporting limb cracked and slowly collapsed, dumping him on his rump in the vegetation 10 feet below. Although gorillas possess good balance, they occasionally fall. Once a displaying male jarred a tree so strongly that an animal lost its grip and fell screaming about 10 feet. On another occasion a juvenile attained too much momentum during a descent, grabbed at a branch which snapped, and tumbled about 8 feet.

Descent from trees with numerous branches is accomplished feet first and with the chest facing the trunk. If the trunk provides no obstacles the animals merely use the soles of the feet as brakes and slide down hand over hand in a shower of bark. A sloping tree is frequently descended sideways, with the rump somewhat lower than the head, by advancing the leg and arm on one side while retaining a secure hold with the other hand above.

It appeared to me that juveniles climbed trees more frequently than adults. To obtain a numerical basis for comparison, all gorillas which under their own power climbed to a height of 10 feet or more were tabulated between March 29 and September 17, 1960, at Kabara (Table 13). In 74 encounters, primarily with groups IV, VI, and VII, 247 gorillas were seen to ascend trees. These animals were quite well habituated to my presence and thus the amount of climbing was not thought to be affected much by my proximity.

Even though silverbacked males are excellent climbers and numerous sturdy trees grow in the Kabara area, they climb infrequently. Only the dominant male in group VII exhibited a predilection for climbing. Blackbacked males climb with a frequency comparable to that of silverbacked males. Females ascend trees twice as often as silverbacked males, and juveniles over four times as often. Infants less than one and a half years old hesitate to climb high into trees, although they frequently ascend into the lower branches. The youngest animal which I observed to reach a height of 10 feet under its own power was eight months old.

It is commonly stated in the scientific and popular literature that all apes are brachiators. Thus Le Gros Clark (1960) writes: "Only the anthropoid apes among the primates are full brachiators, that is, their movements in trees are exclusively arm-swinging movements and never quadrupedal." Inadequate field observation and repetition seems to have led to a widespread misconception. All apes are apparently brachiators on anatomical grounds (Washburn and Avis, 1958), but only the gibbons and siamangs brachiate extensively. Orang-utans do so occasionally (Shaller, 1961), as do chimpanzees, but I have never seen gorillas brachiate in the wild, although I observed them hanging by their arms and reaching for another branch while their legs hung free. All great apes are essentially quadrupedal climbers.

Jumping

Gorillas jump rarely. I observed them to leap from branch to branch or from a tree to the ground only thirteen times and the distance never exceeded 10 feet. Similarly, Cordier (pers. comm.) noted that gorillas refuse to jump from great heights; when tall trees in which gorillas sat were chopped down by him, the animals merely held on and crashed to the ground. Occasionally an animal jumped when excited by my approach or when hurrying after a departing group. Gorillas jump cautiously, frequently hanging first by both arms, then only by one arm as they scan the vegetation below for several seconds before finally releasing their hold. A juvenile squatted on a branch, looked down, extended its arms in front, and jumped 6 feet. A one-and-a-half-year-old infant grabbed the top of a bush while sitting on a branch and leaped forward to land with a crash 7 feet below. A blackbacked male lunged from one branch to another, caught himself with his hands alone and swung back and forth. However, the branches were so low that his feet nearly touched the ground.

Animals land either on all fours, or on their hindlimbs but with the torso bent forward so that they quickly regain their balance.

Gorilla Spoor

A group of gorillas moving through the forest leaves behind numerous signs which persist for several days and, occasionally, even for months. Recognition of such signs is essential in the study of gorillas, for they are needed in tracking the animals and they aid in estimating relative population densities. They are also invaluable in giving a rough idea of group size and composition, of food habits, of daily travel distances, and of other aspects of behavior.

Nests

The crude, roughly circular structures of herbs and shrubs on the ground, and the platforms of branches in trees, resembling hawk eyries, are the most frequently seen and most lasting evidence of gorillas in their habitat. Nest construction and nesting behavior are discussed fully in another section.

Nests remain clearly visible for months. At Kabara, nests on the ground were often readily discernible after four months even though numerous young herbs had grown up around and in them. One ground nest of bamboo near Rweru was first observed in August, 1959, and was still in relatively good condition in September, 1960. Tree nests may remain visible for a similar length of time.

Trails

When traveling, gorillas tend to progress in single file, but when feeding they spread out and make numerous trails. Their activity

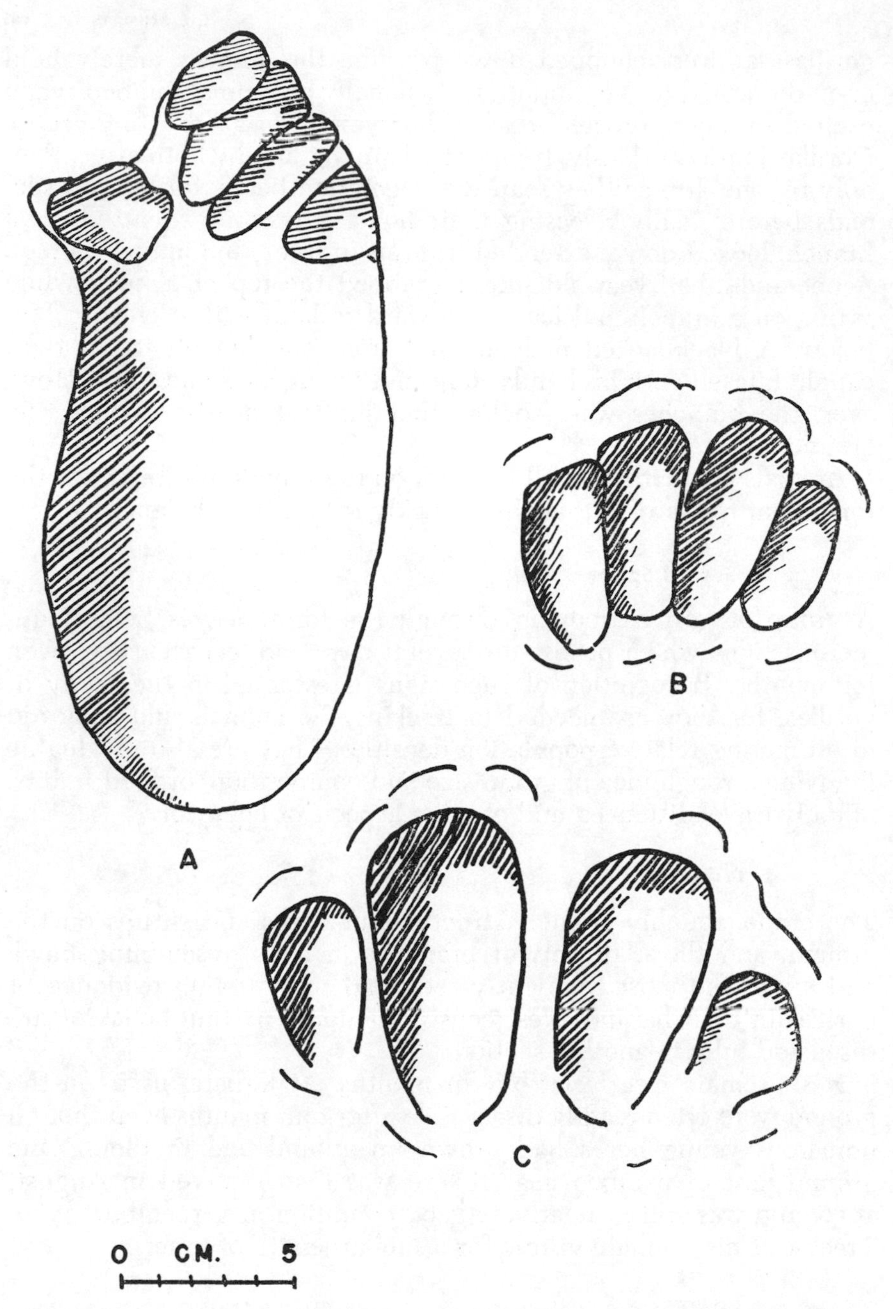

FIG. 14.—Foot- and knuckle prints of gorillas at Kabara. (Redrawn from plaster of Paris casts.) *A*, footprint of a female(?); *B*, knuckle print of a juvenile; *C*, knuckle print of a silverbacked male.

always disturbs the vegetation and where undergrowth is dense their route usually remains conspicuous for several days. Flattened herbs, broken branches, and partially eaten food plants mark the trail. Gorillas congregate intermittently and rest for periods ranging from several minutes to several hours. Such resting sites are easily recognized for the vegetation is characteristically packed down over an area of 20 × 30 feet or more. At least one and as many as four or five such rest stops were made at Kabara by each group during a typical day. Nests are sometimes found in such rest areas, and frequently heart-shaped rump outlines mark the sitting places.

Tracks

Foot and knuckle imprints are frequently discernible in soft sand or earth. Knuckle imprints of a silverbacked male and of a juvenile are illustrated in Figure 14B, C. All sketches were made from plaster casts of tracks which I collected in the wild. Five additional knuckle imprints of silverbacked males measured 14.5 to 16.4 cm. in width;

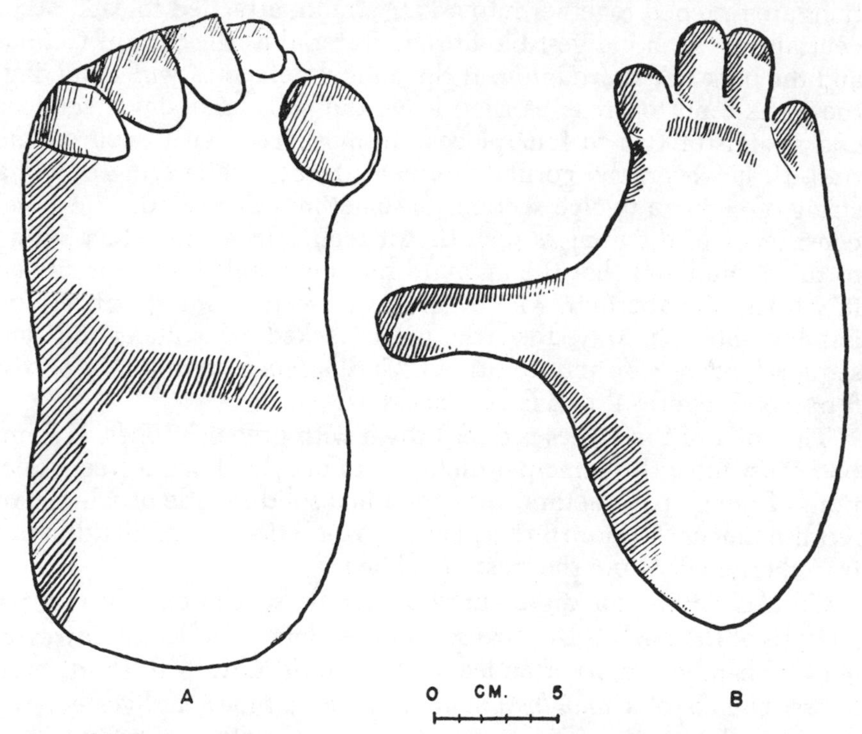

FIG. 15.—Footprints of (*A*) a silverbacked male and (*B*) a blackbacked male at Kabara. (Redrawn from plaster of Paris casts.)

one blackbacked male 13.5 cm.; two females 12 and 14 cm.; and one juvenile 9.5 cm. My own knuckle width is 8.5 cm.

Only the broad heel of footprints is usually visible, although occasionally the toes too are clearly evident. Figure 15 illustrates the prints of a silverbacked and a blackbacked male. Figure 14A probably represents a female. The sketches show that gorillas place their foot either with the big toe abducted or adducted and that the four other toes are either extended (Fig. 15B) or curled under the sole (Fig. 14A, 15A). Measurements of foot length must take into account the position of the toes. With the toes curled under the sole, one foot each of three silverbacked males measured 24 to 26.5 cm. in length, but with them extended the figure was raised to 29 cm. in one animal.

Dung

The dung of gorillas cannot be confused with that of other animals. The boluses or sections are three-lobed, and frequently several sections are fastened together into a long strand, attached to each other centrally by such indigestible fibrous material as the pith of *Galium* and the bark of *Peucedanum.* If the animal defecates while walking, the fibers tend to break as they leave the anus, and single sections and strands of two to four pieces are most frequently seen on the trail. If, however, the gorilla eliminates faeces while lying down, a string of eight to twelve sections is sometimes deposited. The usual consistency of the dung is such that it retains its shape when falling to the ground and that the animal's fur is not soiled when it sits on it. At Mt. Tshiaberimu, where gorillas subsisted almost entirely on bamboo shoots in May, 1959, the faeces lacked gross fibers and consequently were soft and unformed. Gorilla faeces frequently persist for several months if rain is not excessive.

The color of the faeces is dark brown with greenish tones. If bamboo constitutes the principal diet, the color is yellowish-green. Sections of dung are sometimes wrapped in a solid mantle of whitish or reddish mucus. Figure 16 represents cross-sections of gorilla dung as traced originally from the fresh specimen.

Gorillas chew and digest most materials so thoroughly that an analysis of food habits by means of faeces alone is difficult. However, chewed bamboo stems often leave characteristic wads of short, hard fibers; chunks of *Cynoglossum* roots pass at times undigested; the spongy, white pith of *Senecio erici-rosenii* is often apparent; and the seeds of *Stephania, Rubus, Pygeum, Myrianthus,* and *Xymalos*

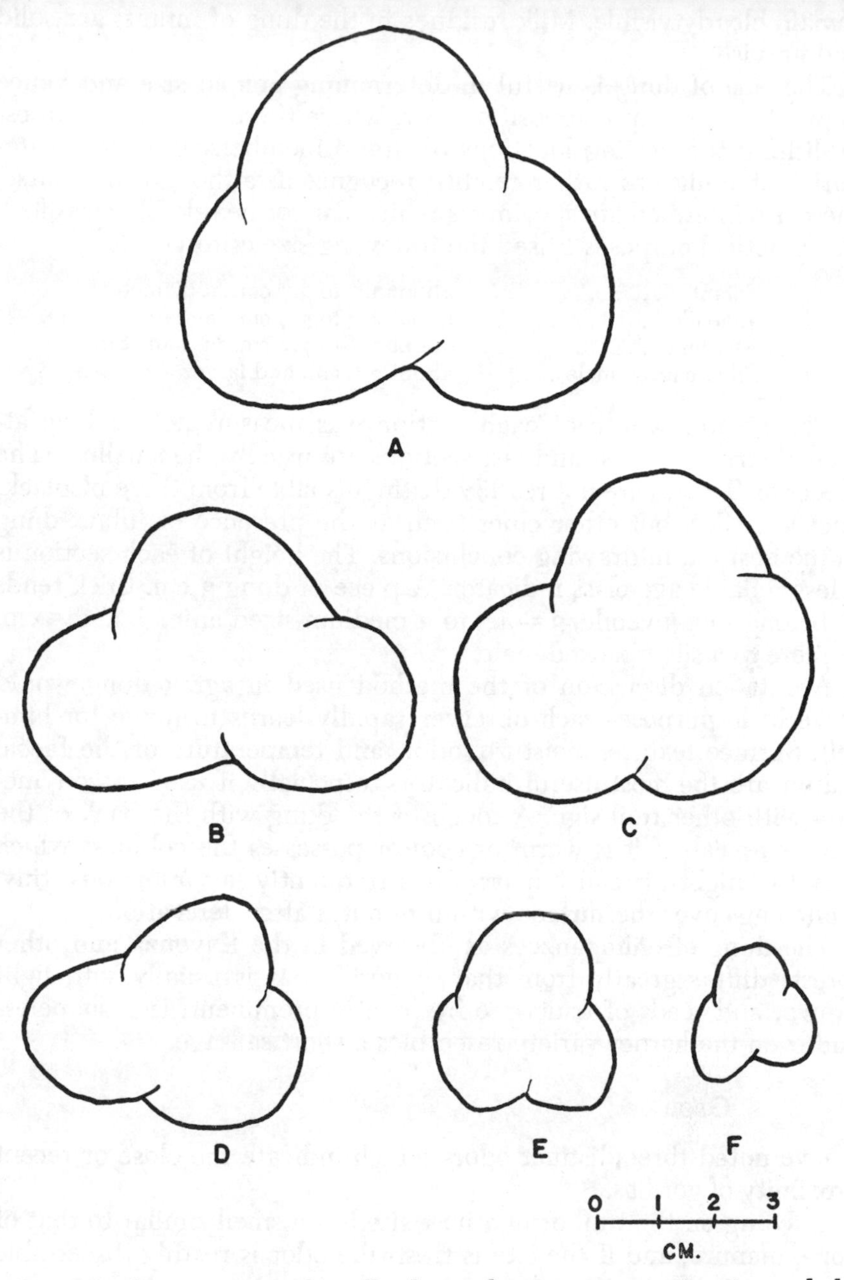

FIG. 16.—Cross-sections of gorilla dung, showing comparative size and the characteristic three-lobed shape. *A*, silverbacked male; *B* and *C*, females; *D*, juvenile; *E* and *F*, infants.

remain clearly visible. Milk residues in the dung of infants are solid and grayish.

The size of dung is useful in determining animal size and hence approximate group compositions, or, where these are known, in establishing the nesting locations of group members. Four size categories of dung are rather readily recognized, although, of course, there are intermediate specimens which cannot be reliably classified. For practical purposes I used the following size criteria:

Infant	about 1.0 to 3.9 cm. in diameter
Juvenile	about 4.0 to 5.4 cm. in diameter
Medium	about 5.5 to 7.2 cm. in diameter
Silverbacked male	about 7.3 cm. and larger

The greatest width of each section was measured. In a long attached string the first and last sections are usually the smallest. The faeces of females are not readily distinguishable from those of black-backed males, but other clues such as the presence of infant dung in the nest aid in drawing conclusions. The height of each section is a less reliable age-class indicator. A piece of dung 3 cm. thick tends to belong to a juvenile, 3.5–4.5 to a medium-sized animal, and 5 cm. or more to a silverbacked male.

A detailed discussion of the method used in aging dung would serve little purpose—each observer rapidly learns to judge for himself. Surface texture, moisture, odor, and temperature of the faecal matter are the most useful indicators, especially if used in conjunction with other trail sign. A touch of the dung with the back of the fingers reveals if it is warm or cool or possesses the coldness which only the night air can impart. Flies frequently lay numerous, tiny, white eggs over the surface within minutes after defecation.

The dung of chimpanzees as observed in the Kayonza and other forests differs greatly from that of gorillas. It is usually soft, light brown, and seeds of fruits are frequently prominent; it is shapeless and even the harder variety resembles a short sausage.

Odor

I have noted three distinct odors which indicate the close or recent proximity of gorillas.

1. Dung on the trail or in a nest site has a smell similar to that of horse manure, and if the site is fresh the odor is readily discernible at 50 or more feet.

2. The hair and skin of gorillas smells slightly sweet. I noted this

in captive animals and sometimes also in wild ones when less than 10 feet separated me from them, especially on wet days.

3. The most potent and unmistakable odor of gorillas "is rather like a combination of very pungent human sweat, manure, and charred wood" (Donisthorpe, 1958). I compared it to the smell of distant burning rubber in my notes. When this odor was discerned I knew that gorillas were within 100 feet of me, although on occasion I noted it as far as 150 feet. I think the strong odor is emitted primarily by silverbacked males, but I was unable to prove this point to my satisfaction. The odor becomes especially noticeable when the group is excited. On some days I smelled it when the group fed or rested leisurely, but on others I was unable to perceive it even though a breeze blew in my direction.

I was unable to determine the source of the odor. Although gorillas were observed to sweat while lying peacefully in the sun, the odor was not apparent at that time, suggesting that sweat alone is not the cause.

CHAPTER 4

POPULATION DENSITY, STRUCTURE, AND BEHAVIOR

Population Densities and Numerical Estimates

Some of the difficulties in estimating densities and numbers of gorilla populations were discussed by Emlen and Schaller (1960*a*):

> Intense surveys of local gorilla populations were conducted on five plots in three areas in the northeastern part of the range. It was hoped that these surveys would provide indications of population densities in these areas. We found, however, that, because of the wandering habits of the gorilla, a survey conducted on a plot of 5 to 10 square miles over a period of a few weeks may be quite atypical of even local situations. On one 10-square-mile plot, for instance, the population was known to have nearly quadrupled through immigration shortly after our twenty-day survey was completed; on another, six days of intense work over four of the six square miles revealed no fresh sign; then two troops were discovered together at the western end of the plot.

Further difficulties are encountered when several groups of approximately the same size frequent the same area, for recognition of each group is essential in determining the number of groups, and hence individuals, involved. Even reasonably accurate estimates can only be based on long residence in one location.

Counts of nests, dung, and other indirect evidence can, however, furnish indices of population density for comparative purposes. Although such figures convey little information concerning the actual number of animals in a locality, they are, nevertheless, useful in determining if gorillas are more abundant in one area than in another. Counts of trail sign involve some bias, for nests are more easily seen in open habitats than in dense ones, and the human element further intrudes when the observer tends to choose the easiest transect route. In the Kayonza Forest, for example, travel on the ridges is more pleasant than along the tangled slopes. Our first transect routes at Kabara inadvertently fell in the area least frequented by gorillas.

Table 14 presents comparative densities of nest sites in several areas, obtained by hiking cross-country and counting all sites, regardless of the number of individual nests, in the course of travel. The highest population was noted at Kabara, the lowest in primary and old secondary forest of the Utu region. These figures were generally confirmed by the number of direct encounters with gorillas and by the frequency of other trail sign such as dung and discarded food plants.

The number of gorillas per square mile also varies from area to area. Table 15 summarizes data for three isolated populations for which it was possible to obtain a fairly reliable numerical estimate. The Virunga Volcanoes support about twice as many animals per square mile as the other regions surveyed. The extensive mountain forests of the Mwenga-Fizi region, and the expanses of lowland forest in the Utu region support a population which is probably somewhat lower than that indicated for the Kayonza Forest and Mt. Tshiaberimu. This would give a rough figure of one gorilla per square mile for the subspecies as a whole.

The literature contains several estimates of the number of mountain gorillas in existence, but no bases for the statements are given. The entire population has been estimated to number in the "thousands" (Gatti, 1936), 2,000 (Burbridge, 1928), 1,000 to 1,500 (Blower, 1956), and 5,000 to 6,000 (Schäfer, 1960). After a six-month survey of gorilla distribution we tentatively presented a figure of 3,000 to 15,000 animals (Emlen and Schaller, 1960*a*).

Several guesses have been made for specific areas. Akeley (1923) placed the number in the Virunga Volcanoes at 50 to 100 animals, Mathis (1954) at "très peu nombreux," Derscheid (1927) at 600 to 850, Hoier (1955*b*) at 700 to 800, and Johnson (1931) at 2,000. The Annual Report of the Game Department, Uganda, for 1925 (1927) estimated the entire Uganda population at 100 individuals. Pitman (1942) placed the number in the Kayonza Forest at 80. Blower (1956) judged that about 50 animals frequented the Virunga Volcanoes on the Uganda side. Johnson (1931) guessed that 20,000 gorillas occurred in the Mt. Tshiaberimu region.

As I pointed out earlier, the number of gorillas per square mile varies from area to area, and population estimates are extremely difficult to make. Even after spending one year in the Virunga Volcanoes, I am able to present only a rough figure for the total population in these mountains. However, there is a legitimate demand for estimates, especially for the small isolated pockets which are threatened with destruction.

Virunga Volcanoes

The difficulty of estimating populations is well illustrated in the data from the Virunga Volcanoes, where guesses have ranged from 50 to 2,000. As a tentative figure I placed the number in January, 1960, at "at least 350" (Schaller, 1960). Further work suggests a total population of about 400 to 500 animals, distributed as outlined below.

Eastern sector Through constant tracking over a period of five years several groups of gorillas have become well known between Mts. Sabinio and Muhavura and these range freely between Ruanda and Uganda. Kawai and Mizuhara (1959*a*) noted four groups comprising thirty-nine animals, the highest number ever recorded in a short period in the 25-square-mile area. The narrowness of the forest habitat precludes the possibility that several groups remain only in Ruanda without venturing into Uganda. On the basis of known animals a population of forty to fifty seemed indicated as of 1959.

The broad saddle between Mts. Sabinio and Visoke supports a low population of gorillas. I obtained evidence of about four small groups in July, 1959. It is doubtful if more than fifty to seventy-five animals frequent the saddle regularly.

Central sector During my work at Kabara I identified about two hundred gorillas in eleven groups. The total area covered by their home ranges is probably on the order of 30 square miles. In addition, good habitat, which was not visited, occurs on the slopes of Mt. Karisimbi in Ruanda, on the western slopes of Mt. Mikeno, and in the flats north of Bishitsi. Probably at least one hundred additional animals can be added to the known two hundred, bringing the total for this area to three hundred and perhaps more gorillas.

Kayonza Forest

I spent two days in the western sector of the forest and found evidence for only one group. Itani (pers. comm.) saw very few signs of gorillas in the southern part of the Kayonza Forest. This suggests that the central, northern, and eastern portions harbor the main concentration of the animals. In 1959 Emlen and I transected about 15 square miles of terrain in the central and northern sectors and contacted or found fresh evidence for four groups comprising about thirty animals. Work in the eastern portion for one day revealed fresh sign of one large group. Thus, work in about one-third of the total forest area showed a population of at least fifty animals. For the whole forest I tentatively suggest 120 to 180 animals.

Mt. Tshiaberimu

In two weeks of intensive field work I obtained definite evidence for about four groups comprising roughly twenty animals. On this basis the total population which frequents the Tshiaberimu massif does not seem to exceed thirty to forty animals.

An estimate of the total population of mountain gorillas cannot be made with any degree of accuracy at this stage of knowledge. Earlier I tentatively derived an average population density of one animal per square mile. If this estimate is acceptable, then about 8,000 mountain gorillas exist in the 8,000 odd square miles of forest inhabited by this subspecies. However, until further work clarifies the abundance in more regions, attempts at precise estimates are of little value. Our work suggests that at least 5,000 but no more than 15,000 mountain gorillas inhabit the forests of Central Africa.

Whether the population is increasing, decreasing, or maintaining a certain level is a moot point, for there are no accurate historical data upon which a comparison can be based. A high population density in the Utu region appears to be dependent on the availability of secondary forest which is primarily created by man. Human activity was more extensive fifty years ago than it is now and it is possible that the gorilla population has decreased somewhat as the forests have been allowed to mature. The gorillas in the Mt. Tshiaberimu region are probably decreasing with the destruction of their habitat and with extensive hunting. The game department of Uganda estimated in 1949 that there had probably been a 50 per cent increase in the Kayonza Forest gorilla population in twenty years (Anon., 1951), but no basis for the statement is given. The animals in the Virunga Volcanoes are presumably increasing under the protection given them.

Population Dynamics

This section contains the meager information which I was able to obtain on the factors governing the turnover of gorilla populations. Some of the data were not readily obtainable in the wild. Thus the age at sexual maturity and longevity figures could only be noted with precision in captive gorillas. Nearly all the data from the wild are based on a cross-sectional study of the Kabara groups during one year; birth rates, mortality rates, and population composition as presented reflect only the characteristics of that population. It should be remembered that the groups at Kabara live in a highly favorable environment with a bountiful supply of forage, limited predation by man, and in a temperate zone where numerous diseases characteristic of warmer climates are absent.

Age at Sexual Maturity

Three pairs of lowland gorillas have reproduced in zoological gardens, and these furnish the only definite data on age at sexual maturity. In addition several females have shown behavior which is indicative of their reproductive state. Haddow (1952) has pointed out that primates in captivity tend to mature somewhat more rapidly than those in the wild. This should be considered in evaluating the data below. Table 16 lists the approximate ages and dates in the development of sexual behavior in the three pairs which produced offspring. The ages are probably accurate to within half a year. The results show that two males were about nine years old when they impregnated the females even though they had copulated for some time previously; the third was only about seven years old. The females at the Columbus and Washington zoos were seven to seven and a quarter years old at conception; the female at the Basel zoo conceived at the age of ten years.

Cyclic sexual receptivity displayed toward the keeper was first shown by Sumaili, a mountain gorilla at the New York (Bronx) zoo, at the age of about seven years (Quinn, pers. comm.). Congo, a mountain gorilla studied by Yerkes (1927), behaved similarly between the ages of six to seven years (the age is my estimation) although she performed thrusting movements against a dog one year earlier. A male, Mambo, in the New York (Bronx) zoo showed no copulatory behavior in response to frequent invitations by a female until the age of nine years.

This evidence suggests that, on the average, females reach sexual maturity between the ages of six to seven years and males between the ages of nine to ten years. Data from the Washington zoo show, however, that viable sperm in captive males may be present by the age of seven years. These ages are considerably lower than some previous estimates. Burbridge (1928), for example, maintained that females reached adulthood at twelve to fifteen years, and Dekeyser (1955) claimed that males were not sexually mature until eighteen years of age.

The data from zoological gardens find general confirmation in the wild. Females do not remain without child long after they outgrow the size of a typical juvenile at about six years of age. The onset of sexual maturity in males could not be determined in the wild, but two records from zoos indicate that it might be correlated with accelerated growth and a development of such secondary sexual characteristics as the angular body build. This occurs in nature when blackbacked males begin to grow silvery hairs on their backs at about nine years of age, a process which continues for at least two to three years. Thus silverbacked males, which are fully adult both physically and sexually and are the leaders of groups, are probably at least twelve years old.

Birth Rates

No female was seen to carry more than one dependent infant. In the few instances when more than one infant associated with a female, the extra one had left its own mother for a short time. Infants remain with their mothers for about three years, and accordingly at least that amount of time elapses between births. Several juveniles about four years old associated extensively with infant-less females, presumably their mothers. A few juveniles of similar ages approached females with very small infants. This suggests that females give birth about every three and a half to four and a half years. Of twenty-seven females in four groups whose status was traced for at least eight consecutive months, only two females out of seven which had no infant at the beginning still had not given birth at the end of the study. These two females still lacked infants after eleven and twelve months, respectively. Both were elderly and physically below par: one had a cancerous-looking eye, the other a skin rash. If a female produces an offspring about once every four years, her life total is probably on the order of four or five young, unless several die in infancy.

Mortality Factors

Evidence for three general causes of mortality was obtained.

1. Injuries, such as those enumerated earlier, probably result in the death of some animals. Most injuries appeared to be mechani-

cally produced, like falls from trees and collisions with pointed branches, but some others seemed to be bites.

2. Mortality through predation by leopards and other carnivores is probably negligible. Man, however, is undoubtedly an important predator on gorillas, hunting the animals primarily for food. Snares, pitfalls, nets, guns, spears—all are used in hunting even though the animal is legally protected throughout its range. Emlen obtained thirteen fresh gorilla skulls from one area south of Mt. Tshiaberimu, and a missionary in the same region remembered that in 1928 natives killed and ate eleven gorillas in one day. By his own admission, one Belgian mine official in the Utu region shot nine gorillas for sport.

3. Diseases—especially viruses, bacteria, and various blood and intestinal parasites—are probably the major cause of death in gorillas. In West Africa gorillas are infected with the tick *Rhipicephalus appendiculatus* which carries tickbite fever in man (Dekeyser, 1955). Photographs published by Schultz (1950) show lowland gorillas with what appear to be yaws. Rousselot and Pellissier (quoted in Ruch, 1959) noted that gorillas suffer from a bacterial skin infection which resembles leprosy. Ruch (1959) records flukes (*Eurytrema brumpti*) from one lowland gorilla, the malaria-causing *Plasmodium reichenowi*, the intestinal protozoan *Troglodytella gorillae*, and the nematode *Oesophagostomum stephanostomum* from others. A captive infant died of an inflammation of the appendix (Grzimek, 1957*b*).

Van den Berghe and Chardome (1949) noted *Microfilaria gorillae* in the blood of a mountain gorilla. Mountain gorillas from Mt. Sabinio and Mt. Kahuzi were infected with the cestode *Anoplocephala gorillae* (Gyldenstolpe, 1928; Sandground, 1930). A male from Kisoro contained *Ascaris* and the ova of *Anaglostoma* (Wilson, 1958); the veterinary report (Anon., 1960) of another male from Kisoro read: "Hookworm in numbers which *may* be significant." The crablouse, *Phthirus*, has been collected on a mountain gorilla (Zuckerman, 1933). Peden (1960) diagnosed the cause of death in a male at Kisoro as gastroenteritis; the stomach and intestine showed inflammation and petechial hemorrhages on the mucous membrane. An adult male mountain gorilla at I.R.S.A.C., Lwiro, died of cirrhosis of the liver after about four years in captivity; one female died of pneumonia soon after capture; and one old female with missing canines also died soon after capture, revealing a "spongy heart" when autopsied (Cordier, pers. comm.).

Fox (1939) found arthritic conditions in fifteen out of eighty-nine gorilla skeletons examined, and Stecker (1958) reported similar pathology. Alveolar abscesses are common in old gorillas (Schultz, 1950), and Colyer (1936) found carious cavities in five adults out of a total of 689 museum specimens checked. Gaffikin (1949) noted a periodontal abscess around the lower left lateral incisor of a black-backed male shot at Kisoro.

On the whole, the Kabara population appeared robust and healthy, except for a few physical injuries and for some animals which showed symptoms resembling those of the common cold. I examined numerous pieces of dung but observed no helminths. However, I preserved forty-five dung samples from four groups and a lone male in alcohol and examined them under the microscope. The ova of a nematode, resembling those of the human hookworm, were found in 53.3 per cent of the samples, suggesting that the whole population may have been infested with the parasite (Table 17). Several pieces of dung were encased in a white or reddish mucous cover, which suggests an enteritic condition perhaps attributable to this nematode parasite.

Although this list of gorilla diseases is incomplete, it is sufficiently long to illustrate that the animals suffer from a wide variety of illnesses. Even minor ailments, such as the nematode infestation at Kabara, can lower the resistance of an animal, especially when it is old, and make it susceptible to more serious disorders.

Longevity

Only a few animals have lived longer than twenty years in captivity. The data on most of these animals as of December, 1961, are listed below:

Name	Zoo	Sex	Present Age	Age at Death	Source
Bushman	Chicago	♂	..	21	Yerkes, 1951
Bamboo	Philadelphia	♂	..	34½	Philadelphia zoo, pers. comm.
Massa	Philadelphia	♂	31	...	Philadelphia zoo, pers. comm.
Susie	Cincinnati	♀	..	22	Walker, 1954
Oka	New York	♀	22	...	New York (Bronx) zoo, pers. comm.

The ages of wild adult gorillas are impossible to estimate except in very general terms. Since males do not reach full adulthood until about twelve years of age they probably live at least twenty years. According to Emlen, Bamboo at the Philadelphia zoo looked definitely old at the age of about thirty-one years. Seemingly very old males and females in the wild are uncommon. On this basis I suggest that the longevity of gorillas is generally about twenty-five to thirty years in the wild.

Population Structure

Only the ten groups in the Kabara area were studied sufficiently to permit a detailed analysis of their population structure. Table 18 presents (a) a summary of the population in groups at the beginning of the study, and (b) the total known Kabara population, including all lone males, births, and animals which had joined or left groups by the end of the study. Two infants which died within three months after birth have been deleted. The number of silverbacked males is probably somewhat low for not all lone animals were individually recognized by me.

Male-female ratios The disparate ratio of males to females (the socionomic sex ratio of Carpenter, 1958) is characteristic of numerous organized primate groupings including the gorilla. In Table 18A, males, silverbacked and blackbacked combined, constitute 18.9 per cent of the population and females 36.7 per cent, a ratio of roughly two females to one male. If, however, the ratio of all males, including the lone individuals (Table 18B), to all females in the population is considered (the tertiary sex ratio of Carpenter), the numbers and percentages change to 43 (22.5%) males and 65 (34.1%) females, or about one and a half females to one male. Although additional lone males undoubtedly exist around Kabara, their number is clearly not sufficient to raise the figure to the expected 1:1 ratio. No clear explanation for the inequality in the sex ratio is available.

Unfortunately, infants cannot be sexed with certainty in the wild. Juveniles, however, behaved and looked sufficiently different on occasion for me to be reasonably certain of their sex. Thus, two of the four juveniles in group VII appeared to be males; at least one of the three in group IV was a male; and group VIII contained a very small blackbacked male, and what seemed to be one male and two female juveniles. My impression was that even if a difference in sex ratio existed in juveniles, it was not sufficiently large to account for the inequality among adults.

On this basis I suggest that the difference in the sex ratio of adults is due to a somewhat higher post-juvenile mortality rate among males than among females.

Female-young ratios About 45 per cent of the Kabara population was either juvenile or infant. Although thirteen females are recorded as not having infants (Table 18B), eight of these were not observed for longer than three months, one for six months only, and two had infants which died. Only two females out of twenty-seven whose history was traced for eight months or longer remained without infant and both still lacked them after eleven months.

Of the fifty-four infants present at some time in the population, about twenty-seven young were born between July, 1957, and December, 1958, and twenty-seven between January, 1959, and September, 1960 (see Table 59), thus showing a steady increment. Using the natality statistic births/1,000 as applied to humans, these Kabara gorillas showed a birth rate of about 17/193 or 90/1,000 per year.

Juvenile-infant ratios Infants outnumber juveniles fifty-two (27.2%) to thirty-one (16.2%). Even if three very small blackbacked males should perhaps have been classed as juveniles, the ratio remains at roughly one and a half infants to one juvenile, or about 30 per cent fewer juveniles than infants.

Infant and Juvenile Mortality Rates

My specific data on mortality are limited to a few isolated instances. Of the thirteen infants which were born between August, 1959, and August, 1960, one died, one disappeared and was undoubtedly dead, and one was so seriously wounded that it probably died shortly after I last encountered the group. Thus in this twelve-month sample period the mortality of infants during the first year of life was 23 per cent.

The following computations based on Table 18B give an indication of the mortality rate of gorillas during the combined infant and juvenile stages. The data are based on males, because females cannot be aged after they pass the juvenile stage.

a) Number of young born per year = about 17.
   If the sex ratio of males to females is 50:50 at birth, then 8.5 males are born per year.
b) Number of males which move from the juvenile to the blackbacked stage = 18 in 4 years or 4.5 males per year.

Therefore, between birth and the age of six years males show about a 47 per cent decline in number.

Although crude, these figures on mortality suggest that 40 to 50 per cent of the gorillas in the Kabara area die during the first six years of their life, and that the mortality rate is highest during the first year of the infant's life.

Group Dynamics

Accurate tallies of group sizes and compositions have rarely been reported, for observers usually obtained only fleeting glimpses of gorillas in dense vegetation. Donisthorpe (1958), for example, spent eight months at Kisoro in an area of less than 10 square miles without obtaining precise information on group dynamics. Fortunately efforts initiated by Osborn and Donisthorpe and continued over a period of three years by Baumgartel, his guides, and visitors to the Kisoro area have yielded some valuable data. In the Utu region Cordier has trapped whole groups. These two sources provide useful additions to the information obtained in this study.

In most areas both direct observation and such indirect means as careful scrutiny of nests and trails were employed in collecting data on group dynamics. At Kabara my best information was based on individual recognition of all members of groups II, IV, V, VI, VII, VIII, several members of groups I, III, IX, XI, and seven lone males.

Group Size

The literature contains remarkably little precise information on group size. In West Africa, von Oertzen (1913) once counted an aggregation of sixteen nests and Reichenow (1920) of thirteen nests. Trouessant (1920) claimed that groups of four to five animals were the rule. More recently Sabater Pi (1960) noted groups of five and eleven animals, Geddes (1955) of ten animals, Sabater Pi and de Lassaletta (1958) of twenty-three animals, and Merfield and Miller (1956) of two, four, nine, and ten animals.

Several estimates of group size are available for the mountain gorilla. Barns (1923) maintained that they usually traveled in groups of six to eight. Lönnberg (1917) noted groups of twenty to thirty, Derscheid (1927) groups of seven to forty-three, and Blower (1956) groups of five to fifteen.

Gyldenstolpe (in Prince Wilhelm of Sweden, 1923), Bingham (1932), and Johnson (1931) all recorded aggregations of over thirty nests. It is likely that such large sites either represent nests of two groups or that nest sites of two different ages overlapped.

At Kabara the smallest group consisted of five animals, the largest of twenty-seven (Table 19), but changes occurred in one group which brought its count temporarily up to thirty. The mean of ten groups in the Kabara area was 16.9 animals.

At Kisoro the smallest group noted by various investigators was three, the largest eighteen. Counts recorded by several observers in the area include the following: Osborn (1957) three, four, five, and about twelve; Donisthorpe (1958) three, five, six, and eleven or twelve; Bolwig (pers. comm.) five and six; Emlen and Schaller, five and seven; Kawai and Mizuhara (1959*b*) five, seven, nine, and eighteen. Most of these represented counts of the same groups, and variations in number were due to births, deaths, and, occasionally, ingress and egress. From the evidence of the various observers it appears that about five or six groups frequented the general area at one time or another. If a tentative mean is taken of the six groups listed in Table 20, the average group size is between seven and eight animals or only half that of Kabara.

Pitman (1942) saw groups of five and eight animals in the Kayonza Forest and he was told of one containing over twenty. Kawai and Mizuhara (1959*b*) noted groups of five, thirteen, and seventeen animals. I obtained counts of about two, five, five, seven, twelve, fourteen, and fifteen animals per group. The groups with twelve and fifteen animals were contacted in the same area one year apart and perhaps represented the same gorillas. A miner saw a group of eight animals several times. The average size of the six groups listed in Table 20 is seven animals.

At Mt. Tshiaberimu the largest nest site contained eight nests, and most others varied from two to six. Freshkop (1953) saw groups of seven and twelve in the mountains to the southwest of Mt. Tshiaberimu. Near Mt. Kahuzi, Emlen and I were shown a site with seven nests. In the Fizi-Mwenga region nests and dung revealed one group consisting of at least thirteen animals. Cordier (pers. comm.) has trapped or counted groups of four, thirteen, fourteen, fifteen, nineteen, and twenty-five in the Utu region; I examined a fresh nest site with only three nests and two others which contained six nests each.

Thus, the size of gorilla groups varies from two to thirty animals. Average group size probably lies between six and seventeen animals, but the number varies from area to area. At Kabara, where food is plentiful and the animals are not hunted, the average size of the groups is considerably higher than in other locations. However, there is no difference between populations, for groups in the Utu region, where gorillas are heavily hunted, occasionally are as large as those at Kabara. Hunting, however, is not the only factor limiting group size. At Kisoro and in the Kayonza Forest gorillas are not hunted much, yet the size of groups is relatively small if compared to those at Kabara. This suggests that perhaps factors such as availability and type of forage may regulate the size of groups.

The social and environmental factors which regulate group size over long periods of time are little understood. However, it seems likely that maximum social stability is found when aggregations are neither so large that cohesiveness is lost in the forest nor so small that social intercourse is limited.

Group Composition

Yerkes and Yerkes (1929) summarized the existing gorilla information without being able to present a single precise group composition. With the exception of Merfield and Miller (1956), publications after 1929 added little until work began on the mountain gorilla at Kisoro. Both Osborn (1957) and Donisthorpe (1958) presented some tentative figures on compositions, and later other investigators, including ourselves, added to the knowledge of that particular area.

Table 19 presents the composition of ten groups at Kabara at the time of the first accurate count of each group. The counts are complete, with the possible exception of group IX which may have contained seven infants rather than six.

Group compositions obtained in other study areas are frequently not exact. All are listed in Table 20, but group counts which may be incomplete are marked with an asterisk. Because of rapid and imperfectly understood changes at Kisoro, I have not recorded compositions from this area after 1959. Some of the data in Table 20 are from nest sites alone, and all undetermined animals are placed in the "unidentified" column.

From Tables 19 and 20 it is evident that all groups contained at least one silverbacked male, one or more females, and a variable number of young. More than one silverbacked male was found in five of the twenty-seven mountain gorilla groups listed, and all five of these groups contained more than ten animals each. At Kabara only the two smallest groups lacked blackbacked males. Females usually outnumbered the combined silverbacked and blackbacked males in the larger groups, and juveniles and infants made up about 45 per cent of the Kabara population.

The average group at Kabara consisted of 1.7 silverbacked males, 1.5 blackbacked males, 6.2 females, 2.9 juveniles, and 4.6 infants. My data are too scanty to treat groups from other regions in this manner. However, if all twenty mountain gorilla groups for which compositions were established are taken from Tables 19 and 20, the average group shows the following approximate composition: 1.5 silverbacked males, 0.9 blackbacked males, 4.1 females, 1.6 juveniles, and 3.0 infants.

Changes in Group Composition

Only at Kabara and Kisoro have groups been followed over a sufficiently long period of time to make it possible to trace changes in the composition and to determine the fate of individuals. At Kabara I was able to check several groups at frequent intervals over a period of one year. At Kisoro, Baumgartel and his guides have been able to follow the changes in one group from 1957 to 1961.

Kabara For the sake of completeness I have recorded all changes in the composition of groups II, IV, V, VI, VII, and VIII below.

GROUP II

Months in which checked: August, October, November, 1959.

Composition on August 25, 1959: 1 silverbacked male, 3 blackbacked males, 6 females, 5 juveniles, 4 infants = 19.

On the night of August 26–27, 1959, groups I and II nested 150 feet apart, and apparently one animal from II, probably a blackbacked male, nested at the periphery of group I. However, no changes in composition occurred.

GROUP IV

Months in which checked: March, August, September, October, 1959; January, March, April, May, August, 1960.

Composition on September 1, 1959: 4 silverbacked males, 1 blackbacked male, 10 females, 3 juveniles, 6 infants = 24.

Two additional animals were present in September than in March, 1959, but this very likely reflects an incomplete count during our brief first encounter. Four silverbacked males were noted in March, 1959. Although at least two of these, including the dominant animal, appeared to be the same ones seen later in September, I was not completely certain of the identification.

March 12, 1959. An infant was born.

August 28 to August 30. A peripheral silverbacked male (The Outsider) apparently left, then rejoined the group.

September 9. A new silverback (The Newcomer) joined the group. The Outsider left the group but rejoined it by September 22.

September 18 to 20. An infant was born.

Between October 2, 1959, and January 11, 1960. The Outsider (peripheral) and D. J. (the No. 2 male in the dominance hierarchy) left. A new silverbacked male (New No. 2) and two females, both with infants, were added, probably representing a small group which joined. An infant was born in late December.

April 24, 1960. A silverbacked male (Splitnose), who had been with the group at least since August, 1959, left.

April 25, 1960. An infant was born but died two days later.

Between May 1 and 15. A new silverbacked male (The Lone Stranger) joined the group and remained at least until May 24.

Between May 24 and August 12, 1960. The Lone Stranger left.

Final composition on August 13, 1960: 3 silverbacked males, 1 blackbacked male, 12 females, 3 juveniles, 10 infants = 29. Thus in the course of 12 months, from August, 1959, to August, 1960, at least 7 different silverbacked males associated with the group (see Fig. 17), 3 infants were born, and 2 females with infants were added.

GROUP V

Months in which checked: September, October, November, 1959; January, April, May, August, 1960.

Composition on September 18, 1959: 2 silverbacked males, 2 blackbacked males, 3 females, 2 juveniles, 2 infants = 11.

Between September 20 and October 30. Two silverbacked males and 2 juveniles joined the group.

Final composition on August 12, 1960: 4 silverbacked males, 2 blackbacked males, 3 females, 4 juveniles, 2 infants = 15.

GROUP VI

Months in which checked: October, November, December, 1959; January, February, March, April, May, June, 1960.

Composition on October 2, 1959: 1 silverbacked male, 1 blackbacked male, 9 females, 2 juveniles, 7 infants = 20.

Between October 3 and November 14, 1959. A distinctive female without infant disappeared and was never encountered again.

November 18, 1959. A silverbacked male (The Lone Stranger) was at the periphery of the group, but he was seen alone the following day.

January 9, 1960. A silverbacked male and a blackbacked male were at the periphery of the group. Both were with the group on January 10, but neither was present on the following day.

First half of January. An infant was born.

February 26, 1960. An unknown silverbacked male was seen and heard at the periphery or near the group for at least 2 days.

Between March 9 and April 9. The infant born in January disappeared from the group and presumably died, for its mother remained in the group.

Final composition on June 1, 1960: 1 silverbacked male, 1 blackbacked male, 8 females, 2 juveniles, 7 infants = 19.

GROUP VII

Months in which checked: October, November, December, 1959; January, February, March, April, May, August, September, 1960.

Composition on October 10, 1959: 1 silverbacked male, 2 blackbacked males, 6 females, 4 juveniles, 5 infants = 18.

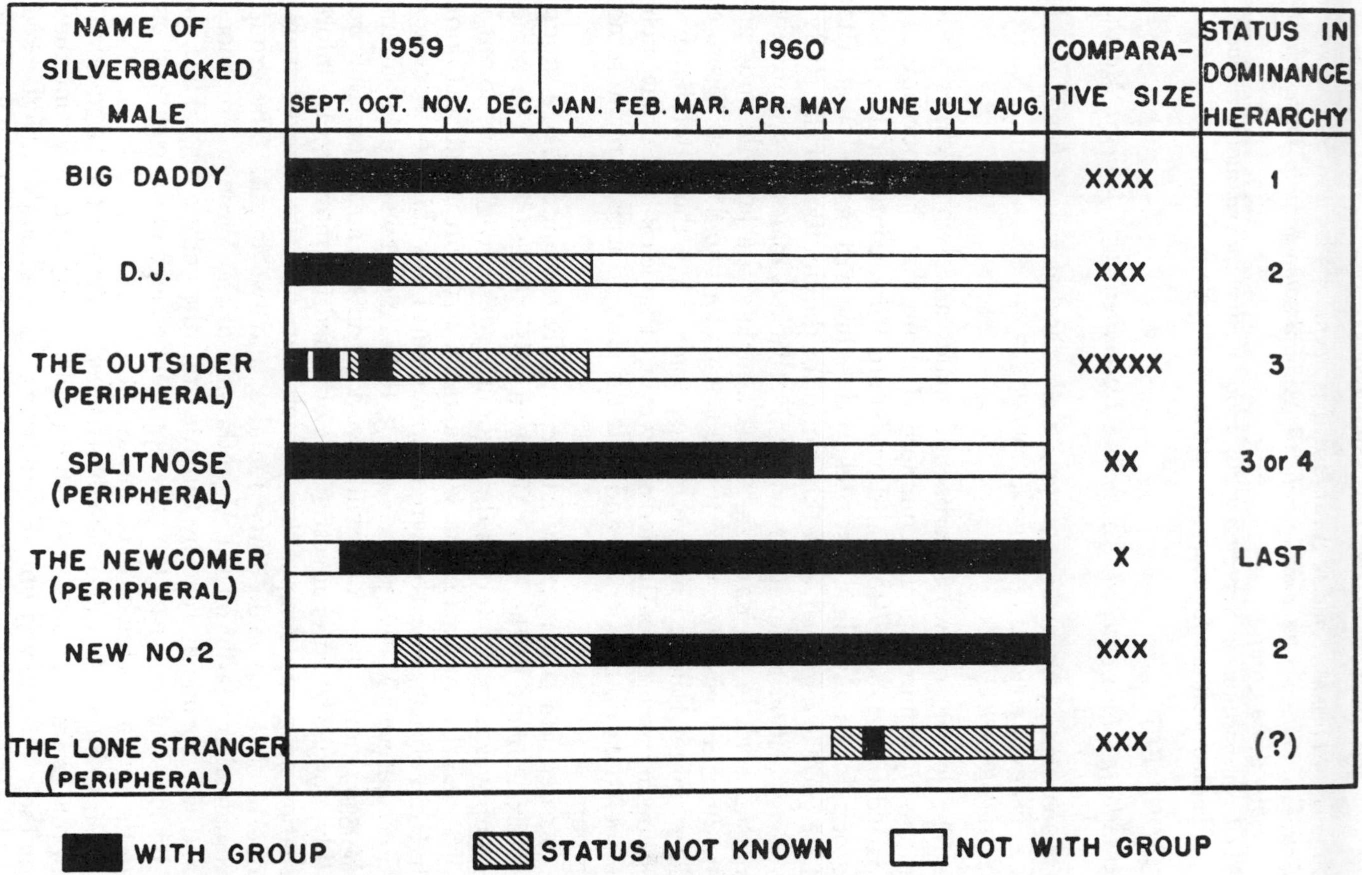

NAME OF SILVERBACKED MALE	1959 SEPT. OCT. NOV. DEC. — 1960 JAN. FEB. MAR. APR. MAY JUNE JULY AUG.	COMPARATIVE SIZE	STATUS IN DOMINANCE HIERARCHY
BIG DADDY		XXXX	1
D. J.		XXX	2
THE OUTSIDER (PERIPHERAL)		XXXXX	3
SPLITNOSE (PERIPHERAL)		XX	3 or 4
THE NEWCOMER (PERIPHERAL)		X	LAST
NEW NO. 2		XXX	2
THE LONE STRANGER (PERIPHERAL)		XXX	(?)

WITH GROUP — STATUS NOT KNOWN — NOT WITH GROUP

FIG. 17.—The joining and parting of silverbacked males in group IV at Kabara

Between February 1 and 6. An infant was born.

Between February 14 and March 16. An unknown female with infant joined the group.

Final composition on September 16, 1960: 1 silverbacked male, 2 blackbacked males, 7 females, 4 juveniles, 7 infants = 21.

GROUP VIII

Months in which checked: November, December, 1959; January, March, May, 1960.

Composition on November 19, 1959: 1 silverbacked male, 2 blackbacked males, 8 females, 3 juveniles, 7 infants = 21.

No changes occurred.

The joining of two silverbacked males and two juveniles with group V requires special comment. I think that the four animals parted from group V several days before I first contacted the main group, for I found their nest sites in the same part of the forest. The four gorillas remained with the group for the duration of the study. An association of juveniles with silverbacked males is apparently not unusual. In the Kayonza Forest a male and a juvenile, judging by dung and nest size, were tracked by us for two days; at Kisoro a large infant followed a silverbacked male when the latter left his group and subsequently died. However, it appears likely that such associations do not persist, and that the animals usually rejoin some group.

KISORO In late 1957 tourists repeatedly encountered one group of six animals in the Mts. Muhavura-Gahinga saddle. The same group was seen at intervals throughout 1958. Bolwig (pers. comm.) observed it between February 11 and 15, 1959. An infant was born in late February or early March. Emlen and I saw this group several times between March 30 and April 5, 1959. And Kawai and Mizuhara (1959*b*) noted the group while working in the area from May to July, 1959. The latter investigators obtained the most precise information on group composition, which was: one silverbacked male, one blackbacked male, three females, and two infants. The group remained primarily on the Uganda side of the Virunga Volcanoes and was commonly seen by tourists until the end of 1959. Thus, in the course of two years the only apparent change in the composition of the group was the birth of one infant.

On February 6, 1960, the silverbacked male and the large infant of this group were contacted alone (Mills, 1960), and they continued on their separate way until February 23, 1960, when the silverbacked male was found dead and the infant was captured. The rest of the group, consisting now of one blackbacked male, three females, and one infant, traveled rapidly and moved into the Ruanda sector of the volcanoes. They reappeared several days later after apparently having joined a group of three animals consisting of one silverbacked male and two females. The combined group of eight (one silverbacked male, one blackbacked male, five females, one infant) was frequently seen by tourists during the succeeding months. An infant was born in late April, 1960. No further changes occurred until September, 1960, when a female, probably of this group, was found dead. In February, 1961, a leopard killed the silverbacked male and later one of the females, after which the survivors moved off, presumably to the Ruanda side of the volcanoes.

On March 17, 1961, a group, comprising one silverbacked male, four females, and one infant, appeared on the Uganda side but then left. The silverbacked male was very distinctive. The same group returned and was encountered on April 20, 1961, but this time it numbered eleven to twelve animals, and the guides felt certain that the additional animals formerly belonged to the group in which the silverbacked male had been killed.

In summary, although my personal data from Kabara are based on observations of only one year, they show that gorilla groups tend to remain quite stable in composition over periods of many months. Most of the changes recorded at Kabara concern the coming and going of males and the births and deaths of infants. One female disappeared from a group and her fate remained unknown; another joined a group, probably by switching over when two groups were near each other. A small group apparently joined group IV.

The interesting observations from Kisoro should be treated with caution, but they are probably indicative of the natural changes in composition occurring in groups over the years and of the response of the animals to these changes.*

Subgroups

For purposes of this discussion I use the term subgroup in the sense of two or more units splitting and remaining completely separated from an established group for various lengths of time before rejoining. This is in contrast to Altmann (1959) and Washburn and DeVore (1961) who consider temporary aggregations of animals within the group as subgroups. My observations on subgrouping are few and most are mentioned below.

* See p. 405, n. 1.

When Emlen and I contacted group IV on March 12, 1959, a silverbacked male and nine other animals were observed for several hours. At 1540 hours we saw one animal about 600–700 feet beyond the group under observation. Contact with the animals on the following day indicated that two groups or subgroups joined after we left at 1600 hours, nested together, and apparently remained as one unit. Although we did not know the previous history of the group, observations during subsequent months suggest that it was an instance of subgroups rejoining rather than of separate groups uniting.

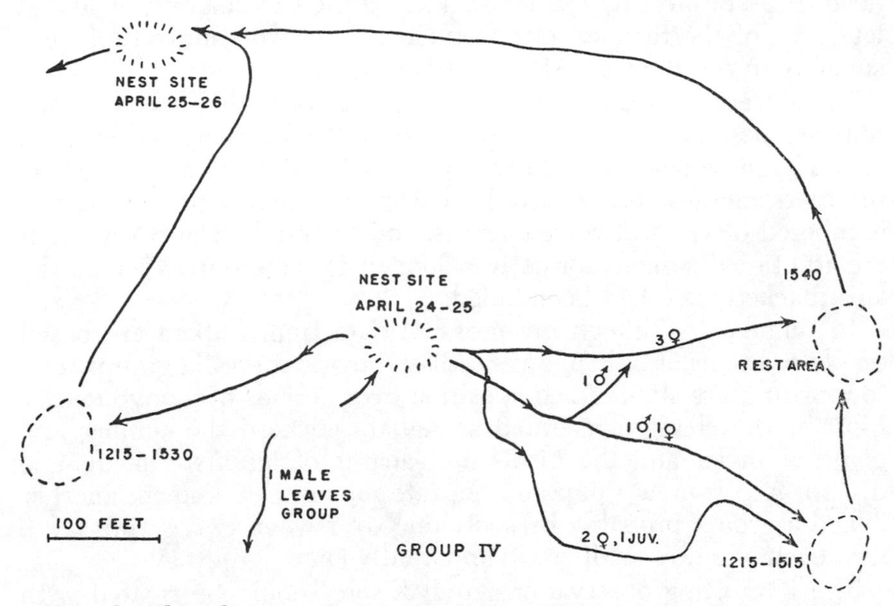

FIG. 18.—The splitting and rejoining of group IV on April 25, 1960, at Kabara. The time of day is indicated near the rest areas.

On April 25, 1960, I encountered this same group when it was split into two units with about 600 feet separating them. Two silverbacked males, six females with infants, and one juvenile had moved in one direction from the nest site; the dominant male and the rest of the group had wandered in the opposite direction (Fig. 18). However, both subgroups rejoined the same evening and remained together on the following days.

These two instances of subgrouping in group IV were the only ones of this type observed during the entire study, and it is perhaps significant that an infant was born each time on the day such behavior occurred. In the first instance the female with the newborn stayed and associated closely with the dominant male; however, in the second instance the female with the newborn moved with the subgroup that did not contain the dominant male.

On one occasion, when I suddenly came upon group XI, the animals fled in two different directions. A blackbacked male, two females with infants, and one unidentified animal ran in one direction and the rest of the group in the other. After a brief rest 150 feet apart, the two subgroups moved independently in parallel directions and rejoined after about 600 feet and two hours.

In the Kayonza Forest I once watched four gorillas for four hours without hearing or seeing others. However, the following day, trails and nests revealed that the group contained at least twelve gorillas. Whether the animals were merely spread out while feeding or if actual subgrouping occurred could not be determined in the dense vegetation.

Occasionally groups in the Kabara area split up at night and slept 50 to 300 feet apart only to rejoin after rising. This seemed to be done more or less inadvertently, and I did not consider the separate sleeping aggregations to be subgroups.

The sum total of evidence suggests that subgrouping in gorillas is a fairly rare phenomenon, occurring primarily in groups where more than one silverbacked male is present.

Contacts between Separate Groups

Since several gorilla groups often occupy the same section of the forest, two or more will sooner or later wander near to each other. At such times gorillas frequently showed no overt response. On at least twelve occasions a group heard but could not see another group only 600 to 900 feet away, yet seemingly ignored the presence of the animals. However, groups approached each other occasionally. Since the contacts were usually of brief duration, it was primarily a matter of luck that I was present to observe the interactions. I obtained notes on the sequence of events at four such meetings. Two times both groups were unaware of my presence for all or most of the encounter, and on the other two occasions only one group gave any evidence that it knew of my proximity. In addition, evidence from trails and contact with the animals after the event confirmed close association between groups eight more times. The twelve instances are summarized in Table 21.

Most interactions involved only two groups. However, groups I, II, and III were very near each other for at least one day when I encountered two of them and heard the third. Once, when groups

VII and V approached each other closely, group II was only about 800 feet away.

The behavior of the groups during periods of contact is best illustrated by quoting extensively from my field notes:

GROUPS VI AND VIII (Figs. 19 and 20)

November 22. 0945. All members of group VIII feed slowly except the silverbacked male (VIII Dominant), who stares uphill, and then beats his chest. About 300 feet uphill sits the silverbacked male (VI Dominant) of group VI looking downhill.

1025. Group VIII is still feeding slowly uphill toward group VI, which also forages. Only VI Dominant retains his seat and remains motionless. When the two groups are 100 feet apart, VI Dominant rises, walks 3 to 4 steps toward group VIII and sits again. But almost immediately afterward he rises on his hindlegs, beats his chest, and then stands quadrupedally for about 5 seconds, staring at Group VIII, before walking about 10 feet uphill. Suddenly VI Dominant moves quite rapidly at an angle downhill followed by his group until they are opposite group VIII and only 60 feet apart. Both VI Dominant and the blackbacked male of group VI beat their chests once, facing group VIII. Then both groups feed intensively, acting as if the other does not exist. VI Dominant and VIII Dominant are now only 35 feet apart, their closest approach to each other. VIII Dominant beats his chest.

1110. Several members of group VIII move uphill and VIII Dominant follows. The groups rest 100 feet apart.

1250–1315. Both VI Dominant and VIII Dominant beat their chests several times, but the other group members pay no obvious attention.

1325. Group VI moves out of sight feeding, and group VIII leaves also, at about right angles to the former.

Apparently the two groups fed and traveled close to each other late in the afternoon and nested in adjacent sites. Figure 20 shows that the beds of the males were only about 30 feet apart. The dung of the two males was soft, about the consistency of mashed potatoes, suggesting that they were excited; the dung of the other gorillas was normal. Perhaps some mingling between groups occurred before bedding and after rising but each animal appeared to have slept with its own group.

November 23. 0945. The two groups feed 120 feet apart after having left the nest site over separate trails. Group VIII is out of sight in a shallow ravine. The animals continue to feed slowly and rest. A blackbacked male of group VIII appears at the periphery of group VI, climbs a tree, descends, and returns to his group.

1240. VI Dominant rises and moves downhill at an angle to within 60 feet of group VIII at the edge of the ravine.

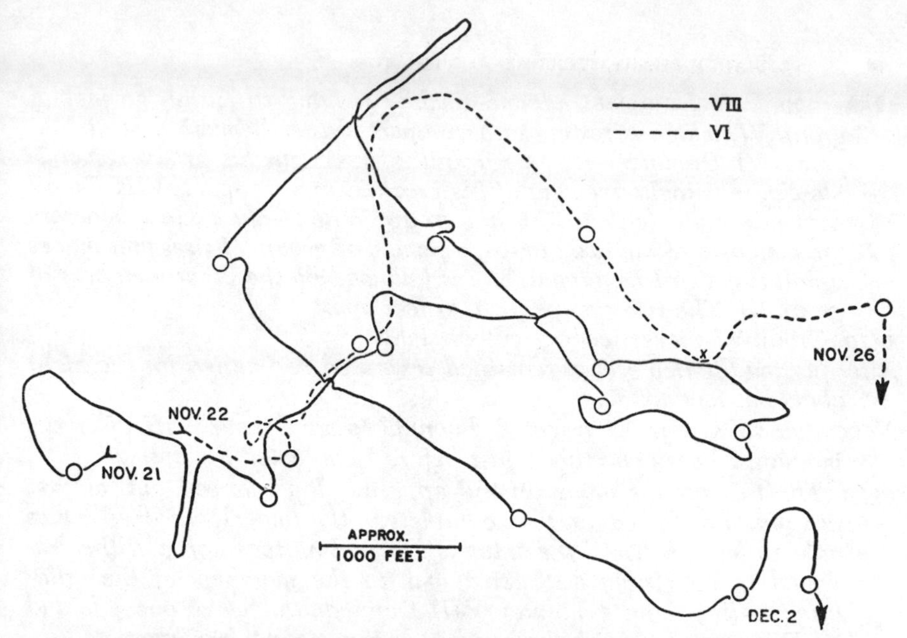

FIG. 19.—The close proximity, joining, and parting of groups VI and VIII between November 22 and 25, 1959, at Kabara.

○ Night nest site
× Place of observed joining

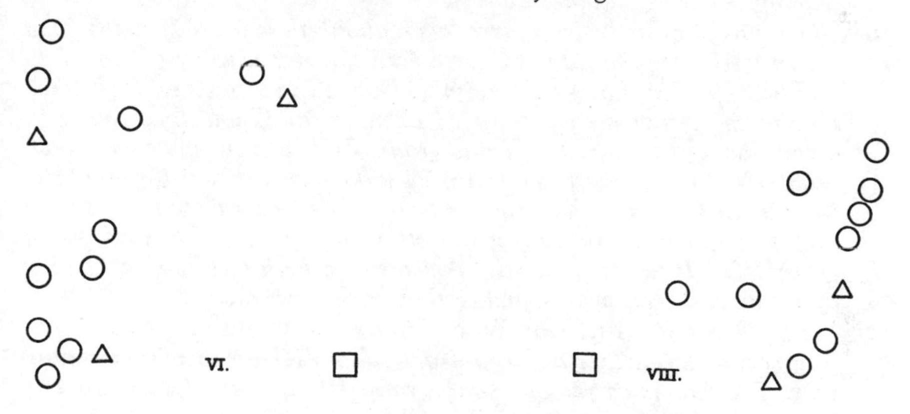

FIG. 20.—Adjacent nest sites of groups VI and VIII during the night of November 23–24, 1959, at Kabara. Only 30 feet separate the nearest animals of the respective groups.

□ Silverbacked male
○ Blackbacked male or female
△ Juvenile

1250–1330. VIII Dominant advances while feeding to within 30 feet of group VI, which is resting in a compact cluster beneath a large Hagenia. *VI Dominant retreats uphill 40 feet, but his group remains. Slowly, seemingly unaware of his proximity to group VI, VIII Dominant advances, followed by his group. When only 15 feet separate the members of the two groups, a female of group VI rises and moves uphill to join VI Dominant. She is followed by the other members of group VI. The two groups feed 40 feet apart.*

1420. Group VI moves feeding out of sight.

After parting the two groups remained separated and nested for the night about 800 feet apart.

November 24. Group VI traveled downhill toward group VIII after rising, and I heard chestbeats just before I contacted the animals.

0930. The two groups move up the opposite slope, spread out, at first just mixed at the edges, then completely, the two silverbacked males only 10 feet apart. All the animals are feeding together, and they respond in no visibly distinctive way to the members of the other group. After about 5 minutes, VIII Dominant angles to one side and VI Dominant sits, with the result that the groups drift apart.

0955. The two groups rest completely separated, with the males about 60 feet from each other and separated by an earth ridge about 8 feet high. The animals are not visibly excited, and feeding, resting, and playing occur as on other days.

1025. Two juveniles and two infants, each about 1½ years old, climb from group VIII over the ridge of earth into the rest area of group VI to within 20 feet of the nearest animal. They sit in a row and look. Two infants and a juvenile of group VI clamber uphill and they too sit in a row facing the intruders from group VIII at a distance of 6 feet. Suddenly the juvenile from group VI jerks forward as if to run at the juvenile of VIII, and the latter turns and flees several steps. Then the infant of group VI sidles step by step toward the other juvenile of group VIII. It jumps forward, takes a swipe with one hand at the leg of the juvenile but misses, and dashes back to its own side.

1055–1115. Dense fog settles in. When it lifts, the infant of group VI has returned to its mother and those of group VIII do so within 5 minutes. The juveniles remain near group VI and are joined by one blackbacked male of group VIII. All rest.

1255. VIII Dominant rises and he and his group move slowly away from group VI. One juvenile of VIII hurries after its group; the other two animals remain with group VI. The blackbacked male of group VIII thumps the ground with the palm of his hand several times as he faces two females with infants of group VI which are only 15 feet away. Suddenly one of the females rises and makes a slight forward motion as if to charge. The blackbacked male flees 30 feet, and then he, followed by the juvenile, rejoin group VIII.

1330. Groups VI and VIII feed and move in opposite directions.
Both groups remained on the same slope at least one more day, but did not rejoin.

GROUPS VII AND XI (Fig. 21)

April 18. 1250. The silverbacked males of the two groups are only 20 feet from each other. The members of group VII sit clustered around their dominant male (VII Dominant), but those of group XI are scattered. XI Dominant is greatly excited: he beats his chest, thumps the ground with one or both hands, emits a long series of hoots, and climbs a log only to jump with a crash into the vegetation below. He continues this for the next 1½ hours. VII Dominant, on the other hand, sits hunched over, his back to XI Dominant most of the time.

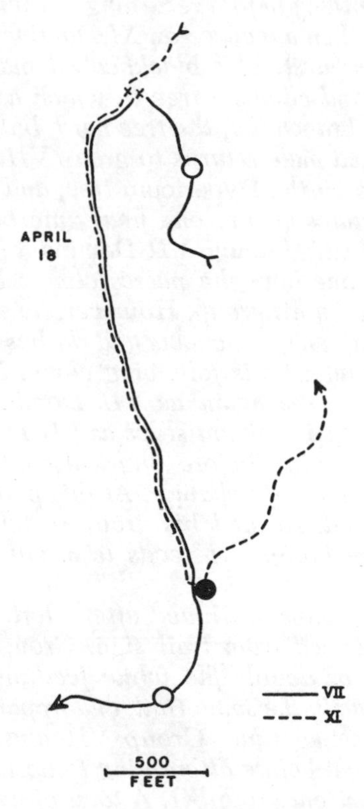

FIG. 21.—The routes of groups VII and XI before, during, and after meeting on April 18, 1960.

○ Night nest sites of group VII
● Night nest site of Group XI
× Place of closest observed contact

Suddenly VII Dominant rises and walks rapidly toward the male of group XI. They stare at each other, their faces but one foot apart, for 20 to 30 seconds. Then VII Dominant returns to his seat. This is repeated two more times. The males stand quadrupedally as they stare at each other. A Hypericum *tree appears to be an arbitrary boundary past which neither animal advances. The other members of the two groups are either watching the proceedings or sitting quietly with apparent indifference.*

1400. As XI Dominant beats his chest, VII Dominant rips off an herb, throws it into the air with an underhand motion and runs at the other male. They face each other silently with browridges nearly touching. They part. VII Dominant moves away and begins to feed.

A blackbacked male and a juvenile of group VII amble within two feet of XI Dominant, pass him, and enter his group. They walk among the other animals briefly before returning to their own side. The same blackbacked male then advances on XI Dominant, stands 6 feet from him, and beats his chest. This blackbacked male continues into the area of group XI and climbs a tree in which a juvenile of group XI sits. He shakes the branches of the tree hard, but the juvenile remains, and the blackbacked male returns to group VII.

1415. XI Dominant sits by the Hypericum *tree, and VII Dominant walks toward him and stares at him one final time before moving toward the tree in which I sit. Although VII Dominant has been aware of my presence for over one hour, he passes almost beneath the tree, followed in single file by his group. However, he stops, and the animals move past him. Suddenly he wheels and dashes back 20 feet over the trail at XI Dominant who is following alone. XI Dominant retreats one step, then advances again as VII Dominant hurries after his group. Once more XI Dominant stops as VII Dominant rushes silently at him and brakes just before physical contact. The other group members show no unusual behavior. At this point XI Dominant spots me and is distracted. He and his group watch me until I leave at 1440. At that time group VII feeds leisurely about 200 feet from group XI.*

The association of the groups continued after I left. The following course of events was deduced from trail sign. Group VII moved steadily ahead in a single or double file while feeding. Group XI followed group VII over exactly the same trail, and apparently approached the latter at about nesting time. Group VII abandoned its nests and moved some 200 feet before attempting to nest again, only to be disturbed by the persistent group XI. A total of 27 nests were built and abandoned and these were scattered over a distance of 400 feet. Several tufts of long hair torn out by the roots lay on the trail suggesting that a tussle may have occurred. Group VII finally moved 500–600 feet in single file and nested, and group XI simply slept on the trail. The following morning the groups parted in opposite directions without further contact on subsequent days.

GROUPS VII AND V (Fig. 22)

November 1. 0835–1245. The two groups nested 500 to 600 feet apart. When I contacted them, both sat and fed slowly, most of them out of sight of each other. Group VII ignored the proximity of group V, and only when several animals in the latter group quarreled loudly did several members of group VII lift their heads to look in the direction of the sound. After resting, both groups fed toward each other.

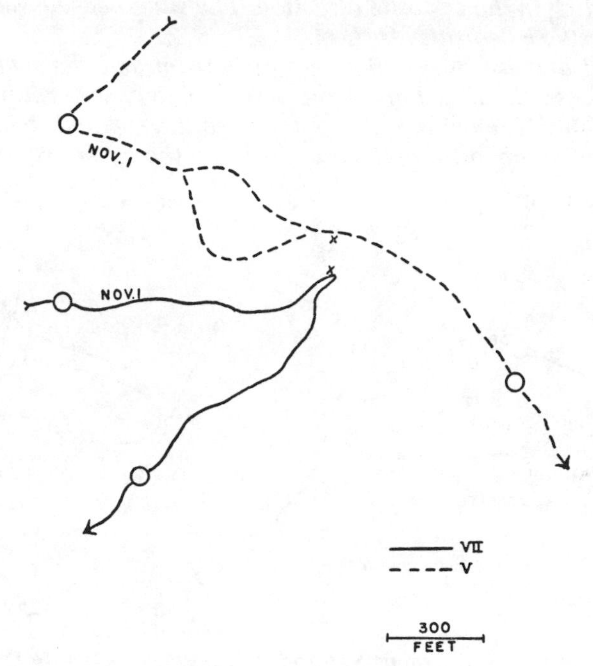

FIG. 22.—The routes of groups V and VII before, during, and after meeting on November 1, 1959.

× Place of closest contact
O Night nest sites

1245. A blackbacked male and 3 juveniles of group V move into the open and sit watching group VII at 300 feet. A large infant of group VII climbs a tree and looks at group V, and a juvenile of group VII approaches the other group to within 150 feet. The other animals continue to feed.

1430. A blackbacked male of group V just lies on a log, head propped on his hand, observing group VII. Two silverbacks of group V jerk up their heads intermittently from their feeding to glance at group VII, which is spread over a slope 200 feet away. The silverbacked male of group VII thumps the ground and beats his chest several times; only one male in group V beats his chest once.

1600. Both groups move slowly in opposite directions and part without further contact.

GROUPS VII AND ? (Fig. 23)

March 25. The two groups nested 250 feet apart. Trails indicated that some animals approached each other to within 100 feet after rising in the morning.

0955. Group VII is near its nest site but the other group is 400 feet away and feeding in the opposite direction. The silverbacked male of group VII is restless; he beats his chest.

1030. Several animals in the other group scream and the male in group VII looks in the direction of the sound. The other group moves off, and group VII remains in the rest area until 1505 and then departs in the direction opposite to the one taken by the other group.

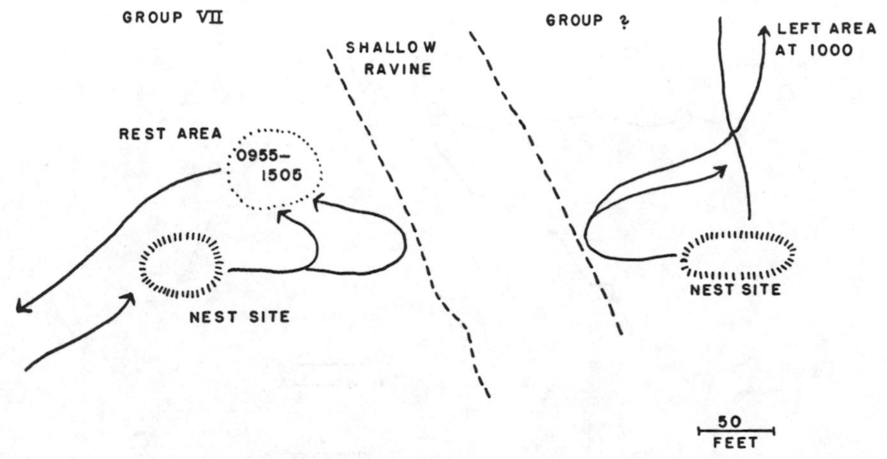

Fig. 23.—The routes of groups VII and ? during a meeting in the evening of March 24 and the morning of March 25, 1960.

GROUPS VII AND III (Fig. 24)

October 19. At 0815 I contacted group III, which then moved onto the slopes of Mt. Mikeno and was last seen at 1150. At 1215 I found group VII within 750 feet of group III. When I left group VII at 1345 I had not detected any sign of group III in the vicinity. However, the two groups joined later in the day, nested together for one night, and had parted by 1000 hours the following morning when group VII was recontacted.

These few observations show that the responses of groups to contact with each other vary considerably, including seeming obliviousness to the other's presence, close approach, and brief mingling. On the whole, the interactions were peaceful. Definite aggressive bluff-charges were made only by one silverbacked male, and weak aggressiveness was observed on the part of a female, a juvenile, and an infant toward intruders from another group. Except for one instance, displays were not prominent, and the animals gave the impression of being only slightly excited. They obviously did not anticipate trouble, for infants were permitted to wander into the adjoining group and the leader allowed another silverbacked male to approach his females closely without obvious concern.

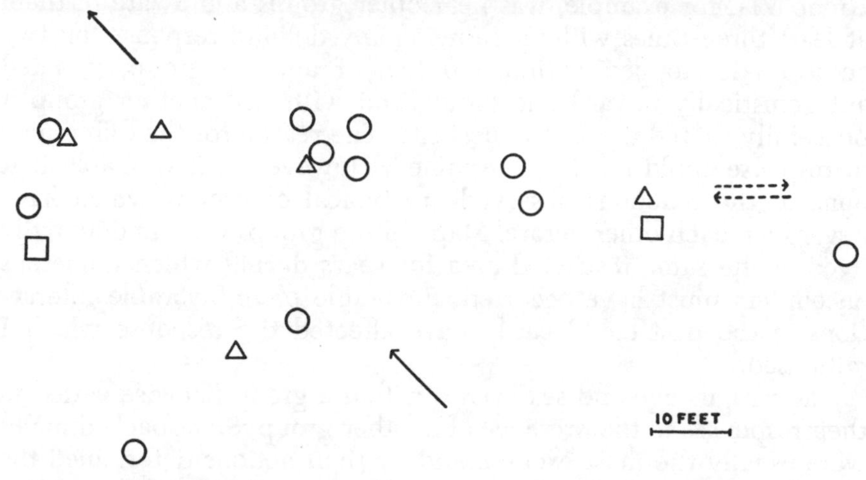

FIG. 24.—A combined nest site of groups VII and III during the night of October 19–20, 1959.

□ Silverbacked male
○ Blackbacked male or female
△ Juvenile
→ Evening and morning direction of travel by group VII
⇢ Evening and morning direction of travel by group III

I never observed serious quarreling or fighting between individuals of different groups. The only seemingly reliable record of fighting between the dominant males of two groups occurred at Kisoro where, according to Reuben Rwanzagire, the African guide, two animals apparently fought several times in July, 1958, and on August 11, 1958. One of the males was found dead on August 13, 1958, of unknown causes. The only evidence of fighting on his body was a bruised right eye and full-depth skin injuries on the backs of both hands (Wilson, 1958). The reason for these fights was not determined. Barns (1922) also obtained evidence for possible fighting: "I found open wounds on the crest of only one of the old males

I shot; they were apparently teeth marks, and this same animal, by the way, had several big boils in different parts of the body." These reports, as well as some of the wounds and scars which I noted in gorillas (see section on injuries) suggest that fighting occurs occasionally. However, such behavior is undoubtedly infrequent and the reasons for it remain obscure.

One striking aspect of intergroup behavior was the highly variable response which a given group showed in the presence of others. Group VII, for example, was near other groups and aware of them at least three times without showing any definite response; on two occasions it stopped within 100 feet of another group; it acted antagonistically toward one group; and with still another group it peacefully shared a common nest site. The reason for the differences in response could not be determined. However, it is probable that none of the reactions observed are typical of groups which have never seen each other before. Many of the groups have undoubtedly lived in the same restricted area for years during which numerous encounters must have occurred. Favorable or unfavorable interactions in the past could easily have affected the response which I witnessed.

The various age and sex classes within a group likewise varied in their responses to the presence of another group. Silverbacked males were usually the most excited, and by their actions determined the closeness and duration of meetings. Some evidence of dominance was seen when the male of group VI retreated at the approach of the male of group VIII. On the other hand, the male of group XI was not intimidated by bluff charges on the part of the male in group VII.

Females showed the least amount of emotion and they commonly appeared oblivious to the behavior of another group. However, one female retreated from the advance of the silverbacked male of the other group, and one made an incipient bluff-charge at an intruding blackbacked male. Juvenile and blackbacked males were by far the most venturesome and some readily sat at the periphery or penetrated into another group.

The gorillas exhibited strong attachments to members of their own group with the result that mingling and subsequent parting of two groups usually had no effect on group stability. Apparent exceptions were noted, however. One female with infant probably switched from one group to another. Work at Kisoro showed that groups which have lost their leader through death may integrate with another group. And the fact that a new male and two females with infants suddenly appeared in group IV perhaps indicates that small groups may occasionally join other groups.

The peaceful nature of most interactions between gorilla groups find some parallels among other primates studied so far. Nissen (1931) presented observations on chimpanzees which suggest that groups may join, and I obtained similar data on this ape in the Budongo Forest. Haddow (1952) and Carpenter (1958) both mentioned aggregations of monkey groups at localized food supplies. Washburn and DeVore (1961) observed baboon groups drinking at waterholes within a few feet of each other without mixing and without fighting; only once did a juvenile briefly cross into the next group.

Lone Males

Many primate societies have an excess of males, some of which may be encountered leading a lone life away from organized groups. It is a popular supposition that such males have been ejected forcefully from groups by younger, stronger rivals. Carpenter (1958) has challenged this concept, presenting evidence from several primate species which shows that lone males may be of different age classes and that they probably do not remain completely isolated from groups. The work on gorillas supports his evidence.

For purposes of discussion I apply the term "lone male" to those animals which are not associated with an organized group at the time of observation. Such males apparently occur in all gorilla habitats, in both Central and West Africa. Lone males frequently wander considerably, and records of individuals 20 or more miles from the nearest known gorilla group are not uncommon. Female gorillas apparently do not leave groups to lead a lone life. At Kabara I identified four silverbacked and three blackbacked males which were lone animals during all or most of the twelve months of the study; perhaps two or three other silverbacked males occurred also, but these were not identified individually during my brief encounters with them.

Most of my detailed observations were made on group IV. Figure 17 summarizes the joining and parting of the various silverbacked males in this group, as well as presents the comparative sizes of the animals and their status in the dominance hierarchy. Males which tend to remain at the outer edge of the group are designated as "peripheral." Of the seven silverbacked males which associated with the group at one time or another during the twelve months of observation, only the dominant animal remained the whole time.

Some joined and stayed for several months; one animal left after less than one month; and one male parted and rejoined twice within a period of about twenty days. Most of the changes occurred in peripheral males (e.g., The Outsider, Splitnose), but the number two male in the dominance hierarchy (D. J.) likewise disappeared from the group and was replaced by another. With all these changes, the number of silverbacked males was in a constant state of flux, varying from three to five.

Figure 17 also shows that the dominant animal is not necessarily the largest member of the group. The Outsider was the largest silverbacked male in the area, yet he occupied third place in the hierarchy.

The lone males fell into several age classes. Of the three blackbacked males seen, one was small, one medium, and one large. At least five of the silverbacked males appeared to be in the prime of life, and only one seemed old.

Several observations were made on a large silverbacked male, dubbed The Lone Stranger, who lived alone most of the time. His center of activity was on the slopes of Mt. Mikeno near Kabara. I tracked him first on November 17, 1959, and observed him at the periphery of group VI. He apparently left the group the same day. On November 23, I flushed him from the one-day-old trail of group VI. On December 3, I again encountered him alone on a fresh trail of group VI, and the following morning he sat on a two-day-old rest area of group VI. He appeared to be attracted by the fresh gorilla spoor. In May, 1960, when group IV moved onto the slopes of Mt. Mikeno near Kabara, this same animal joined the group and remained at least one week.

Some lone males were never observed with a group. For example, one blackbacked male visited the slope behind the Kabara cabin on September 19, 1959. Two months later, on November 19, I ran into him on the same slope. He revisited the area on April 12, 1960; I met him on a fresh trail of group VI on April 18; and between May 20 and May 25 he again remained near our cabin.

Some lone males appeared to wander considerably for I observed them only once or twice during our year at Kabara.

Lone males associated with some groups but not with others. All interactions of lone males with groups were with groups IV and VI. On the other hand, groups II, V, VII, and VIII never established contact with lone males in my presence. This suggests some form of selectiveness on the part of the lone males. There is no correlation with the number of silverbacked males in a group and the joining of lone males, for group IV always contained several silverbacked males while group VI usually had only one; conversely, group V retained four silverbacked males without apparent change for nearly one year and no other males were seen to associate with it. It is likely that through previous contacts lone males have learned which group will accept them and which will not. Acceptance or rejection probably depends on the temperament of the dominant male in the group as well as the type of interaction between the group and the lone male during past encounters.

I observed the response of groups to lone males three times:

1. Group IV and a lone silverbacked male (The Newcomer)

I saw the trail of a lone silverbacked male within about 1,000 feet of group IV on the morning of September 9, 1959.

1100. Most members of the group sit and watch me when suddenly the lone male appears at the edge of the group within 30 feet of the nearest animal. He sees me, lets out a roar, and runs out of sight. However, about 10 minutes later he reappears and walks without hesitation toward the 3 silverbacked males of group IV. All 4 males then sit in an area of about 30 feet in diameter. None of the group members react to the new male.

2. Group VI and two "lone" males

January 9, 1960. 0940. All members of group VI except one feed behind a slight rise. About 100 feet from them is the small blackbacked male of group VI, and with him is a strange silverbacked male and a strange blackbacked male. The silverbacked male of group VI and the new one beat their chests at least 3 times each even though they are out of sight of each other. Suddenly group VI, led by a female, marches across the slope in single file with the dominant male in the middle. Both blackbacked males join the group, but the new silverbacked male remains 100 feet away. All animals rest for one hour before moving slowly downhill, with the lone silverbacked male tagging 100 feet behind.

1125. Dense fog rolls in, making further observations impossible.

January 10. 1230. The group rests. The dominant male of the group lies out of sight, but the new silverbacked male sits on the slope surrounded by at least 3 females and a juvenile. The new blackbacked male remains at the periphery of the group.

1435. All animals except the new silverbacked male follow the dominant male when he begins to feed. I leave after the whole group has moved out of sight.

Both lone males apparently left the group later in day, and they did not rejoin on subsequent days.

This was my only observation in which two "lone" males associated and seemingly traveled together, and it was the only time that I saw a "lone" blackbacked male join a group.

3. Group VI and a lone silverbacked male (The Lone Stranger)

On November 17, 1959, I found the fresh trail of a lone male near group VI.

November 18. 1420. The group sits compactly in a small area and pays no obvious attention to a new silverbacked male lying 30 feet away. As the animals rise and slowly move up the slope at 1435, the dominant male remains behind and silently faces the lone male. The lone male also rises and the two stare at each other from a distance of 30 feet for about 10 seconds. The lone one unfortunately spots me and beats his chest. The dominant male then joins his group.

Apparently the lone male and the group established no further contact, for on the following day I found the night nest and saw The Lone Stranger about ½ mile from group VI. (This lone male also joined group IV for a brief period in May, 1960.)

The picture which emerges is one in which males apparently have the choice either of joining with and parting from some groups, or of leading a lone life in the forest. Considered from the viewpoint of genetics and evolutionary trends, the movements and contacts of lone males are important in reducing inbreeding.

Formation of New Groups

Although I was not able to observe the process of group formation, the various topics on group dynamics discussed in this chapter suggest two ways in which it could occur:

1. A lone male joins a group and when he leaves one or two females follow him, thus forming the nucleus of a new group.
2. Large groups occasionally form subgroups, and, if the two units fail to rejoin, distinct groups are created.

Group Ranges and Movements

Gorillas lend themselves well to a study of movements over long periods of time. The trails made by groups are sufficiently distinct to enable the observer to trace precise routes from day to day. At Kabara, groups were frequently tracked daily for fifteen to twenty days and on occasion as long as twenty-five to thirty days if they remained in the study area.

Home Range and Territory

In discussing home range and territory Burt (1943) stated: "Every kind of mammal may be said to have a home range, stationary or shifting. Only those that protect some part of the home range, by fighting or aggressive gestures, from others of their kind, during some phase of their lives, may be said to have territories." This distinction between home range and territory has not been clearly recognized in several publications on primates (Carpenter, 1958; Haddow, 1952) with the result that data from some of the species studied are not directly comparable. Territorialism cannot be determined merely by noting a certain group in the same locality over a long period of time, but interactions between groups must also be repeatedly observed.

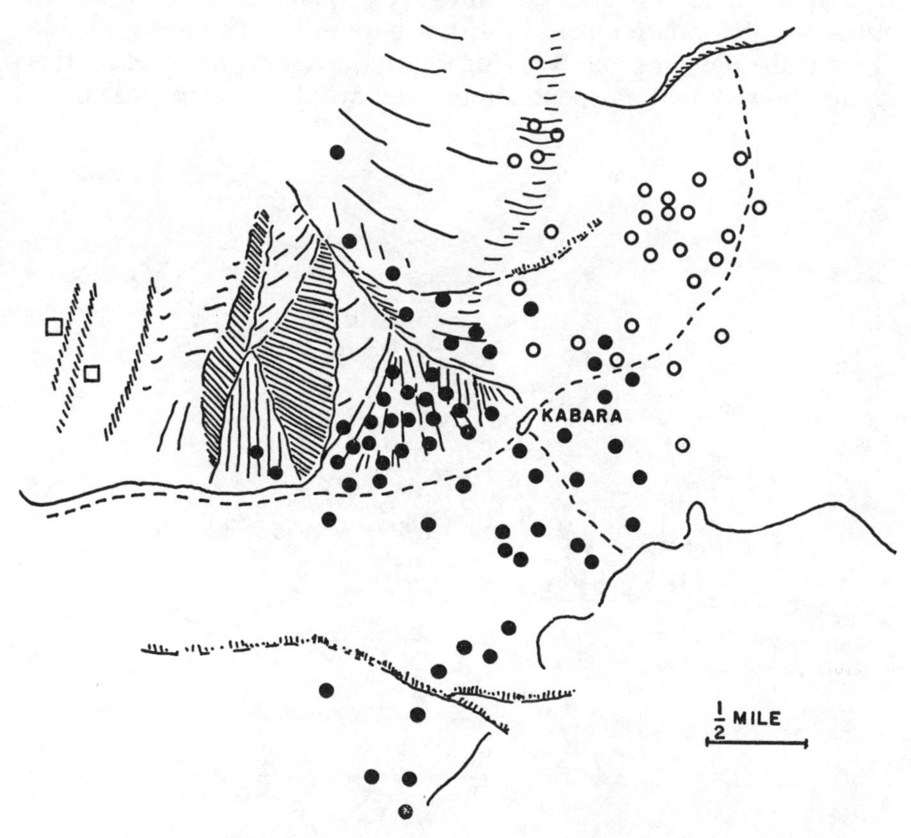

FIG. 25.—The extent of home ranges of gorilla groups in the Kabara study area. Each symbol represents one encounter with, or definite sign of, a group. The scale of the drawing is the same as that of the aerial photograph in Figure 9.

- - - Main foot trails
○ Group II
● Group VI
□ Group IX

HOME RANGE The extent of home ranges of gorilla groups at Kabara during the year of study was plotted by marking each separate observation of a group on a map. In Figures 25, 26, and 27 each symbol represents one observation; nest sites or trails are also marked at times to denote the periphery of the range. Some groups, like IX and XI, barely entered the study area at its edges, and others, like I and III, were not seen frequently enough to determine the extent of their range. Groups which were often observed spread their activity over the following approximate number of square miles: II, 4 square miles; IV, 6 square miles; V, 4 square miles; VI, 8 square miles; VII, 8½ square miles; VIII, 5 square miles. The most reliable figure is the one for group VII, for my frequent encounters with this group showed that it spent almost the whole year near Kabara.

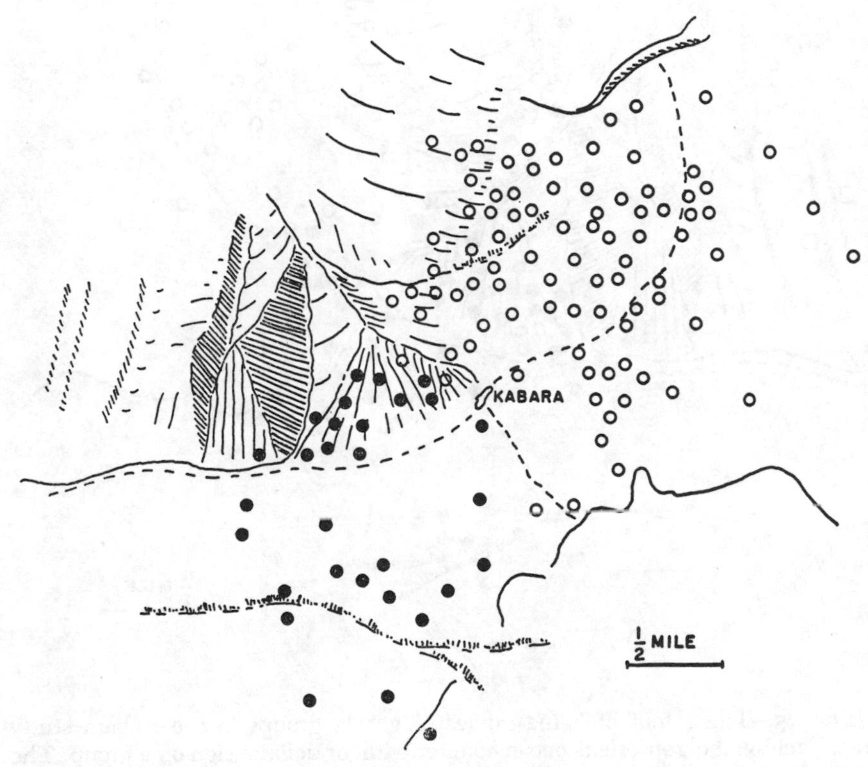

FIG. 26.—The extent of home ranges of gorilla groups in the Kabara study area. Each symbol represents one encounter with or definite sign of a group.

- - - Main foot trails
○ Group VII
● Group VIII

Most of its activity centered in one area of 4 square miles, and its absence from the study area was of such short duration that it precluded wandering much beyond the area of 8½ square miles noted in Figure 26. Groups IV, VI, and VIII utilized the Kabara area for roughly half of my study period, which suggests that it constituted a substantial part of their range and that their entire range was, at most, no more than twice the figure stated above. Based on this evidence, the home ranges of gorilla groups at Kabara were of the order of about 10 to 15 square miles each.

At Kisoro, one group spent the major portion of its time from 1957 to 1960 in about 4½ squares miles of the Uganda sector of the volcanoes, and probably extended its activities over an area of similar

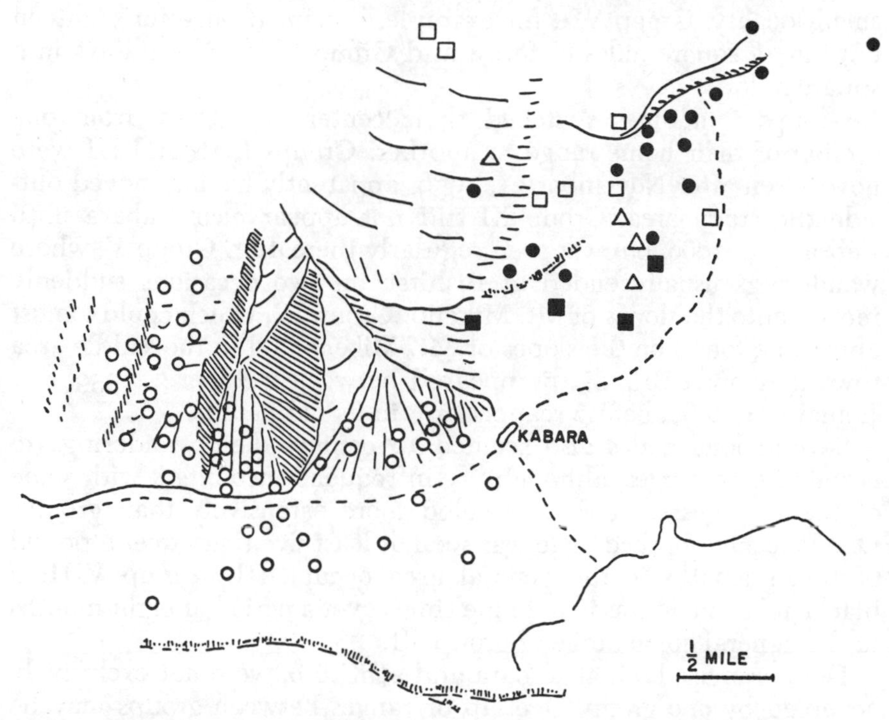

FIG. 27.—The extent of home ranges of gorilla groups in the Kabara study area. Each symbol represents one encounter with or definite sign of a group.

- - - Main foot trails
△ Group I
■ Group III
○ Group IV
● Group V
□ Group XI

size on the Ruanda side. Sabater Pi (1960), on the basis of rather meager data, maintained that one group in West Africa limited its activity to about 3½ square miles of terrain. Gorilla groups thus frequent home ranges somewhat larger than most other primates studied so far. Gibbon groups range over 30 to 100 acres each (Carpenter, 1940), *Colobus guereza* groups over about 15 hectares (Ullrich, 1961), baboons over 3 to 15.5 square miles (DeVore and Washburn, 1963), *Macaca fuscata* over ½ to 5 square miles (Imanishi, 1957), and langurs, *Presbytis entellus,* over ½ to 3 square miles (Jay, 1962).

Although the home range of each gorilla group at Kabara was fairly large, a group occasionally spent considerable time in one small locality. Group VII, for example, remained once for eighteen days in 1½ square miles of forest and Group II for fifteen days in 2 square miles.

Groups sometimes switched their center of activity from one section of their home range to another. Groups I, II, and III were never seen after November 21, 1959, apparently having moved outside the study area. Group XI did not appear near Kabara until February 2, 1960, but was seen regularly thereafter. Group V, whose wanderings usually ended at Bishitsi, on two occasions suddenly moved onto the slopes of Mt. Mikeno. Group IV, which could almost always be found on the slopes of Mt. Mikeno and in the saddle area toward Ruanda-Urundi, disappeared between October 2, 1959, and January 11, 1960, before reappearing in its old haunts.

Several lone males also seemed to confine their wanderings to certain home ranges, although my infrequent encounters with some of them suggest that they traveled more extensively than groups. One lone silverbacked male was seen at least five times over a period of seven months in the general area occupied by group VIII; a blackbacked male was noted nine times over a period of eight months in the general home range of group VI.

Home ranges, both at Kabara and at Kisoro, were not exclusively occupied by one group. Overlap of ranges between groups may be small, such as between groups II and VI; or extensive as between VI and VIII (Figs. 25, 26, and 27). Six groups frequented the same section of the forest near Bishitsi, and three groups periodically occupied the southeast slope of Mt. Mikeno.

In some areas, permanent cultivation, grasslands, and dry scrubby forests, all vegetation types which are little favored by gorillas, determine the boundaries of home ranges. However, most gorillas occupy forests that furnish an unbroken expanse of suitable vegetation in which no physical barriers prevent the animals from wandering freely. Yet groups restrict their activity to definite areas. The recognition of certain physical features, coupled with a preference for familiar terrain, is probably the most important factor which confines the activity of groups to specific ranges. For example, an imaginary line from Rukumi to Kabara and along the crest of a prominent spur of Mt. Mikeno appeared to be a boundary for all the local gorilla groups except one. Neither group IV nor group VIII crossed this boundary although they sometimes approached it closely; group VII skirted over it three times. Only group VI traveled freely back and forth over this line.

Gorillas probably visited most of their home range several times each year, and they appeared to be familiar not only with its grosser aspects but also with many details of its physical features. One ravine on Mt. Mikeno was only about 15 feet wide, but its walls were perpendicular and up to 30 feet high. The most convenient way to reach the other side was via a log that spanned the chasm. Even though the log was well hidden in brush, one group approached this log directly on two separate occasions in order to cross, indicating a prior knowledge of its presence. Another group tended to follow the rim of a shallow ravine before descending at a particular spot.

TERRITORY Direct observations of the actions of groups near each other never revealed the slightest indication that a certain section of forest was defended directly by fighting or indirectly by aggressive displays against intrusion of another group. Although groups frequently hesitated to mingle and individuals occasionally behaved aggressively toward one another (see section on group dynamics), no attempt was made to restrict movements into a certain area. This evidence suggests that territorial behavior is absent in gorillas.

SEASONAL MOVEMENT

The literature contains several statements to the effect that gorillas shift their areas of activity seasonally with the food supply. Sabater Pi and de Lassaletta (1958) observed that in West Africa gorillas visited native fields during the rains from March to mid-May and from September to early December, and retired to the forest for the remainder of the year. Blower (1956) stated that at Kisoro gorillas showed "undoubtedly considerable seasonal movements influenced by food requirements." Donisthorpe (1958) attempted to pinpoint seasonal changes in gorilla distribution at Kisoro but assembled no conclusive evidence. An African and a white planter told us that in the region south of Mt. Tshiaberimu gorillas move closer to the road during the rainy seasons to eat the tender new stems of elephant grass.

Except for bamboo, gorilla habitats throughout the range show no conspicuous seasonal differences in the abundance of forage. An attempt was made to summarize the movements of the Kisoro gorilla groups from all published and unpublished sources over a span of five years. Unfortunately most of the records are too vague as to exactly where the animals were observed. However, gorillas have been recorded in all forest zones during every month of the year. A possible seasonal correlate is found at an altitude of above 11,500 feet, where fresh gorilla sign has been noted during every month from January through August, but only once between September and December. But this correlation is perhaps an artifact produced by the reduction of observer activity at high altitudes during the period of heavy rain.

Striking seasonal changes occur in bamboo, where shoots and tender green stems are generally absent from late June to mid-September. Along the Rweru-Kabara trail, gorillas visited the bamboo zone more consistently when many shoots were present than during periods when few or none were available (Table 22).

Thus, it appears that gorillas can usually obtain some forage in all vegetation zones throughout the year. But some groups appear to utilize bamboo, and perhaps other types, more extensively during certain months.

General Patterns of Group Movement

The patterns of movement of several groups which I followed over extended periods at Kabara were for the most part highly irregular; at no time were movements predictable with any degree of certainty. The only generalizations regarding movement which can safely be made are that gorillas travel continuously within the boundaries of their home range and that they appear to arrive in certain sectors of their range at irregular intervals.

Times of arrival and departure into and from certain sections of the home range were variable for each group. Table 23 lists the dates of arrival of three groups in a certain section of the forest. All three groups spent considerable time outside the study area and the date given represents the earliest known evidence that they had arrived. Group VI made its appearance on the slopes behind Kabara at intervals of about forty days (twenty-four to fifty-seven days), group VIII on the same slope at intervals of about sixty days (fifty-

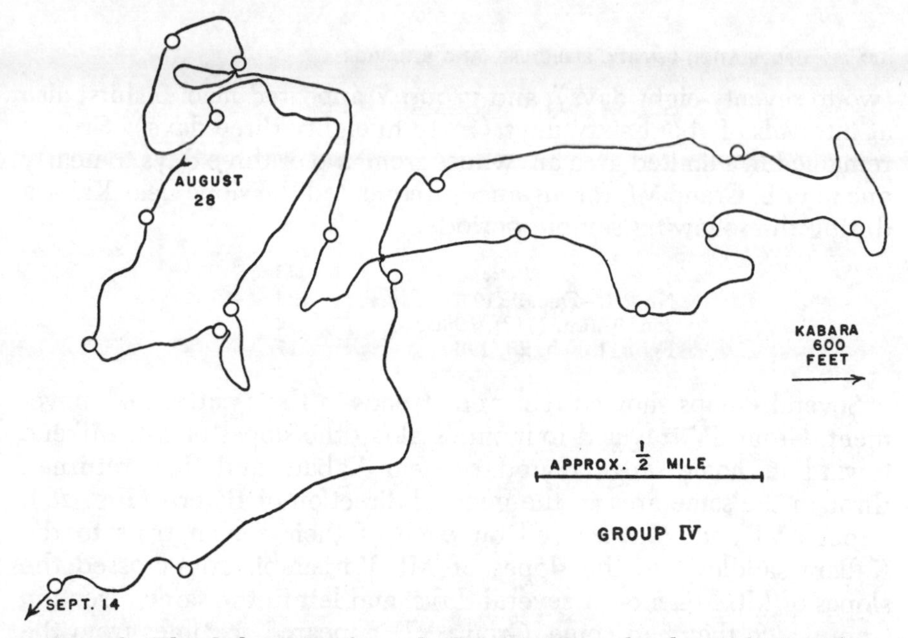

FIG. 28.—The daily route of travel of group IV on the slopes of Mt. Mikeno near Kabara between August 28 and September 14, 1959. Each circle represents one nest site.

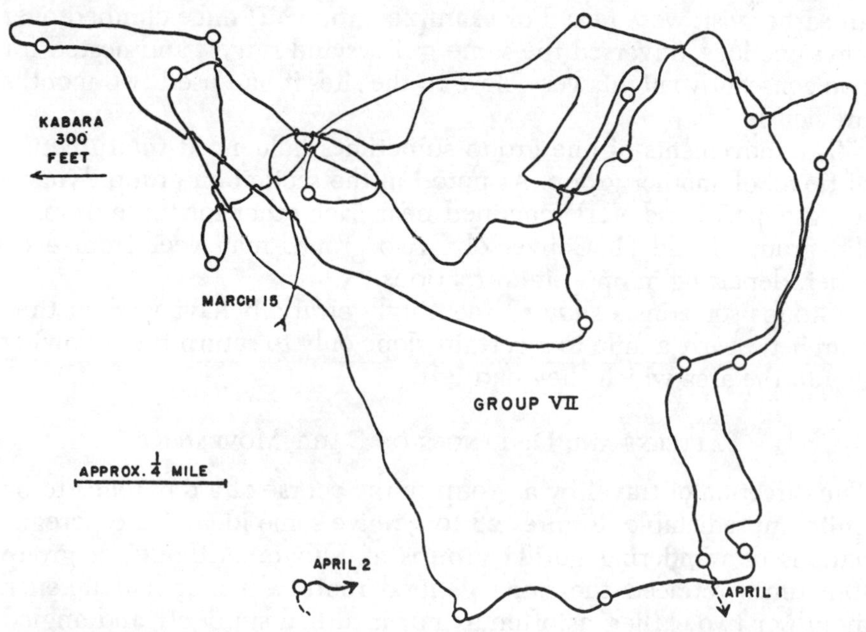

FIG. 29.—The daily route of travel of group VII in the saddle area between Kabara and Bishitsi from March 15 to April 2, 1960. Each circle represents one nest site.

two to seventy-eight days), and group V appeared near Bishitsi also at intervals of about sixty days (forty to eighty-three days). Groups remained in a limited area anywhere from two or three days to nearly one month. Group VI, for instance, frequented the slope near Kabara during the following sample periods:

Dates	No. of Days
Nov. 13–Dec. 9, 1959	26
Jan. 9–Jan. 11(?), 1960	2
Feb. 11–Feb. 28, 1960	17

Several groups showed recurrent trends in their pattern of movement. Group IV traveled four times along the slopes of Mt. Mikeno toward its home range boundary near Kabara and then returned through the same area in the general direction of Rweru (Fig. 28). Groups VI and VIII arrived on each of their eleven trips to the Kabara saddle from the slopes of Mt. Karisimbi, criss-crossed the slopes of Mt. Mikeno for several days, and left in the same direction from which they had come. Group VII appeared six times from the slopes of Mt. Karisimbi, crossed the saddle area, and briefly looped up and down Mt. Mikeno near Kabara (Fig. 29).

Instances in which groups used almost the same path as during an earlier visit were rare. For example, group VII once climbed over the same logs, traversed the same gulleys and ridges, and nested for two consecutive nights very close to the sites it had used two months previously.

The movements of one group sometimes influenced the direction of travel of another group. As noted in the section on group dynamics, groups VI and VIII remained near each other for three days, as did groups I and II. Conversely, two groups may veer from each other, departing in opposite directions.

Groups sometimes showed seemingly erratic behavior when they marched nearly a mile to a certain slope only to return the following day to the area which they had left.

Patterns and Distances of Daily Movement

The direction of travel by a group in the course of a day tends to be quite unpredictable. Figures 28 to 32 give some idea of the peregrinations of wandering gorilla groups at Kabara. Although a group sometimes retraced the same general route which it had taken a month or two earlier, as often as not it turned suddenly and angled up a slope which it had not visited during my previous encounters with it. Groups may circle and in the afternoon cut across their trail

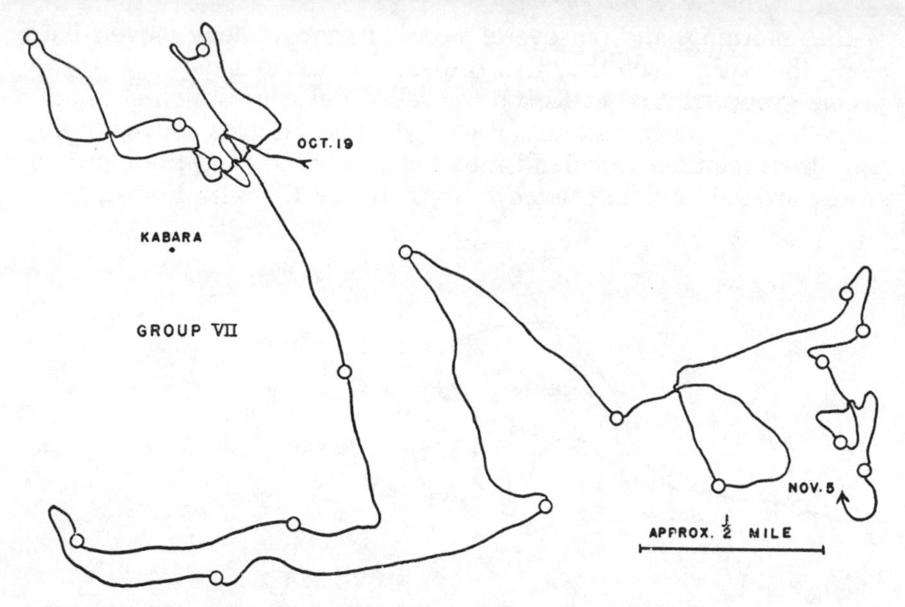

FIG. 30.—The daily route of travel of group VII in the saddle area between Kabara and Bishitsi from October 19 to November 5, 1959. Each circle represents one nest site.

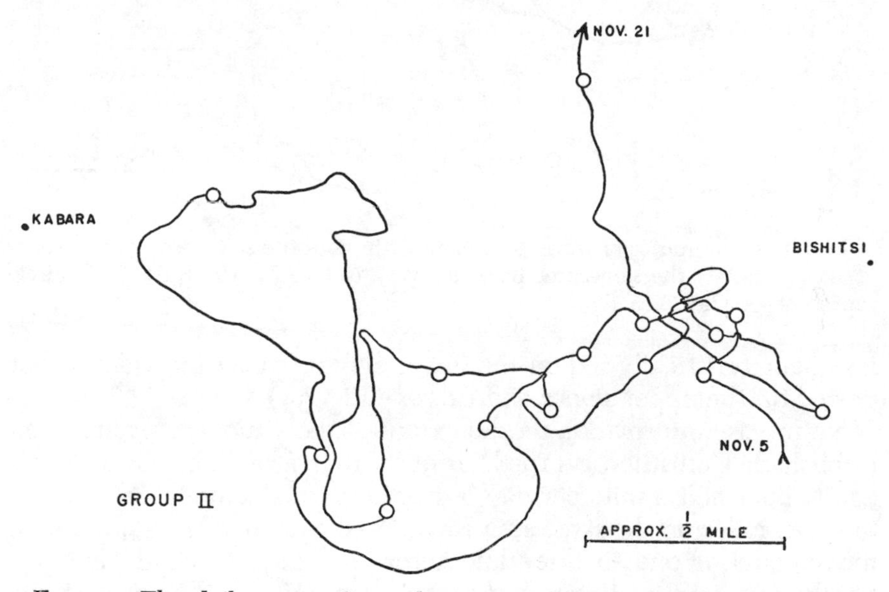

FIG. 31.—The daily route of travel of group II in the saddle area between Kabara and Bishitsi from November 5 to 21, 1959. Each circle represents one nest site.

of that morning, and on several occasions they briefly moved back along the path which they had broken the previous day.

One group tracked at Kisoro for seven days by Emlen showed a fairly regular daily pattern (Fig. 33). The animals moved uphill from the mountain woodland into the bamboo during the morning, swung around, and descended to nest. In the Kayonza Forest, goril-

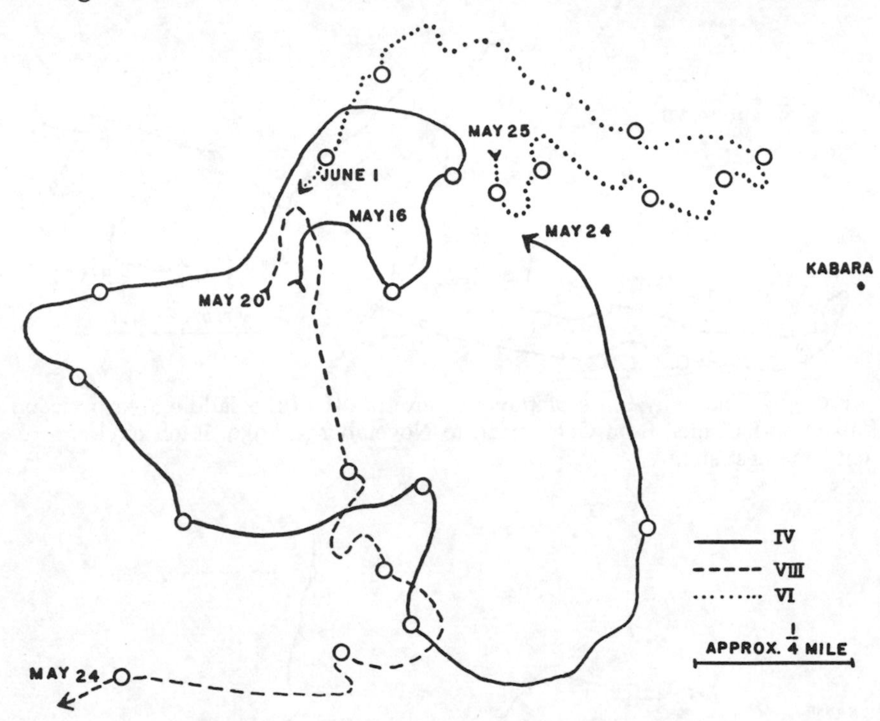

Fig. 32.—The routes of travel of 3 groups through the same area near Kabara at approximately the same time between May 16 and June 1, 1960. Each circle represents one nest site.

las characteristically fed on the lower slopes and in the valleys, but nested on the upper slopes and ridges (Fig. 34).

No precise information on the extent of daily movement has been published. Donisthorpe (1958) reports that groups at Kisoro averaged about half a mile per day but occasionally some traveled three to seven miles; in the same area Kawai and Mizuhara (1959*b*) noted movements of one to three kilometers per day. I paced, or alternately paced and estimated, the complete daily trail from nest site to nest site on 114 occasions at Kabara. Distances traveled by seven groups varied from 300 to 6,000 feet with an average of 1,742 feet.

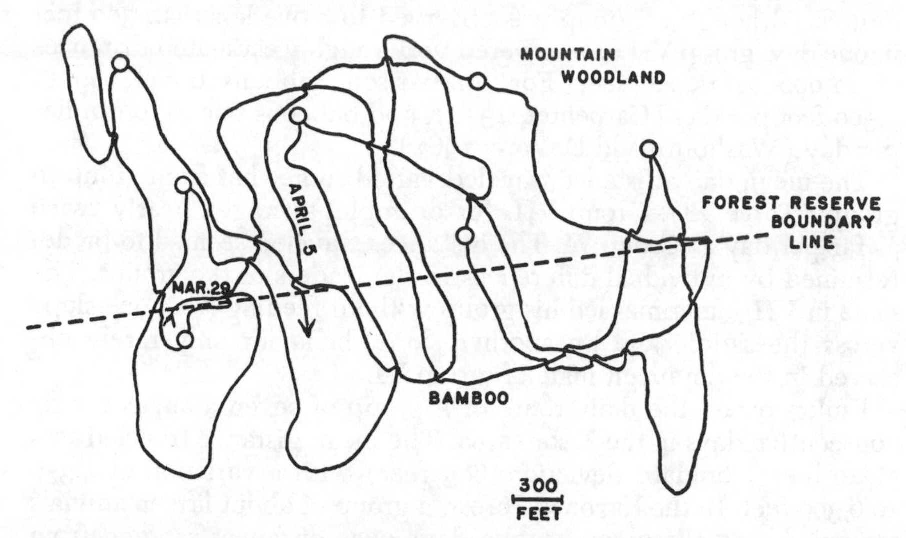

FIG. 33.—The daily route of travel of one group of 7 animals in the Kisoro area from March 29 to April 5, 1959 (after J. T. Emlen). Each circle represents one nest site.

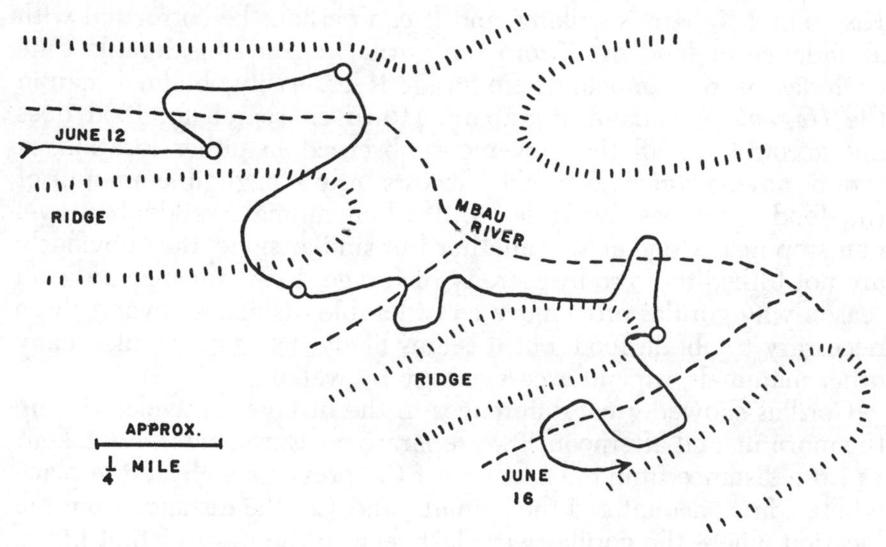

FIG. 34.—The daily route of travel of one group of about 15 animals in the Kayonza Forest from June 12 to 16, 1960. Each circle represents one nest site.

However, the whole range of variation is not represented by these figures. Although no group was observed to move less than 300 feet in one day, group VII once covered very roughly estimated distances of 11,000 to 15,000 feet. For comparison, gibbons travel 600 to 4,500 feet per day (Carpenter, 1940), and baboons two to four miles per day (Washburn and DeVore, 1961).

The mean daily distance traveled varied somewhat from group to group (Table 24). Group VII, for example, averaged nearly twice as far per day as Group VI. The distances traveled seemed to be determined by individual differences in the leaders of the groups. The male in VII sometimes led his group without feeding from one slope across the saddle and up another slope, behavior only rarely observed in the dominant male of group VI.

Emlen paced the daily route of a group of seven animals for six consecutive days in the Kisoro area. The mean distance traveled was 5,310 feet (standard deviation, 685 feet) with a variation of 4,350 to 6,300 feet. In the Kayonza Forest, a group of about fifteen animals moved during three consecutive days over distances ranging from 1,500 to 4,000 feet with an average of 2,767 feet (standard deviation, 1,249 feet).

The variation in the average daily distances traveled between Kisoro and Kabara is striking, and it can perhaps be correlated with abundance of food. At Kisoro the gorillas spent considerable time each day in the bamboo, where forage is less easily obtained than in the *Hagenia* woodland at Kabara. However, search for food does not account for all the movement observed in either location. A group moving some 2,000 feet harvests only a negligible amount of the food resources available. And when animals suddenly travel non-stop nearly one mile to another but similar slope, they obviously are not forced to do so by scarcity of forage. I can advance no good reason why gorillas often move considerable distances beyond those necessary to obtain food, but it seems likely that gorillas, like many other mammals, expend excess energy by walking.

Gorillas showed great differences in the distances traveled during the morning and afternoon. Two separate measurements were taken: (1) the distance from the nest site of the previous night to the place where I first encountered the animals, and (2) the distance from the location where the gorillas were last seen to the nests of that night. The results are presented in Table 25. Many of the data were derived from group VII. During the morning the movements of gorillas were quite leisurely: by mid-morning they were only about 300 feet beyond their nest site, and by mid-afternoon only about 700 feet. About half of the average daily route was covered between 1500 and 1700. After 1700 movement usually slowed down or ceased entirely as the animals built their nests.

Speed of Movement

Undisturbed groups moved slowly and with many pauses to feed and look around. The speed of leisurely moving groups was estimated by noting the time required for one animal to walk between two arbitrary points a certain distance apart. The average walking speed, including the various pauses but not rest periods, in twenty-two observations, was 532.7 feet per hour (standard deviation, 452 feet) with a variation of 100 to 1,800 feet.

Groups moving to another area without pausing traveled steadily at an estimated speed of two to three miles an hour. Excited groups walking rapidly probably proceeded at about four to five miles an hour. Running animals moved so rapidly that they could easily have escaped a pursuing man. For example, I followed a lone blackbacked male through the *Hagenia* forest at my maximum speed of about five miles per hour for nearly one mile without being able to overtake the animal again after the brief initial contact. An author who writes that ". . . the gorilla is a slow moving creature, and on the ground is easily overtaken by man" (Montagu, 1951) has obviously never tried to do so. Although no one has so far sought the opportunity to race a gorilla under controlled conditions, the maximum running speed of the ape is probably on the order of 15 to 20 miles per hour over short distances.

The Daily Activity Cycle

Gorillas are diurnal with nearly all their activity falling between the rather abrupt daybreak at 0600 and the equally abrupt nightfall at 1800 (Appendix D, Table 74). Within this twelve-hour period they follow a pattern which, though highly variable, is usually consistent enough for the major forms of activity to be predicted. Roughly their day consists of a morning feeding period, a long midday rest, and an afternoon feeding period. This pattern is similar to that described for gibbons (Carpenter, 1940), chimpanzees (Nissen, 1931), howler monkeys (Carpenter, 1934), and red-tailed monkeys (Haddow, 1952).

Rising Time and Activity

In West Africa one group of gorillas observed in an enclosure by Geddes (1955) rose at "daybreak." Merfield and Miller (1956) found that gorillas rose "before dawn." Osborn (1957) once encountered gorillas in the nest about twenty minutes after sunrise. I obtained direct observations at Kabara on eight occasions:

Date	Weather	Group	Rising Time
Sept. 8, 1959......	Cloudy	IV	Some of the animals apparently awakened to the distant chest-beating of a lone silverbacked male at 0530. The dominant silverbacked male was the first gorilla to leave its nest at 0600. The last animal left its nest at 0625.
Sept. 26, 1959.....	Partly cloudy	IV	A peripheral silverbacked male was the first animal to leave its nest at 0650. The last animal left the nest at 0710.
Feb. 8, 1960.......	Clear	VII	All animals left their nests between 0700 and 0705.
Mar. 24, 1960.....	Partly cloudy	VII	Nearly all animals were still in their nests at 0840 and did not leave the nesting site until 0950.
Mar. 10, 1960.....	Partly cloudy	VIII	The first animal sat up in its nest at 0635 and the last animal left the nest at 0700.
Oct. 6, 1959.......	Partly cloudy	Lone silver-backed ♂	I contacted the animal in the nest at 0805.

Two further examples, condensed from my field notes, illustrate typical rising behavior.

1. Group VII. January 4, 1960. Cloudy sky

0550. All animals sleep.

0630. The silverbacked male sits up, stretches one arm, yawns, and lies back down.

0645. The silverbacked male rises and moves about 5 feet from the nest and sits. Most of the other animals sit in their nests; several move slowly near their nests.

0650. The male leaves the nest site and walks 25 feet into the vegetation and begins to feed. The other group members follow him immediately.

0715. All feed. The group is spread over a diameter of 150 feet.

2. Group VII. April 2, 1960. Clear sky

0615. All animals sleep.

0640. The first sun rays creep over the sleeping forms. Everything is quiet except for the rumbling of stomachs and wind-breaking.

0650. A male of group V, which was nesting some 900 feet away, beats his chest twice but elicits no response from group VII.

0652. A female sits up in her nest, but lies down again at 0700.

0715. *Another female sits up and pulls in some* Galium *and* Peucedanum *which she eats in the nest. Two other animals sit up.*
0730. *Three gorillas have left the nest and feed slowly. Others remain in bed.*
0745. *The dominant male, whose nest is out of sight behind a low ridge, apparently has risen, for suddenly all animals move in the direction of his nest and a few seconds later reappear with the male, feeding.*

From these data and from nine other rising periods which I could not record as precisely, it appears that, in general, gorillas begin to stir in their nests during the hour after sunrise, and leave the nest site one-half to one hour later. The rising activity is characteristically slow. Some animals sit up, yawn, stretch, and recline again. A few feed near the nest while others are still in bed. As a whole, group activity usually does not commence until the dominant male leaves the site to feed.

The data are too limited to attempt a correlation between rising time and weather. Merfield and Miller (1956) noted that lowland gorillas remained longer in their nests on cold, damp mornings. However, in all my observations, groups which remained in bed past 0730 did so on sunny mornings.

Daylight Schedule of Feeding, Moving, and Resting

In order to determine the daily schedule of groups not influenced by my presence, I recorded their activity at the time of contact. I divided the activity into (1) feeding and/or moving along, (2) resting, and (3) feeding slowly and resting, with about half the animals occupied on one project and half on the other. The results of 279 encounters between 0700 and 1800 hours at Kabara are presented in Table 26 and Figure 41. Groups whose rising and bedding times were observed directly are not included in the tabulation. The data show that considerable variation in the time of certain activities exists, but that the tendency is toward a generalized pattern. Between the time of rising from the nest to roughly 0900, groups usually feed intensively. The hour between 0900 and 1000 is one of flux, in which about half the groups still feed but the other half are already resting. The major rest period falls between 1000 and 1400. After 1400, feeding and moving is again the primary activity (Figs. 36, 37, and 38).

Several factors influence the daily cycle. Repeated observations revealed that a gorilla requires about two hours of fairly intensive feeding after rising from the nest to become satiated. Thus, when gorillas rise late from their beds, they feed later into the morning than when they rise early. Any interruption in the feeding period affects the subsequent routine on that day. A sudden burst of sunshine during a spell of rainy weather tends to curtail feeding as the animals sunbathe. Heavy rains inhibit foraging activity. My arrival

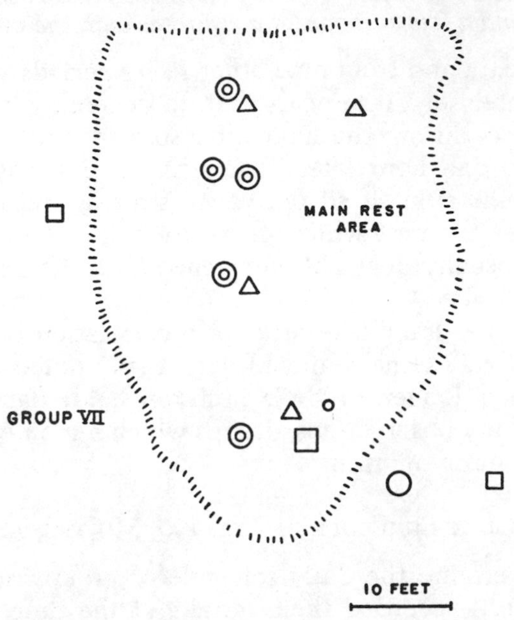

FIG. 35.—A typical scattering of animals in the midday resting area (group VII, February 14, 1960). The dashed line encircles a heavily trampled and packed-down area.

□ Silverbacked male
□ Blackbacked male
○ Female
◎ Female with infant
△ Juvenile
∘ Infant

tended to interrupt the routine of groups not habituated to my presence. Throughout the study I preferred to contact the animals after 0830, for if they ceased feeding before that time they soon began again and moved from my sight.

Every group rests at some time between mid-morning and mid-afternoon for a period lasting usually one to three hours. This rest period may be briefly interrupted while the group forages or moves to a new rest area. The midday rest period does not commence suddenly. Every animal feeds until satiated with the result that a few are still foraging actively while others are already resting. Sometimes one hour or more elapses between the time that the first and last animals in a group settle down to rest.

During the midday rest, most members of the group congregate near the dominant male. Figure 35 represents a typical distribution

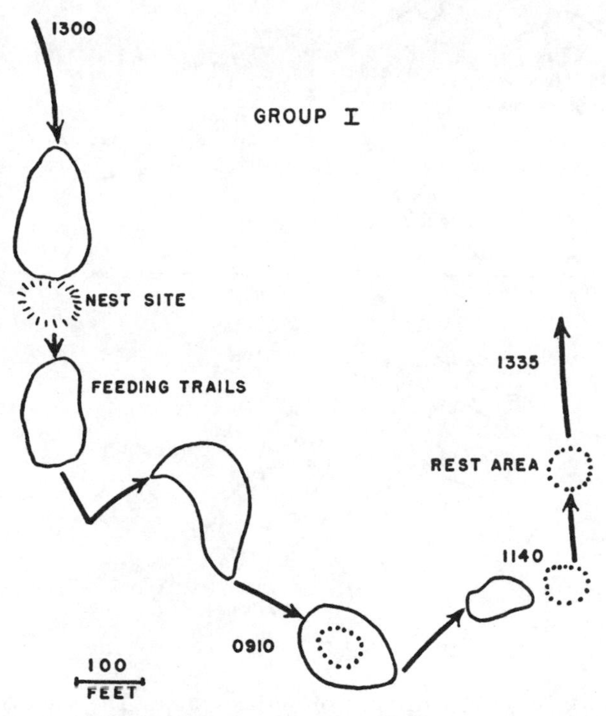

FIG. 36.—A typical daily route of a gorilla group, showing feeding areas, rest areas, and a nesting site (group I, August 23–24, 1959). The numbers refer to the time of day. The arrows indicate group movement with little or no feeding as well as the direction of travel.

of animals in a rest area. Sometimes the group rests very compactly, in an area about 20 feet in diameter, with bodies occasionally touching each other. At other times the animals are spread over a hundred feet of terrain.

Sleeping, dozing, and simply sitting are the main occupations of the animals during the rest period (Figs. 39 and 40). Some infants and juveniles play; both self- and mutual grooming occurs. Resting animals are essentially silent except for occasional wind-breaking and stomach-rumbling.

The animals usually show little preference as to near whom they rest. In general, females, juveniles, and infants tend to concentrate in a central area, but the silverbacked male rests as readily at the periphery as in the center of the group. Blackbacked males frequently stay at the periphery (Fig. 35).

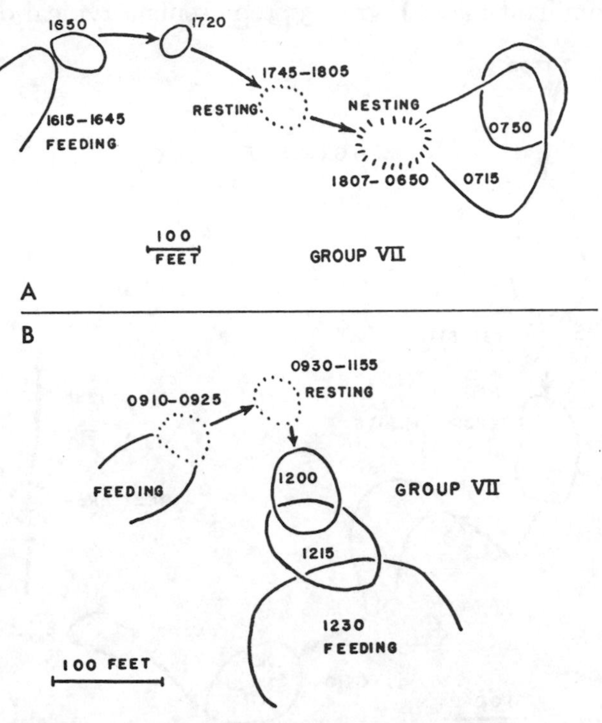

FIG. 37.—Daily activity patterns of gorilla groups. The solid circles indicate the approximate shapes and sizes of feeding areas; the dotted circles represent resting areas. The arrows mark group movement with little or no feeding as well as the direction of travel. The numbers indicate the time of day. *A*, a typical pattern of activity by a group during the late afternoon and early morning (group VII, January 3–4, 1960). *B*, a typical pattern of activity by a group preceding and following the midday rest period (group VII, March 27, 1960).

Weather influences the length of the rest period: on warm, sunny days the animals sleep longer than on cool, cloudy ones. One group sat in the rain from 1030 to 1430 hours. My presence, especially during the early weeks of the study, sometimes caused the animals to move to another resting site.

As the rest period draws to an end, more and more animals feed or move slowly in the vicinity of the rest site. The rest period is over

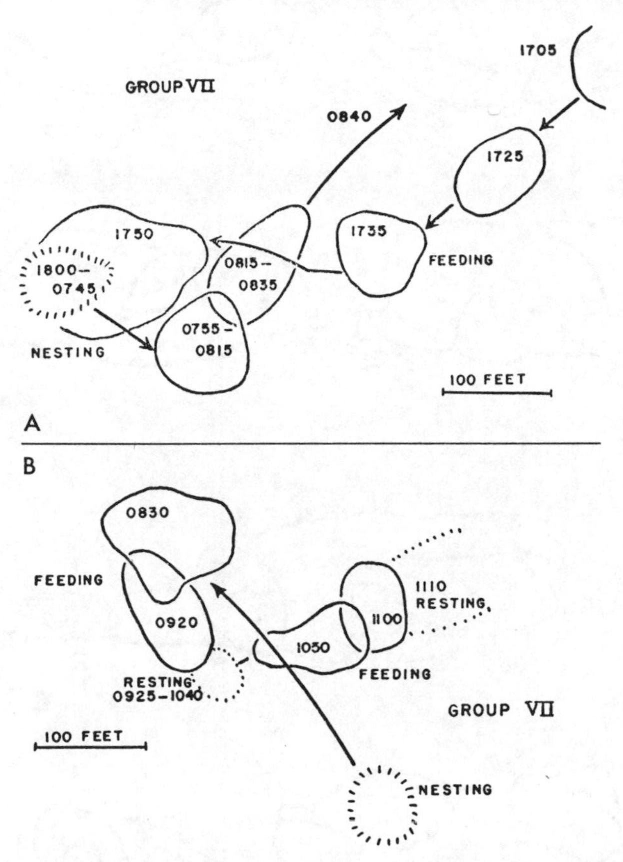

FIG. 38.—Daily activity patterns of gorilla groups. The solid circles indicate the approximate shapes and sizes of feeding areas; the dotted circles represent resting areas. The arrows mark group movement with little or no feeding as well as the direction of travel. The numbers indicate the time of day. *A*, a typical pattern of activity by a group during the late afternoon and early morning (group VII, April 1–2, 1960). *B*, a typical pattern of activity by a group from the time of rising to the midday rest period (group VII, April 20, 1960).

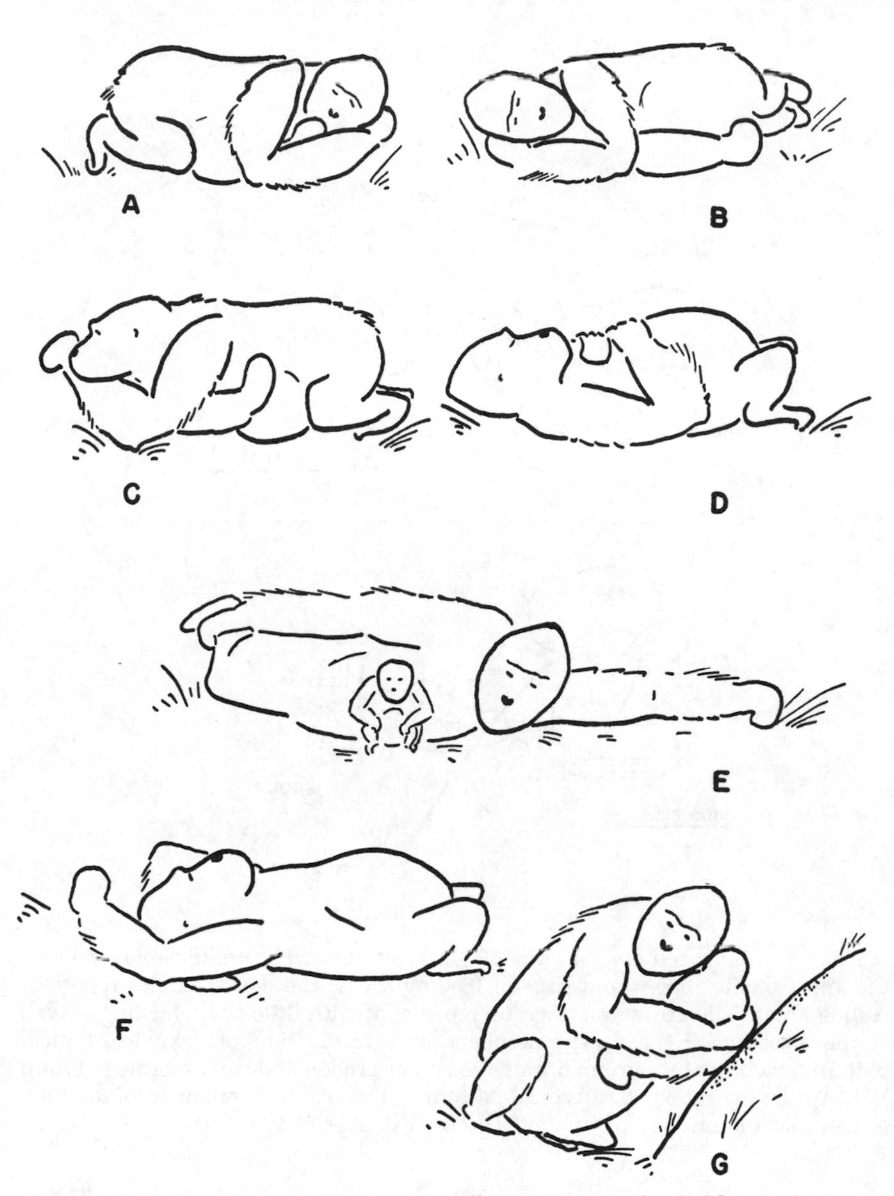

FIG. 39.—Body positions of resting gorillas. *A, G,* females holding infants; *B, C, D, E, F,* females.

FIG. 40.—Body positions of resting gorillas. *A*, juvenile; *B*, *F*, females holding infants; *D*, *E*, *G*, females; *C*, blackbacked male.

when the dominant male moves out to feed. The sequence of events transcribed below is typical:

Group IV

0900. Most of the animals are sitting; several feed slowly or climb around in trees.
1000–1130. All are either lying or sitting, and a few groom.
1130. First one female, then several others, start to feed.
1150. The dominant male moves out from the rest area and all feed except one of the peripheral silverbacked males, a blackbacked male, and a female with infant, who remain on the rest site.
1200. All are foraging.

The activity during the hours between leaving the rest area and nesting for the night is highly variable. Movement is usually more extensive (see Table 25), and the feeding period is more prolonged than the one in the morning, a fact which has also been noted for the red-tailed monkey (Haddow, 1952). Groups occasionally travel several hundred feet with little or no feeding. Since the afternoon feeding period is less intensive than the morning one, foraging generally continues intermittently until dusk. Group activity after 1700 usually slows down considerably and much time is spent in sitting, although the animals tend to accelerate their speed of movement just before nesting.

Bedding Time and Activity

In West Africa, Geddes (1955) noted that a group in a large enclosure bedded down at 1700 hours. The precise time at which gorillas built their night nests at Kabara was difficult to determine, for the gorillas usually delayed construction in the presence of an observer. Group VIII, for example, remained until 1840 to watch me on March 9, 1960, before moving about 400 feet in darkness to nest. The following exact nesting times were obtained:

Date	Weather	Group	Nesting Time
Sept. 25, 1959.....	Cloudy (rain)	IV	The animals sat on the nest site at 1620. By 1700 several had built nests, but at least 3 animals continued to move around until 1745.
Sept. 7, 1959......	Cloudy (rain)	IV	The group fed slowly at 1625. At 1650 the dominant male built the first nest, and the other animals also constructed theirs during the following half-hour. All was quiet at 1725.
Nov. 17, 1959.....	Cloudy	VI	All nests were built between 1805 and 1810.
Feb. 7, 1960.......	Cloudy	VII	The male built the first nest at 1805, but the group became aware of me after that.
Jan. 3, 1960.......	Partly cloudy	VII	The whole group built its nests between 1800 and 1810, with the male constructing his first. The group was aware of my presence but seemed undisturbed.

One further example illustrates nesting behavior:

Group VII. April 1, 1960. Cloudy sky

1700–1800. The group alternately feeds and travels.

1807. The male is some forty feet from his group and partially hidden by a low ridge. The movement and snapping of the vegetation indicate that the male is building his nest. At the same time a juvenile constructs a nest at the base of a tree. The other gorillas continue to feed.

1810. The group is about 40 feet beyond the male. The juvenile abandons its nest and joins the main group.

1820. It is too dark to distinguish the animals clearly. Quite suddenly they all start to break in the vegetation to construct their nests.

1825. All is quiet except for one juvenile, which is still moving about.

The dominant male was the first animal to build its nest in all four instances in which the behavior was observed. Nesting tended to be leisurely, with some gorillas moving and feeding fifteen or more minutes after the first animal had constructed its nest. The

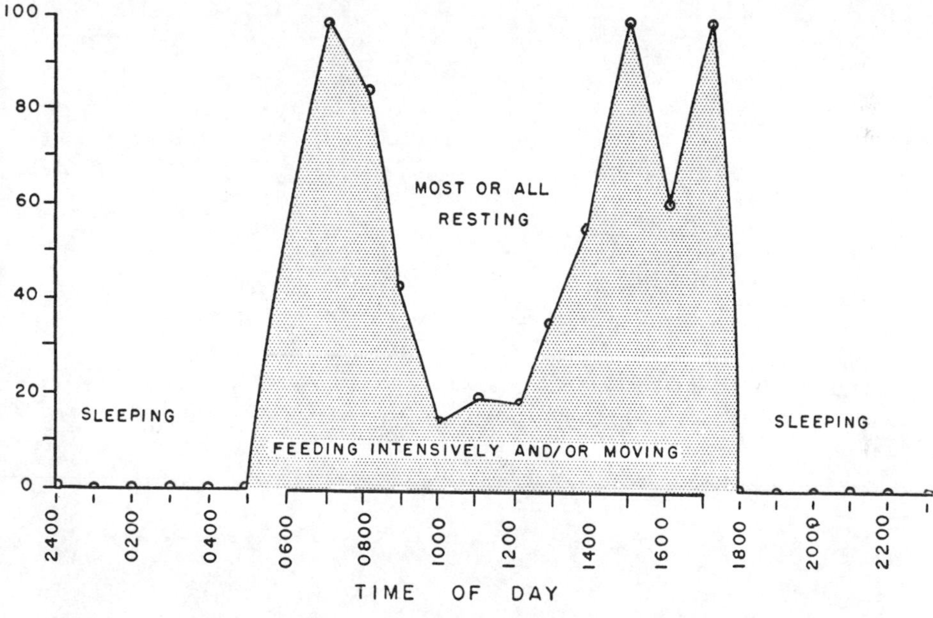

FIG. 41.—The daily activity cycle of gorilla groups. The data are taken from Table 26 to illustrate the peaks of morning and afternoon feeding separated by the midday rest period. (The daily variations in rising and bedding times, and periods when half the group is feeding and the other half resting, are not indicated.) The figure shows that feeding activity may start by 0600 or somewhat earlier, and that by 0700 all groups which have left their nests are actively foraging. The main resting period falls between 1000 and 1400. Activity usually ceases by 1800.

average time of day for the construction of nests appeared to be just before dark. Weather perhaps affected the time of construction: both instances of early nesting occurred on rainy afternoons.

Activity ceases as soon as the animals have built their nests. I obtained no evidence that gorillas ever left their nests at night. During eight nights when I slept close to a group, no snoring was heard. Occasionally a silverbacked male beat his chest intermittently throughout the night, behavior apparently elicited by the proximity of another group or a lone male. For example, a peripheral male in group IV beat his chest at about 2300, 0200, and 0530 hours in response to the sound of chest-beating by a lone male a few hundred feet away.

CHAPTER 5

INDIVIDUAL ACTIVITIES AND BEHAVIOR

Food Habits and Feeding Behavior

The publications of Akeley (1923), Maxwell (1928), Bingham (1932), Donisthorpe (1958), and others contain some information on the food habits of gorillas. I collected food plants and took notes on their utilization by gorillas in all study areas. The plant specimens were later identified by the East African Herbarium, the Uganda Forest Department, and the I.R.S.A.C. station at Lwiro.

Species of Plants Eaten and Food Preferences

In order to obtain a fairly complete list of plants eaten in a specific region, collections must be made over a long period of time, for gorillas utilize several species only incidentally and a few are seasonal. When gorillas consume an entire plant, direct observation is essential since no visible remains litter the trail. For example, the abundant vine *Galium* has never been recorded as a forage species by numerous observers in the Virunga Volcanoes, yet direct observations of feeding animals demonstrate that this plant is a major food item in these mountains.

Information on the forage species utilized by gorillas is based on several sources. At Kabara all data were obtained through direct observation, in the Kayonza Forest about half. In the other areas trail signs such as tooth marks on bark, leafless petioles, and shredded and discarded stems usually provided clues to the species of plant eaten. I have also included some records from other investigators who worked in the area, from natives, and from the literature.

KABARA Forage is abundant in the *Hagenia* woodland—the gorilla can sit down anywhere and be within reach of food. The following four plants, listed in the order of frequency with which they were eaten, probably furnished at least 80 per cent of the daily food supply in the *Hagenia* woodland: (1) *Galium simense,* (2) *Peucedanum linderi,* (3) *Carduus afromontanus,* and (4) *Laportea alatipes. Rumex, Vernonia, Pygeum, Polypodium,* and *Cynoglossum* also contributed considerably to the diet at times. *Rubus* was the only fruit eaten throughout the year; *Pygeum* ripened in April and was then consumed extensively.

In the bamboo zone the shoots of *Arundinaria alpina* were commonly eaten during the wetter seasons, and above 11,000 feet *Lobelia wollastonii, Peucedanum kerstenii, Helichrysum* sp., *Senecio erici-rosenii* and *Senecio alticola* formed the main bulk of the diet.

A few foods could have furnished only meager nourishment, and I was unable to determine the reason for ingesting them. For example, gorillas commonly ate the dry slabs of *Hagenia* bark and gnawed on the bark and rotten logs of *Hypericum.*

Table 27 presents a list of the food species at Kabara.

KISORO During March–April, 1959, the most commonly eaten foods in the bamboo and mountain woodland of the Kisoro area were *Arundinaria alpina, Cynoglossum amplifolium* and *C. geometricum,* several vines (*Crassocephalum bojeri, Basella alba, Piper capense*), and *Peucedanum linderi* (see Table 7). In the *Hypericum* woodland, *Vernonia adolfi-frederici* was the food plant remnant most often seen. Above 11,000 feet *Senecio erici-rosenii, Senecio alticola, Peucedanum kerstenii,* and *Lobelia wollastonii* provided most of the nourishment. Table 28 lists the species which are eaten at Kisoro. Donisthorpe (1958) published a list of nineteen food plants for the area, and species presented by her but not noted as being utilized during the present study are included in the tabulation.*

KAYONZA FOREST The bark and leaves from various vines were important gorilla food items in the Kayonza Forest. Especially favored was *Momordica foetida,* but the animals also commonly ate *Urera hypselendron, Piper capense, Basella alba,* and *Mikania cordata.* Shredded fronds of *Cyathea deckenii* were abundant on the

* Kawai and Mizuhara (1959*b*, printed in 1962) recently added *Acalypha bipartia* and *Momordica foetida* as food plants to the area.

trails, indicating that this species also constituted a major food plant. Occasionally the gorillas raided cultivated banana patches. Table 29 presents a list of the food plants.

MT. TSHIABERIMU Bamboo shoots furnished the bulk of the gorilla's diet when we visited Mt. Tshiaberimu in the latter part of May, 1959. The animals utilized *Rumex* sp. and *Urera hypselendron* in the more open valleys, and commonly chose bark from the twigs of *Pygeum* and *Maesa* in the stands of mountain forest. Fields were frequently visited southwest of the mountain with peas and maize providing preferred forage. The food plants are listed in Table 30.

MT. KAHUZI The bamboo-covered rift mountains northwest of the town of Bukavu provide a habitat very similar to that on Mt. Tshiaberimu. We spent only one day in the forest, but some of the forage species were kindly pointed out to us by Mr. Christiaensen, a botanist from I.R.S.A.C., Lwiro (Table 31).

MWENGA-FIZI REGION Work in these extensive bamboo and mountain forests was limited to two weeks in June, 1959. We saw no gorillas and the data on food habits are based entirely on information from natives and on trail sign. Specimens and data were collected at Mulenge, which lies on the slopes of the rift between Lakes Kivu and Tanganyika; near Fizi, the most southernly extension of gorilla range; and at Nzombe, just south of Mwenga. Table 32 lists the food species.

UTU REGION For purposes of food habit analysis the vegetation in the lowland rain forest region can be divided into two types: (1) cultivation and young secondary forest, and (2) old secondary and primary forest. Examination of gorilla trails showed that in the former type *Aframomum* and the pith of domesticated banana stems provided the bulk of the diet. *Palisota, Costus, Marantochloa, Ficus, Manihot,* and several others, although frequently consumed, were of lesser importance.

I made observations in old secondary and primary forest during a four-day hike through uninhabited country in late June and early July, 1959. A tabulation of food remnants seen along the trail revealed the following:

Species	No. of Observations
Aframomum	19
Palisota	5
Marattia and *Dryopteris* ferns	4
Pycnanthus and *Uapaca*	2

Aframomum again appeared as a major food species. Table 33 lists the food plants from the Utu region.

WEST AFRICA Unfortunately the literature contains only a few observations on the forage species utilized by gorillas in West Africa. Nearly all authors agree, however, that gorillas favor *Aframomum* and cultivated bananas. Sabater Pi and de Lassaletta (1958) autopsied five gorillas and noted that the marrow of banana stems provided about 80 per cent of the bulk eaten, followed by *Manihot* at 10 per cent, and with the remainder consisting of *Aframomum,* sugar cane, and fruits. A male autopsied by Sabater Pi (1960) contained the following food items:

	Per Cent
Prevostea africana	25
Aframomum sp.	25
Musanga smithii	15
Manihot utilissima	13
Fleurya aestuans	10
Uapaca guineensis	10
Unidentified	2

Table 34 lists those species of plants reported in the literature as being eaten by gorillas.

TOTAL NUMBER OF PLANT SPECIES EATEN BY MOUNTAIN GORILLAS A perusal of Tables 27 through 33 shows that mountain gorillas subsist on a wide variety of plant species, consisting of the following types:

Type of Plant	No. of Species
Grass—sedges	4
Ferns	5
Herbs	29
Vines	22
Shrubs	9
Trees	25
Cultivated	6
Total	100

The total figure for species eaten represents a minimum, for such plants as *Aframomum* and *Palisota* were usually not identified below the genus level, and more intense work in some areas will surely reveal additional food plants. However, of the large number of species eaten, only about a dozen provide most of the forage. *Aframomum, Musa,* and *Urera* are the three major food plants in the mountain gorilla's range as a whole, while *Arundinaria, Galium, Peucedanum, Momordica, Cyathea,* and a few others are locally prominent.

Actually gorillas eat only a small percentage of the total number of plant species available to them, and some of the most abundant plants are not utilized. At Kabara, for example, gorillas do not eat *Senecio trichopterygius* and *Impatiens,* two of the dominant herbs.

Nutritional Value of Gorilla Foods

Very little is known about the nutritional value of the various gorilla food plants. A determination on *Peucedanum linderi* from Kisoro, made at the Veterinary Department in Entebbe for Dr. R. Dart and Mr. W. Baumgartel, is transcribed in Table 35 through the courtesy of the latter. It is interesting to note that the leaves, which are not eaten by gorillas, are, in general, more nutritious than the stem. The animals peel the stem and the high percentage of crude fiber contained in the bark is not ingested.

Similarities and Differences in Food Habits between Study Areas

Gorillas occupy a wide variety of environments and have adjusted their food habits to them. Local adaptation has proceeded so far that I found, for example, no overlap between the plants eaten in the Utu region and in the Virunga Volcanoes, although the animals utilized a total of fifty-five forage species in the two areas. A total of fifty-seven food species have been collected in the Kayonza Forest and in the Virunga Volcanoes (Kabara, Kisoro, and Mts. Visoke-Sabinio saddle combined) but only nine (15.7%) of these were shared between the two areas. On the other hand, in the contiguous and similar environments of Kisoro and Kabara about half (eighteen) of the thirty-eight forage species were shared (Table 36).

The low frequency of shared food plants listed in Table 36 reflects primarily the geographic changes of the vegetation. However, several forage species are widespread and tolerate considerable altitudinal variation. With these I attempted to determine if "cultural" differences in food habits occurred from area to area. In other words, did some gorilla populations eat a certain species which others did not, even though the plant was present in their area? Such local preferences for certain foods were noted by Miyadi (1959) in isolated colonies of *Macaca fuscata.*

Table 37 lists seventeen selected forage plants and shows their occurrence and utilization by gorillas in six study areas. Conclusions drawn must be tentative, especially from those species marked as present but not eaten by gorillas, for more detailed investigations may show that they are used. Gorillas ate *Urera, Mimulopsis, Xymalos, Piper, Myrianthus,* and *Arundinaria* whenever these species occurred within their range. The cultivated banana (*Musa*) was a major food plant in the Utu region, but only a minor one along the edge of the Kayonza Forest and in the Virunga Volcanoes even though the plant was often readily available near the forest.

At Mt. Tshiaberimu gorillas apparently ate neither *Laportea* nor *Cyathea deckenii*, yet they commonly consumed the former in the Virunga Volcanoes and the latter in the Kayonza Forest. The pith of the stem of *Aframomum* was one of the most widely used gorilla foods, but in the Kayonza Forest I found no evidence of its having been eaten, even though the plant was fairly abundant and gorillas fed on the fruit. *Galiniera coffeoides* was very common in the Kayonza Forest, where gorillas apparently did not feed on it; however, the animals utilized the plant in three other areas (Table 37). Both *Marattia fraxinea* and *Palisota* were relatively important food items in the Mwenga-Fizi and Utu regions, but neither were eaten in the Kayonza Forest, although the trails of feeding gorillas were followed to within reach of these species several times; *Pennisetum* in the path of a foraging group also remained untouched. Perhaps the rarity with which these plants were utilized merely reflected preference for other foods, but my limited observations suggested qualitative rather than quantitative differences.

General Feeding Behavior

Gorillas often feed in the morning while still sitting in bed, and they sometimes eat their last snack of the day after having built their night nest. However, the major feeding periods, as discussed more fully in the section on the daily activity cycle, fall between about 0700 to 1000 and 1400 to 1700. Immediately after rising, the animals feed intensively, but activity slows down considerably as the morning progresses, with some feeding leisurely and others sitting. Sometimes the dominant male lies down to rest, while the others continue to forage in the vicinity. Even during the main rest period it is rare that an hour goes by without at least one animal snacking. Any gorilla may rise at any time to feed for a few minutes in the vicinity of a resting group. Once the dominant male of a group remained alone in the rest area for nearly one hour as the other members fed within a 100-foot radius.

After leaving the main resting site, the animals usually feed intensively for about half an hour before continuing at a more leisurely pace. Quite frequently groups interrupt their afternoon foraging to sit or travel with little or no foraging. While feeding on the move, members of the group spread out, making numerous trails. The diameter of feeding groups is usually about 100 to 150 feet, but they sometimes scatter over 200 to 300 feet of terrain. Once I encountered group VII extended over 400 feet, and group IV several times spread itself over a similar distance. A large group in the Kayonza Forest fed scattered along a slope with members strung out over some 300 to 400 feet, and my one observation on gorillas in the Utu region revealed a similar group diameter. Gorillas usually concentrate while resting or nesting but fan out rapidly when they begin to feed again (Figs. 37 and 38).

Apparently gorillas do not share food; at least I did not observe a single instance in which one animal offered vegetable matter to another in the wild. However, I noted that the male at the Columbus zoo transferred a piece of apple through the bars from his mouth into the mouth of the female.

The animals usually forage leisurely, alternately sitting and walking. Characteristically each gorilla sits and reaches for food in all directions. If it eats *Peucedanum* or a similar item, the discarded leaves and peelings litter the lap as well as the area immediately in front of the animal. After most of the forage within easy reach has been consumed, the gorilla moves a few steps, grabs a handful of food in passing, and sits again. The motions are quite unhurried and frequently minutes pass without anything being eaten. The animals are usually silent during feeding except for an occasional grunt or belch, the snapping of branches, and the smacking of lips. They rarely lick their lips after eating. The leisurely pace can best be illustrated by quoting my notes, which describe the actions of a blackbacked male during a typical half-hour of feeding:

Junior sits and peers intently at the vegetation, reaches over, and bends the stalk of a Senecio trichopterygius *to one side. He stretches far out and with a quick twist decapitates a* Helichrysum. *After stuffing the leafy top into his mouth, he looks around and spots two more plants of the same species which he also eats in similar fashion. He then yanks a* Peucedanum kerstenii, *including the root, from the ground, and with rapid sideways and backward jerks of the head bites apart the stalk before gnawing out the pith. The sun appears briefly and Junior rolls onto his back. But soon the sun hides behind a cloud, and Junior changes to his side, holding the sole of the right foot with the right hand. Then he switches and rests on his belly. After about 10 motionless minutes he suddenly sits up, reaches far out, slides his hand up the stalk of a* Carduus afromontanus, *thus collecting the leaves in a bouquet which he pushes with petioles first into his mouth. A leafy thistle top, prickles and all, and a* Helichrysum *follow. Then he leaves his seat, ambles 10 feet, and returns to his former place, carrying a thistle in one hand and a* Helichrysum *in the other. After eating the plants he sits hunched over for 15 minutes. The rest of the group feeds slightly uphill and Junior suddenly rises and moves toward the other members, plucking and eating a* Helichrysum *on the way.* A Senecio erici-rosenii *has been torn down by another gorilla, and Junior stops and rips off a leafy top. From the stem he bites large splinters until only a two-inch section of pith remains in his hand, which he eats. A strand of* Galium *follows, and just before he moves out of sight, a final* Helichrysum.

Movement is characteristically faster during periods of intensive feeding. The animal may stuff food into its mouth with one hand while the other is already reaching for more. A considerable amount is eaten while the gorilla stands on three legs and pulls in food with one hand. In the bamboo zone at Kisoro, and in other vegetation types where food is not as abundant as at Kabara, each animal walks until it finds a palatable plant, feeds briefly, and continues on to the next one.

Regurgitation and subsequent re-ingestion of food, which I saw in the New York, Columbus, and Antwerp zoos, were never noted in the wild.

Feeding Behavior on Various Plant Species

Collecting of food was almost entirely manual. I observed only a few occasions on which a gorilla used its mouth directly to detach a plant, although the mouth and hands not infrequently worked in conjunction to remove a branch. The feet rarely entered directly into foraging activity, and their function appeared to be limited to occasionally carrying or holding a food item.

The amount of preparation necessary before each species became palatable varied from such plants as *Galium,* of which the gorilla ate the whole vine, to *Vernonia,* from which it extracted the pith. Both the hands and the teeth were used about equally in tearing and shredding the plant, with the former prominent in manipulation, holding, and pulling and the latter in ripping, tearing, and gnawing. Gorillas showed great dexterity in combining the use of hands and mouth, with the result that the palatable parts of each plant were rapidly exposed and eaten.

The purpose of this section is to describe the methods, and variations in the methods, of eating some of the major food plants. Most of the descriptions are based on direct observations, but a few rely solely on circumstantial evidence in the form of discarded plant sections and tooth marks.

Pennisetum purpureum Two captive animals at the I.R.S.A.C. station, Lwiro, fed on the stalks of elephant grass by holding the stem in both hands, biting into it, and tearing out a section with a sideways jerk of the head. They then grasped the torn edges, pulled the stalk apart, and ate the tender white pith near the base.

Arundinaria alpina Gorillas eat both bamboo shoots and tender green stems. From indirect evidence collected, it appears that shoots are eaten in several ways. Very young shoots, 5 inches or less in length, are often entirely hidden from sight beneath leaves and humus, but gorillas expose them by digging. I noted that for some reason several very small shoots remained uneaten after having been dug up. Gorillas eat the tender white pith of the shoots and the more solid base. To reach the pith the animals apparently rip the shoot apart, leaving the tough and hairy outer layers in a shredded heap. Sometimes the remains look as if the animal had peeled back the various layers to expose the pith, much as a human prepares a banana.

At Mt. Tshiaberimu, gorillas apparently collected several shoots and ate them at one location, leaving five to eight and even ten to fifteen husks piled in one spot.

Shredded remnants of bamboo shoots eaten by *Cercopithecus mitis* often look similar to those left by gorillas and care must be taken in interpreting trail sign.

Young green stems of bamboo one inch or less in diameter are eaten in a characteristic way. The animals break the stems into sections, snapping them off at the internode. They then chew only the lower portion of each internode, leaving the upper part with the attached bract intact. The internodes are hollow and lined with a soft material one mm. or less thick. One animal held an internode horizontally in both hands, bit into it with one set of canines, and jerked its head back, thus splintering the whole section into two parts. It then gnawed out the thin film of pith with the upper incisors.

Cyathea deckenii Small piles of bark from the fronds of tree ferns lie at frequent intervals along gorilla trails in the Kayonza Forest. The animals apparently tear the fronds from the tree, bite off the bark, and eat the more tender inner part. They also eat the large shoots of young tree fern after discarding the curled top and tough bark.

Marattia fraxinea and *Dryopteris* sp. Gorillas eat the stems of these ferns, first detaching them, and then biting off the bark. In the Utu region, one animal had dug up a *Dryopteris* and eaten the root.

Polypodium sp. Numerous small ferns, rarely more than six inches long and consisting of but a single leaf, hang suspended below *Hagenia* branches from which gorillas pluck them, behavior illustrated by one typical example:

A juvenile lies on its belly on a large horizontal branch 25 feet up. It reaches below the branch and without looking grabs a handful of hanging ferns which it pulls in. After severing and discarding the roots with one bite, it stuffs the greens into its mouth.

Sometimes animals rested on a branch and picked ferns slowly one at a time between thumb and index finger. One female removed a large cushion of moss from a branch, placed it in her lap, and plucked and ate the ferns which grew from it.

Peucedanum linderi Commonly seen along a gorilla trail at Kabara are the small scattered piles of flaccid peelings and leafy tops, which are all that remain of wild celery after the animals have eaten. Gorillas usually detach the stalk near the ground with a jerk, and snap or bite off the upper leafy part and some of the smaller stems so that only the large basal portion of the main stem remains in their hand. One typical excerpt from my field notes illustrates how the gorillas eat this species:

A female climbs onto a mound, a decapitated stalk of celery hanging from the corner of her mouth like a large cigar, and sits. She then holds the stalk vertically in one hand, bites into the edge of the fibrous bark at the basal end and at the same time twists her hand outward and her head slightly backward. A long section of bark is torn loose, and by repeating the motions, she rapidly peels much of the stem. She then holds the stalk horizontally in both hands and rapidly detaches slivers of bark by biting and tearing them off with abrupt sideways jerks of the head. Finally she eats the tender center, both by chewing from the end of the stalk and by biting into the side of it.

Peucedanum kerstenii is eaten in a similar manner except that the upper woody section of the root is also peeled.

Laportea alatipes The virulence of nettles at Kabara was such that they readily burned through two layers of clothing; after a sojourn among them my knees were swollen and red welts covered my face. Yet gorillas handled them without hesitation and fed on stems and leaves that bristled with white hairs—the animals were apparently insensitive to them. Several examples illustrate typical feeding behavior:

1. *A blackbacked male slides the entire stem of a nettle from base to top between his thumb and index finger and thus collects a bouquet of leaves, which he stuffs into his mouth.*
2. *A silverbacked male bends a nettle toward himself with one hand, and with the other pinches off the leafy top, which is eaten.*
3. *An infant, 1¼ years old, breaks off the top of a nettle, after which it sits, holding the stem in one hand and with the other pulling off and eating each cluster of flowers.*
4. *A blackbacked male plucks and eats one nettle leaf at a time from a standing plant.*
5. *A blackbacked male tears off a large nettle stem, carries it in one hand 8 feet to a log, sits, holds it sideways with one hand and bites off the bark, much as if eating* Peucedanum. *He eats some of the center but discards most of it.*

Cynoglossum amplifolium and *C. geometricum* Gorillas feed primarily on the taproots of these herbs, although they also eat the basal part of the stem of *C. amplifolium* after biting off the tough bark. At Kabara the animals simply grabbed the stem and pulled slowly until the whole plant including the root was freed. However, at Kisoro, where the soil was frequently hard, the stem tended to break, leaving the root in the ground. In April, 1959, I noted numerous holes 5 to 12 inches deep and 5 to 10 inches wide made by gorillas digging for the roots of this plant (Pl. 21). The stem usually lay by the side of the hole. The taproots were either peeled or eaten whole. A female provides a typical example:

She sits and with both hands grasps a C. amplifolium *near the base and pulls the whole plant from the ground. With a downward jerk of both hands she breaks the stem in half and with a sideways flick of the wrist discards the leafy portion. She holds the stem sideways with one hand and bites off the rootlets, the rootbark, and the bark from the lower part of the stem. Finally she eats the taproot.*

Carduus afromontanus Although the edges of the thistle leaves are prickly, gorillas either ignore the spines or carefully pick out the leafy portions between them. Gorillas eat most of the plant, and the method of feeding on each part is best illustrated from my field notes:

1. *A 1½-year-old infant breaks off the top of a thistle, holds the larger end between its teeth, and slides its hand down the stem, clearing it of leaves and spines. It then eats the stem as a man eats a stalk of celery.*
2. *A female holds the base of a stalk in one hand and slides the other hand vertically up the stalk thus collecting the leaves. She discards the stem and eats the leaves by stuffing them all at once, petiole first, into her mouth.*

3. *A juvenile holds a thistle stem in one hand, and with the other picks off the dried flowering heads one at a time between thumb and index finger and eats them.*
4. *A silverbacked male slides thumb and index finger up the petiole of a leaf to collect the greens, which he eats. He then slides his hand down the main stem, breaking off all leaves. Finally, he snaps the bare stem off and eats it, large end first.*

Lobelia giberroa Although this giant lobelia is extremely abundant at Kabara, I observed only five gorillas as they briefly fed on the plant. Injured plants exude a sticky white fluid which adheres to skin, tastes extremely bitter, and is painful in the eyes. One female broke off a young green stem and took one single bite from the basal end before discarding it. Three other animals tore and bit apart dried and dead lobelia stems. They appeared to eat something from the center of each stem but the nature of the food could not be determined.

Lobelia wollastonii The animals apparently snap off the head of young plants and eat the soft pulp at the base of the leaves; they also pull up the root, which they peel by biting off the bark.

Rumex sp. Gorillas eat the inside of the lower part of the stem by first biting off the bark to expose the white inner parts, much in the manner of *Peucedanum.*

Helichrysum sp. Animals usually snap off the upper leafy part of the stem and stuff the whole top into their mouths. Occasionally an animal plucks one leaf at a time or slides its hand up the stem to collect the leaves all in one motion.

Aframomum sp. Probably the most characteristic gorilla sign in the Utu region is a broken stalk of *Aframomum* still attached to the ground, but with the center of the stem shredded and the round, white pith extracted. A silverbacked male, trapped and held captive by Mr. Cordier, ate *Aframomum* in the following manner: he held the stem horizontally with both hands, bit into it with one opposing set of canines, pulling his head back at the same time as jerking his hands away from his body. This combined motion exposed the pith, which he either ate directly from the stem or pulled out with one hand to consume.

Trail sign suggests that gorillas either bite or break in half the plum-sized red fruit of *Aframomum* and eat the mass of seeds. According to Cordier (pers. comm.) more than five or six *Aframomum* fruits eaten at a time caused diarrhea in his captive male.

Palisota sp. Gorillas apparently snap the plant off near the ground and bite into the short stem and leaf base.

Galium simense Three small rows of fairly blunt hooks grow on the back of each *Galium* leaf with the result that the vine readily adheres to fur and clothes, and, if pulled along the tender fleshy parts of the body, the leaves are quite abrasive. Gorillas handle the vine so as to neutralize the effect of the hooks, behavior best illustrated by quoting from my field notes:

1. *A sitting female reaches forward and with her right hand bends some* Senecio trichopterygius *toward her, and with the left hand pulls off strands of* Galium. *After examining the vine closely, she removes several dry leaves with her lips. Then she picks out several dry* Galium *stems between thumb and index finger, using first one hand and then the other. Finally she pushes the* Galium *several times against her partly-opened lips while twisting the vegetation around in her hand, thus forming a tight green wad in which all leaves adhere to each other. She stuffs the mass into her mouth and chews.*
2. *A silverbacked male sits and pulls in* Galium *from the surrounding vegetation. Once in a while he picks a dead* Hagenia *leaf from the vines with his lips and discards it. He then pushes the loose handful of* Galium *several times against his closed lips or against the teeth while rotating the vegetation until a tight plug is formed, which he either bites in half or shoves into his mouth whole.*
3. *A 5-month-old infant sits on its mother's back and grabs a strand of* Galium *from a stump in passing. It stuffs the vine laboriously into the mouth without first consolidating the leaves into a tight mass.*

Behavior similar to the last example was noted in other infants up to the age of about one and a half years, which suggests that the infants learn to handle *Galium* with efficiency only after considerable practice.

Droquetia iners This vine was merely pushed into the mouth.

Momordica foetida, Basella alba, Mikania cordata One group in the Kayonza Forest was watched as it fed leisurely on a mixed assortment of vines. A male lay on his back, looked to one side and pulled in a long strand. This he held with both hands and bit off the leaves one by one. Next he gathered in a mass of vines which he held in one hand and with the other picked off several leaves in a row before transferring them to his mouth. A female also either plucked leaves by hand or removed them from the vine directly with her mouth. Once a female held the end of a vine in her mouth, and slid her hand downward and outward along the stem thus collecting all the leaves, which she then ate.

Urera hypselendron I once observed a female feeding on the bark of this vine in the Kayonza Forest. She slid her hand down the vine to remove the leaves, and then bit into the bark, which she detached in long strands by twisting her head sideways and backward and simultaneously moving her hands away from her body.

Fleurya ovalifolia A female in the Kayonza Forest fed on the vine once during my presence. She held the long stem horizontally in her mouth so that only the tip protruded to one side. This she grasped and with a wide sweep of her arm pulled the vine through her mouth. Apparently she had applied pressure to the bark for the whole stem was stripped clean of leaves and bark, which collected in her mouth. These she spat out and then proceeded to eat the stem, end first.

Rubus sp. The thorns on this blackberry are sharp and gorillas handle them gingerly. They eat the ripe fruit, which often pass through the alimentary tract undigested. A lone silverbacked male moved from bush to bush and picked berries by pressing them between the tips of his fingers and the heel of his hand and pulling downward.

An observation made on a female is typical of the way in which gorillas eat leaves and tender stems:

> *She sits, and carefully bends a* Rubus runssorensis *branch toward her, holding it between thumb and index finger only; she then bites off and eats the tip of a branch including the young leaves. Next she breaks off another branch tip, using only thumb and index finger.*

Vernonia adolfi-frederici This shrub furnishes a soft white pith which gorillas obtain by breaking off branches and biting them apart. The branches break more readily if force is applied at the distal end rather than proximally where they join the main trunk. Thus, I observed gorillas frequently pull at a branch near the base, then slide the hand outward a few inches, pull again, and continue this until the branch bends sufficiently to snap. Sometimes the animals bite into an attached branch and jerk backward with the head while at the same time pushing the distal end of the branch with one hand in the opposite direction. The broken limb can then be easily ripped loose. Detached branches are usually shortened to 3-foot lengths or less by breaking off the part with numerous small twigs and leaves. This is accomplished either by breaking the top off with one hand, between both hands, or by biting into the center of the stem and bending both ends outward. Infants often seem to lack the strength to break off branches. I have watched them several times pick up and feed on *Vernonia* discarded by others. The following quote from my notes is typical of the method used in feeding, but it represents the only instance in which an infant took a food plant from the hands of its mother:

A female reaches up and bends the branch of a Vernonia *downward until it snaps. She jerks the three-foot-long segment loose, and, while holding it horizontally in both hands, bites off 6- to 8-inch slivers of bark and wood with rapid sideways and backward twists of the head. Then she gnaws out some of the pith with her upper incisors. Thus, she advances down the branch exposing the pith and eating it until only an 18-inch segment remains, all of which has at least been partially chewed upon. Her infant, 1¼ years old, reaches over her shoulder and takes the branch from her. It settles on the log behind the female and gnaws out the remaining pith in the manner typical of adults.*

On two occasions gorillas ate the blossoms of *Vernonia.* Once a female stood on a sloping trunk, grasped the top of a *Vernonia,* and walked up the log, bending the tree as she went. She then lay on her back, pulled the leafy crown close to her, and held the clusters of flowers in front of her face. One at a time she picked the blossoms between thumb and index finger and ate them.

Senecio erici-rosenii The gnarled branches of this arborescent *Senecio* are surprisingly brittle. The soft pith is much utilized by gorillas, and after a visit by the animals only the skeletons of some trees remain standing. Notes on a female illustrate the typical feeding behavior:

She reaches up, and with a quick twist snaps off a leaf cluster from the terminal end of a stem; she takes only one bite out of the leaf base and drops the rest. Then with one hand she breaks off a piece of stem about one foot long, holds it horizontally to her mouth, and bites off the bark and wood on one side until the pith is exposed. This she gnaws out.

Ficus vallis-choudee A captive adult male, belonging to Cordier, fed on fig leaves from large branches which were daily placed into his enclosure. He merely ripped off one or several leaves at a time with his hand and stuffed them into his mouth.

Pygeum africanum Gorillas readily climb high into the branches of *Pygeum* trees to obtain small twigs from which they eat the bark. Several methods of eating were employed:

A silverbacked male sits beneath a Pygeum *tree and picks up a branch which a female above him has inadvertently dropped. He bites off a twig, and, holding it solely between his teeth, nibbles off the bark. Then he rips off another twig, sticks it end-first several inches into his mouth, closes on it with the incisor teeth, and pulls the branch out, thus neatly stripping off the bark. Finally he holds a larger branch horizontally in both hands and bites off the bark with the incisors—much like a man eats corn-on-the-cob.*

The latter means of detaching bark is also used in feeding on such trees and shrubs as *Pycnostachys, Erica, Rapanea, Myrianthus, Hagenia, Hypericum,* and others.

Pygeum fruit appears to be little chewed, for the soft pulp attached to the large seed passes undigested through the alimentary tract at times. Similarly the cherry-sized seeds of *Xymalos monospora* are frequently noted in the dung with fragments of pulp still adhering.

Myrianthus arboreus The ripe, compound fruits of this plant are yellow and about the size of an orange. I observed one female in the Utu region as she climbed slowly up a tree to a height of 30 feet, squatted on a horizontal branch, and plucked a fruit, which she held in one hand. She bit into the fruit, but at that moment saw me and fled.

Dry bark of Hagenia and Hypericum Gorillas occasionally eat the dry bark of various trees, but the reason for doing so could not be determined. Two examples illustrate this behavior:

1. *A juvenile squats in the crotch of a tree, reaches down and tears off a piece of dry* Hypericum *bark, which it places between its lips. Then it collects another piece which it retains in its hand. After that it settles back and bites off, chews, and swallows pieces of the bark.*
2. *A silverbacked male detaches a dry section of* Hagenia *bark with one hand before biting off pieces and chewing them. Once he turns a sliver of bark over and over between his lips and finally swallows it.*

Hypericum log Gorillas eat the decaying wood of *Hypericum* and perhaps other trees, for logs with tooth marks are not infrequently found in the forest. Once I observed a silverbacked male as he held a section of half-rotten *Hypericum* wood, 18 inches long and about 10 inches in diameter, in his hands. He bit off large pieces, part of which he swallowed, while ejecting the rest between pursed lips. The lips of another male, a female, and a juvenile in the same group were red, indicating that they too had been eating wood.

Fungus? A juvenile at Kabara removed the bark from a dead branch and carefully touched its lips and tongue to some whitish material on the wood. Several times it wiped its finger on the wood and licked it afterwards. Although I later examined the branch, I could not determine the nature of the sticky fluid, which tasted slightly bitter. Earlier Emlen and I had observed the same type of behavior in two juveniles. Bingham (1932) obtained evidence for feeding on a similar whitish material. At Kisoro, Kawai and Mizuhara (1959*b*) noted that gorillas may eat fungi.

Manihot In the Utu region I noted one site where a gorilla had dug up manioc and eaten half of the tuber by merely biting into it.

Musa I observed banana plantations which had been raided by gorillas both at the edge of the Kayonza Forest and in the Utu region. According to the natives the animals only rarely eat the fruit, and I obtained no direct evidence myself. However, the tender white pith of the stems is much sought. Characteristically the whole banana tree is broken 2 to 4 feet above ground, so that the crown rests on the ground but the base of the stem remains upright. The stem is then ripped apart with hands and teeth and the pith extracted (Pl. 22).

A captive silverbacked male at I.R.S.A.C., Lwiro, grabbed the petiole of a leaf and tore it sideways, thus ripping a long piece of bark as well as the leaf from the stem. He then bit deeply into the stem and jerked his head sideways, leaving a large jagged hole. With one hand he grabbed the frayed edges of the hole and ripped the bark and much of the inner soft pulp for over 3 feet down the length of the stem. Finally, using both hands, he pulled the stem apart, thus exposing the large central core. He then tore out a section of the pith with his fingers and ate it, either holding it horizontally with two hands and biting pieces from it, or by holding one end and feeding from the other. I noted similar behavior in a captive male, which was trapped as an adult by Cordier. The animal usually took one bite, then pulled the stem apart with his two hands.

Miscellaneous foods eaten by infants Young gorillas occasionally eat materials which are not ingested by older animals. Such behavior possibly reflects a learning process by which infants attain the ability to distinguish between those plants which are palatable and those which are not.

Approximate Age of Animal	Type of Plant Utilized	Disposition of Material
4 months	Moss	Spit out
5 months	Dried leaf	Spit out
6 months	*Usnea* lichen	Swallowed
7 months	Petiole of *Hagenia* leaf	Taken away by mother
1 year	Humus	Spit out
Juvenile	Moss	Swallowed

Taste of Gorilla Food Plants

In an effort to determine if the plants eaten by gorillas are also palatable to man, I ate some of the species, especially in the Kabara area. Table 39 summarizes the results. I found that many of the food plants were bitter or in other ways somewhat unpleasant to my taste. However, I liked the roots of *Cynoglossum* and the fruits of *Rubus* and *Myrianthus*.

Feeding on Soil

I found three places at which gorillas had eaten soil. In two cases the feeding areas were located on the slopes of Mt. Mikeno. The gorillas ate volcanic soil from small bare patches at the base of bluffs where the substratum was quite dry, crumbly, and of a light brown to gray color. The soil tasted chalky to me. At one site in the Utu region the soil had apparently been sampled but not extensively eaten by gorillas.

1. *I observed a blackbacked male, a silverbacked male, and a female of group IV as they stood on a patch of bare soil about 15 feet long and 6 feet wide. They fed very intently, touching their mouths directly to the ground or picking up small objects between tip of thumb and side of index finger.*

Later investigation showed that they had scraped along the soil with their upper incisors, leaving long grooves on the ground, after which they had fed on the dislodged dirt.

2. *Nearly all animals of group VIII appeared suddenly in a rainstorm with soil-smeared faces resembling gray masks. Inspection of the area the following day revealed that the animals had scraped the soil from a small vertical bluff with their incisors. Similar but older tooth marks pointed to the fact that the soil had also been eaten on previous occasions.*
3. *A gorilla trail near Miya in the Utu region passed through a shallow ravine in which one side was partially bare, exposing a layer of white rather crumbly soil interspersed with small quartz rocks. The trail passed by the soil and a few tooth marks indicated that it had been sampled.*

Soil samples were collected at all three of these sites. The Kawanda Agricultural Research Station, Kampala, Uganda, analyzed Mikeno Sample No. 1, and The State of Wisconsin Soils Department checked samples Nos. 2 and 3 (Table 38).

Sample 3 from Miya shows no concentration of salts. Sample 1 from Mikeno is very high in potassium and quite high in sodium. Sample 2 from Mikeno is also high in potassium and so high in sodium that the locality takes on the nature of a salt lick.

Although gorillas ate soil occasionally, they apparently did not seek out such salt-rich sites habitually. At least three times group VIII passed within 100 feet of the locality from which sample 2 was collected without deviating from its course.

Feeding on Honey

In West Africa, Sabater Pi (1960) observed a gorilla as it chewed wax from the nest of subterranean bees. Both the Batwa in the Kayonza Forest and the Bantu guides at Kisoro maintained that gorillas raid the nests of wild bees to obtain honey.

No evidence for the raiding of bee nests was observed during this study although wild bees were common at Kabara. Several times gorillas passed one bee nest near ground level in a *Hagenia* tree without attempting to enlarge the hole from which bees flew in and out. Group IV nested one night within 30 feet of a hollow log inhabited by bees. Trails indicated that gorillas had climbed on the log, but the combs, easily visible through the large open end, remained undisturbed.

Feeding on Animal Matter

Reliable evidence that gorillas eat animal matter in the wild is lacking. Mathis (1954) maintained that gorillas in West Africa turned over and tore apart termite hills; Urbain (1940) believed that they ate termites and the eggs of birds; and Sharp (1927) thought that they were fond of grubs and caterpillars. These records are not well documented, however, and Yerkes and Yerkes (1929) in their summary of great-ape literature do not list a single unquestionable instance of meat eating in free-living gorillas.

Visual examination of several thousand sections of gorilla dung revealed not a single instance of hair, chitinous material, bone, skin, or other evidence that animal matter had been ingested. Several times gorillas tore off a piece of *Hagenia* bark and intently picked at something on its surface. But just what they found could not be determined, and stripping bark from the same tree after the animals had left never disclosed any particular aggregations of insects or spiders. Once a group traveled directly over the intestinal remains of a recently killed *Dendrohyrax* without handling them. Another time a dead duiker (*Cephalophus*) was ignored in similar manner. One group rested within 12 feet of an incubating olive pigeon (*Columba arquatrix*) without destroying the obvious nest. Thus, I found no evidence that gorillas in the wild eat animal matter.

However, in captivity, gorillas readily eat meat, as pointed out by Yerkes and Yerkes (1929). During my visit to the Columbus zoo, for example, the keeper fed the two adult gorillas a slab of boiled beef nightly.

Acquisition of New Food Habits

Observations on captive animals have revealed that adult gorillas do not readily accept foods with which they are not familiar. Cordier (pers. comm.), for example, trapped an adult male and retained him for several years in a large enclosure. When the animal was offered carrots, palm nuts, and sweet potatoes, he refused to eat them. The only unnatural food which the male accepted after some months was bread. Similarly adult males and females trapped in the Utu area, where bamboo does not occur, refused to eat it in captivity (Rahm, pers. comm.). Osborn (1957), who attempted to bait in gorillas to a particular location with banana fruit, sugar cane, and maize, found that gorillas had handled but not eaten these foods unknown to them. On the other hand, a two-year-old infant which was captured at Kisoro readily ate banana fruit, carrots, bread, pineapple, paw-paw, and elephant grass, all items it had never encountered before (Baumgartel, pers. comm.). This evidence suggests that gorillas are quite adaptable during infancy in accepting new foods but that once they have grown to adulthood and become accustomed to feeding on particular species of plants they show great hesitation in trying something new and different.

In spite of the fact that gorillas do not adapt to new foods readily, they have occupied widely diverse vegetation types which today provide little or no overlap in the forage species. However, several food plants show considerable altitudinal tolerance, occurring in two or more vegetation zones (see Table 37). Such species probably provided the means by which gorillas could have penetrated new habitats without adapting immediately to an entirely foreign assortment of food plants. For example, *Aframomum, Palisota, Myrianthus, Marattia,* and several others occur both in lowland and mountain rain forest. Imanishi (1957) observed that in *Macaca fuscata* new food habits were acquired through trial and error and imitation, with females learning to eat new items from their infants, which were more inclined to try unknown foods than adults. Perhaps gorillas adapt to new food items similarly.

Gorillas also consume several domesticated plants which are not indigenous to the continent. The cultivated banana reached Uganda sometime after A.D. 1000 (Wainwright, 1952). However, the cultivated banana (*Musa*) closely resembles the wild one (*Ensete*) which is widespread. Manioc (*Manihot*) reached the Utu region not before 1750 (Dr. J. Vansina, University of Wisconsin, pers. comm.). Maize (*Zea*) was not grown as a crop in Uganda before the arrival of the Arabs in 1844 (Wright, 1949). It is evident from the accounts of missionaries that most of the land in the Mt. Tshiaberimu region has been cleared within the past forty to fifty years; in fact, cleared lands have increased by about 50 per cent since 1956 (agricultural officer, pers. comm. to Emlen). Gorillas now eat maize and peas in the area, crops with which the animals probably had no contact before 1920. This suggests that a gorilla population may adapt to new food plants within fifty years and perhaps more rapidly.

Drinking Behavior

Except for a small lake, which to my knowledge was never visited by gorillas, no permanent water exists in the Kabara area. Although ample water fills the creek beds after heavy rains, it dries up within a few days. During the dry season little or no water becomes available to gorillas. Conditions are similar at Kisoro, except that an open cement trough relays water from a swamp in the Mts. Muhavura-Gahinga saddle to the cultivation below. In the other areas studied, numerous small streams furnish water throughout the year.

I have never seen gorillas drink in the wild, and, at Kabara at least, they probably do so rarely. Rains are common, the morning dew lies heavy on the vegetation which they eat, and many of their foods are succulent. At Kabara, I have observed a group as it crossed a creek without stopping to drink, and several times tracks indicated that the animals had moved across on a log without venturing to the water's edge. Similarly, I watched a group cross a stream in the Kayonza Forest but no animal drank.

Captive animals drink in several ways. The male and female at the Columbus zoo either placed their lips directly into the water and sucked, or they scooped the fluid up with one cupped hand and drank from the inner side. The animals in the captive group of mountain gorillas photographed for the Belgian movie "Masters of the Congo Jungle" ("Les Seigneurs de la Forêt") either drank the water directly with the mouth, or they dipped hands into the water and sucked the moisture from the long hairs on the back of them. A mountain gorilla in the San Diego zoo filled its cupped hand with water, lifted it high, and poured it into the open mouth (Carpenter, 1937; Benchley, 1942).

Nests and Nesting Behavior

Because nests remain visible for several months after construction and represent the only tangible evidence of gorillas which many observers encounter, the literature on the topic is quite voluminous, if not always accurate. However, all investigators, except Garner (1896), agree that gorillas build crude platforms to which the term "nest" can be applied. Useful descriptions of nests, their probable means of construction, and other general notes pertaining to them are found in Maxwell (1928), Bingham (1932), Donisthorpe (1958), and Bolwig (1959*a*).

Gorillas may build nests at any time of the day. For the sake of convenience I have termed those structures built for the purpose of resting only as "day nests" and those used as sleeping platforms for the whole night as "night nests," a distinction also made by Bingham (1932). In the course of the study I examined 3,052 mountain gorilla night nests as well as numerous day nests in various parts of the range and in all types of habitat. This sample forms the basis for the present discussion.

Day Nests

At the midday rest site most gorillas merely lie down, but about 5 per cent of the animals built a recognizable day nest. Many of these nests tended to be so crude that after only brief occupancy they could barely be distinguished from the surrounding trampled vegetation.

Basically a day nest is similar to a night nest in construction. The most reliable way of telling one from the other is that the vegetation is less flattened in the day nest. Day nests usually contain less dung than night nests. Donisthorpe (1958), Bolwig (1959*a*), and Kawai and Mizuhara (1959*b*) never found dung in the day nests examined by them, but I noted many which contained it.

Day nests were usually built at the beginning of the long morning rest period. Twenty-two nests, whose entire construction was observed, were built at the following times:

Time	No. of Nests	Per Cent
0901–1000	8	36.0
1001–1100	10	45.5
1101–1200	3	14.0
1201–1300	1	4.5
Total	22	100.0

The nests varied considerably, from simple ones using one or two handfuls of herbs to large elaborate platforms in trees. In one case, a group at Kisoro, which took refuge under a rock ledge during a rainstorm, had covered the large sharp pebbles on the ground with soil scraped from the adjacent turf. Nest construction is discussed more fully in conjunction with the night nests.

Night Nests

Adult gorillas usually built a recognizable nest in which they spent the night. Although such nests were often as crude as day nests in construction, the impression made by the body readily testified to long occupancy. In all my observations, night nests were used for one night only. The same nest location was occasionally re-used by gorillas on a subsequent visit to the area a month or more later, but no nest was ever occupied for two consecutive nights. Sabater Pi (1960) claimed, however, that in West Africa, one group used the same nests two times in a row.

Careful study of nests and nest sites can provide much useful information about group behavior, but extreme caution must be exercised in aging nests and interpreting the results. Emlen and I found that the extent of growth of the bent herbaceous ground cover provided a useful general index to the time elapsed since gorillas disturbed the vegetation. The tips of herbs, which are bent but not broken off, grow skyward at right angles to the main stem. Growth rate varies, however, from species to species and with the weather. Even after examination of the vegetation, the dung, and other sign I was not always able to determine with certainty the age of a nest more than four days old. One example suffices to show the ease with which wrong deductions can be drawn: group IV traveled downhill from its nest site, fed, and returned uphill to nest in the same general location as on the previous night. The two nest sites thus overlapped (Fig. 42).

Location of Nest Sites The prime requisite for the location of a nest site is the presence of suitable and adequate vegetation in the form of herbs, shrubs, or the branches of trees. In the mature forest of the Utu region, where groundcover is sparse, gorillas nested primarily in the more open valleys. At Kisoro a group, which was tracked by Emlen for one week, chose the tangled mountain woodland below the bamboo zone for nesting (see Fig. 33). In the Kayonza Forest, gorillas favored the upper half of the slopes rather than the valleys (see Fig. 34). Since both valleys and slopes contained ample vegetation, the explanation for the nesting preference may lie in the relative dryness of the soil at the higher elevations.

I have observed nest sites near streams (Fig. 43A), beneath the canopy of trees and under the open sky, on 45° slopes and on flat ground. In the Utu region and at Kisoro gorillas sometimes nested within 100 to 200 feet of human habitations.

At Kabara ground cover was abundant and the animals apparently nested wherever they happened to be. In the giant senecio zone, where materials for construction grew sparsely, clumps of *Senecio erici-rosenii* were favored. Groups sometimes seemed to exhibit a consistent preference in nest location over several consecutive nights. On several occasions group VII utilized *Lobelia giberroa* stands for two to four nights in a row. Once this group slept on turf

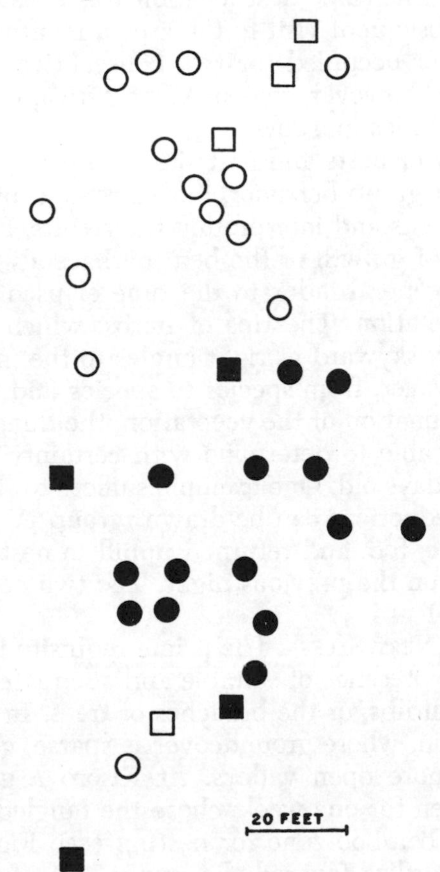

Fig. 42.—Two consecutive nest sites of group IV, showing overlap of nesting locations. Slope of terrain, about 30°.

white symbols Nests of night of August 30–31, 1959
black symbols Nests of night of August 31–September 1, 1959
□ Silverbacked male
○ Blackbacked male, female, or juvenile

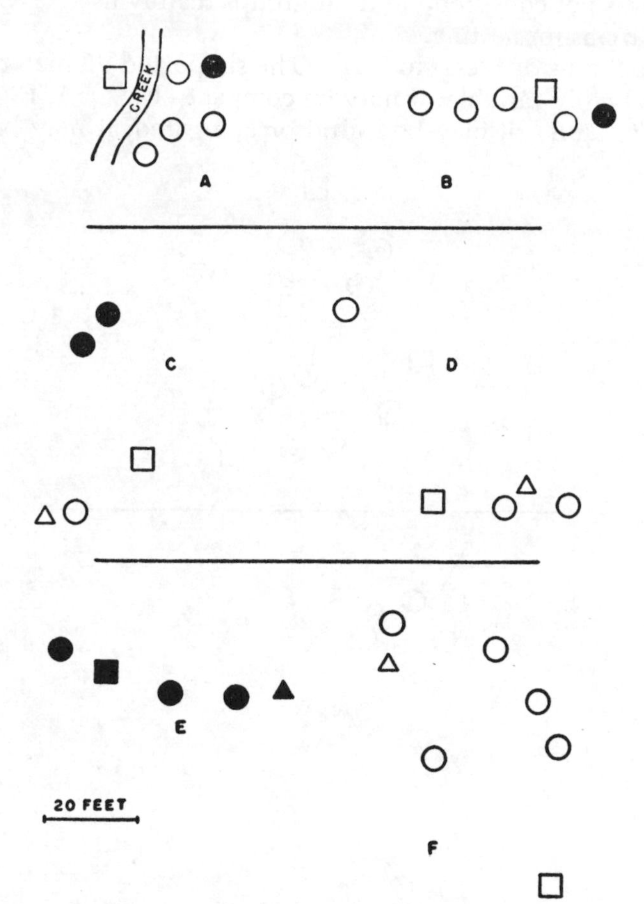

FIG. 43.—Typical nest sites of small gorilla groups. *A, B,* Utu region, between Utu and Niabembe. July, 1959. Slope of terrain, flat to 10°. *C, D, E,* Kisoro, Virunga Volcanoes. March–April, 1959. Slope of terrain, about 40°. *F,* Kisoro, Virunga Volcanoes. March–April, 1959. Slope of terrain, about 15°.

white symbols Ground nests
black symbols Nests elevated two or more feet above ground
□ Silverbacked male
○ Blackbacked male or female
△ Juvenile

and soil on an open slope for two consecutive nights. However, such behavior was not consistent, and all groups readily used a wide variety of situations for nesting.

SIZE AND SHAPE OF NESTING SITE The shape and diameter of each nest site is highly variable: it may be compact (Fig. 44A, Fig. 45) or diffuse (Fig. 46B); it may be round or elongated; it may be in one

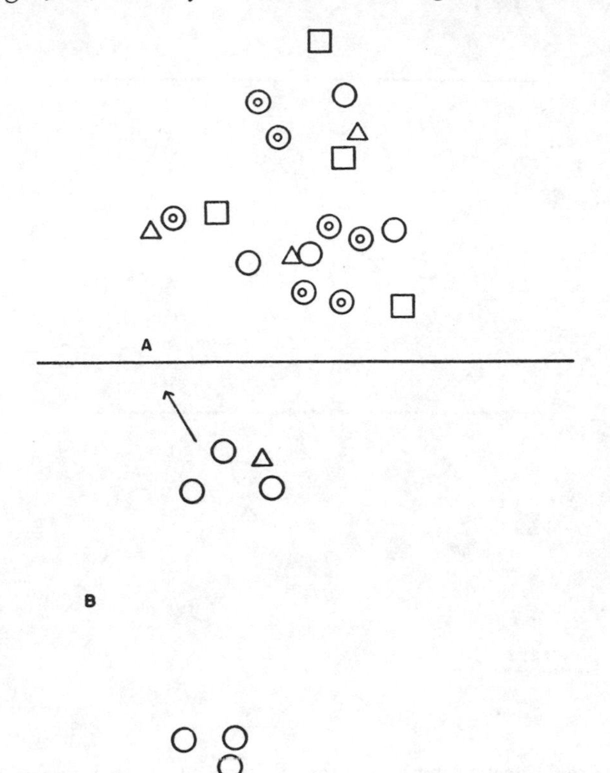

FIG. 44.—Two Kabara nest sites, one compact, the other split. *A*, group IV. September 25–26, 1959. Slope of terrain, about 30°. Data are based on direct observations of the animals in the nest. *B*, group VII. May 3–4, 1960. Slope of terrain, flat.

□ Silverbacked male
○ Blackbacked male or female
◎ Female with infant
△ Juvenile
→ Direction of travel

unit or split into two (Figs. 44B, 47B). The nest site diameter at Kabara varied from about 30 to 100 feet, if split sites and extreme peripheral animals are not included.

Figures 43, 48, and 49 represent some nest sites from other areas for comparison with those from Kabara. Most of them are small, and the nests of several animals are above ground. The nests in Figure 43C, D, and E belong to a group in the Kisoro area, illustrating well the variability in the placing of nests from night to night. The distance between ground nests and tree nests is actually greater than indicated on the maps, for height adds a third dimension to the dispersal pattern represented in two dimensions.

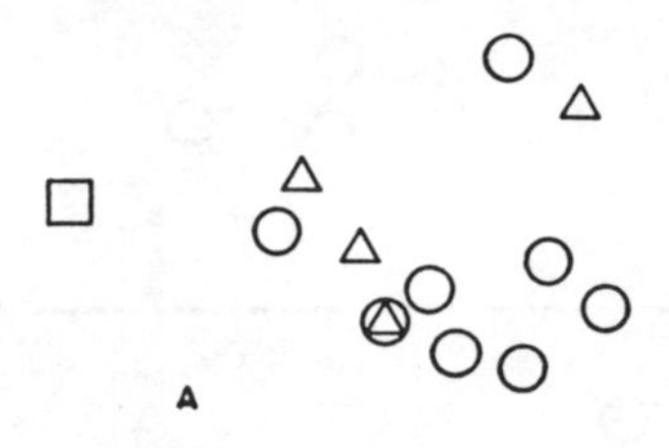

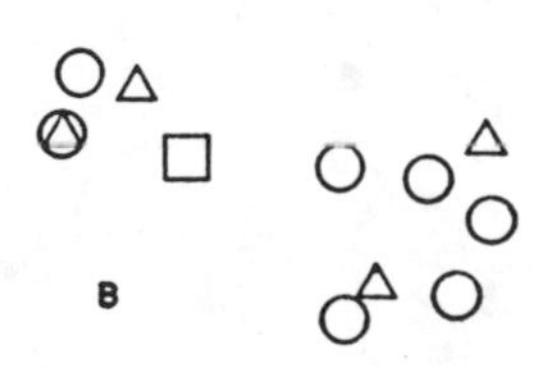

20 FEET

FIG. 45.—Two typical nest sites of group VII at Kabara. *A*, October 15–16, 1959. Slope of terrain, flat. *B*, December 16–17, 1959. Slope of terrain, flat to 10°.

□ Silverbacked male

○ Blackbacked male or female

⊛ Female with juvenile

△ Juvenile

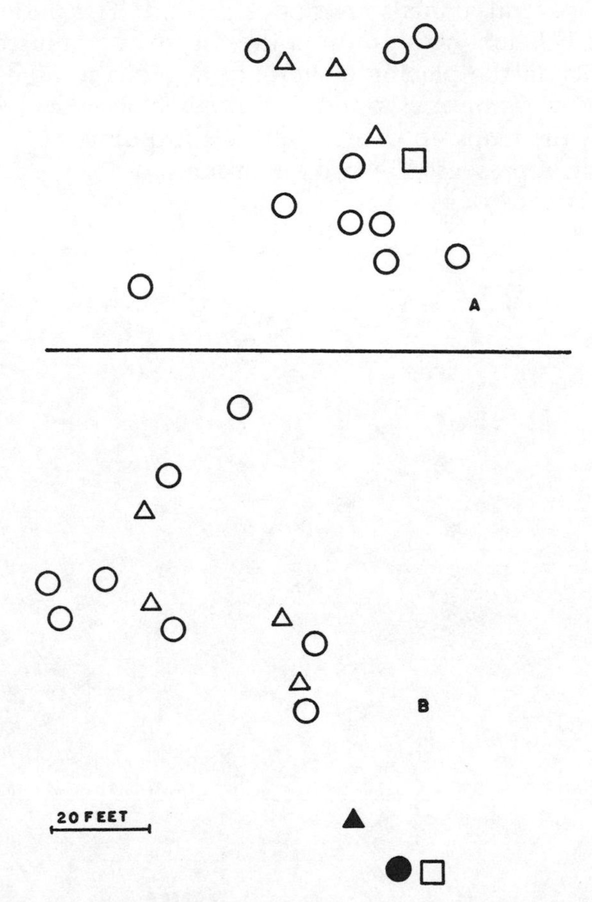

FIG. 46.—Two typical Kabara nest sites. *A*, group VIII. November 28–29, 1959. Slope of terrain, about 20°. *B*, group VII. April 1–2, 1960. Slope of terrain, flat to 10°. (One nest was not found.) Data are based on direct observations of the animals in the nest.

white symbols Nests used throughout the night
black symbols Nests built but not slept in
□ Silverbacked male
○ Blackbacked male or female
△ Juvenile

GROUND AND TREE NEST POSITIONS The literature contains numerous generalizations but limited specific data on nests with respect to their position on the ground or in trees. Savage and Wyman (1847) stated that in West Africa gorillas nested in trees; Du Chaillu (1861) and von Oertzen (1913) maintained that the animals nested both on the ground and in trees; Reichenow (1920) noted that gorillas rarely nested higher than 5 feet from the ground; and finally Merfield and Miller (1956) and Sabater Pi (1960) found that all nests were located on the ground. Maxwell (1928), Barns (1922), Akeley (1923), Bingham (1932), and others all found mountain go-

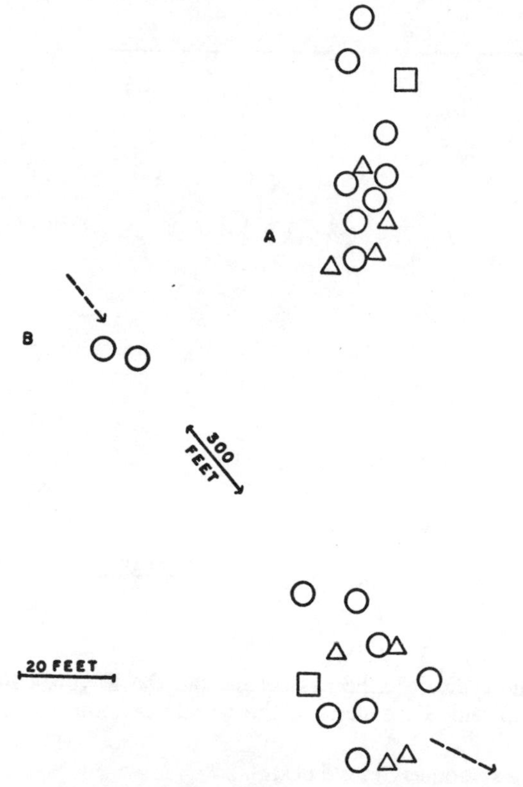

FIG. 47.—Two nest sites of group VII, one compact, the other split. *A*, December 25–26, 1959. Slope of terrain, about 25°. *B*, March 25–26, 1960. Slope of terrain, flat to 25°.

□ Silverbacked male
○ Blackbacked male or female
△ Juvenile
--→ Direction of travel

rillas nesting both on the ground and in trees. Blower (1956) recorded nests almost invariably on the ground at Kisoro, an observation not confirmed by Donisthorpe (1958). Of 225 nests examined by her, 104 were above ground. In the same area, Kawai and Mizuhara (1959*b*) found 210 out of 365 nests above ground; in the Kayonza Forest they noted 13 out of 48 nests above ground. Pitman

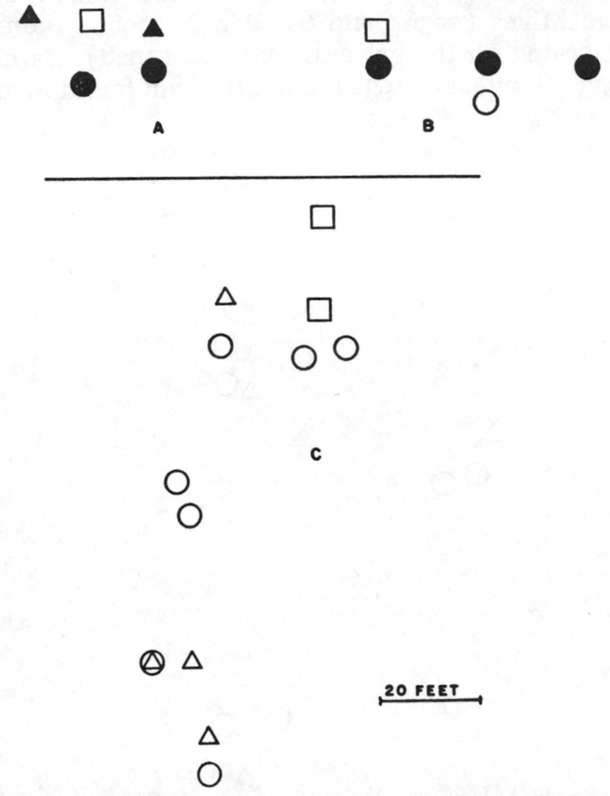

FIG. 48.—Some typical gorilla nest sites from the Kayonza Forest as noted in April, 1959. In *A* and *B* the slope of the terrain is about 35°; in *C* the slope is about 10°.

white symbols Ground nests

black symbols Nests elevated two or more feet above ground

□ Silverbacked male

○ Blackbacked male or female

◬ Female with juvenile

△ Juvenile

(1935) maintained that tree nests in the Kayonza Forest were the rule.

I recorded the location of a total of 3,012 nests in the various study areas (Table 40), measuring the height of nests above ground either directly, by estimating it, or by computing it with a camera range finder.

Table 40 illustrates the great variation in nest locations from area

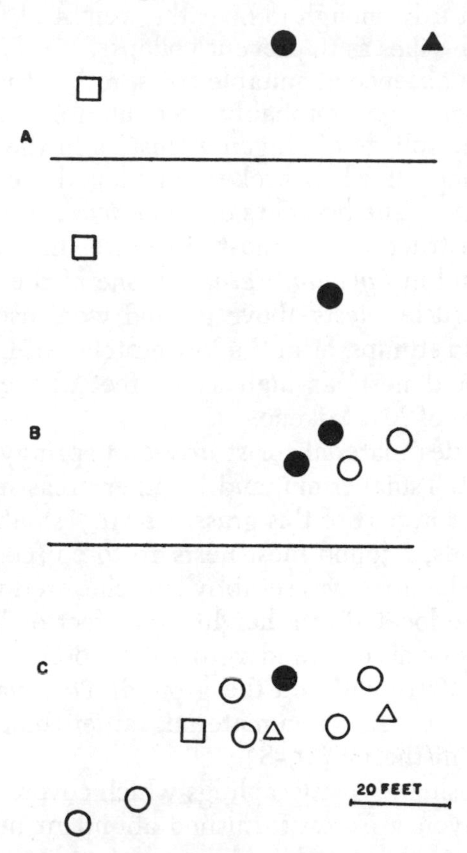

FIG. 49.—Some typical gorilla nest sites. *A,* Utu region near Kasese. September, 1960. Slope of terrain, flat. *B,* Mt. Tshiaberimu. May, 1959. Slope of terrain, about 15°. *C,* Mwenga-Fizi region near Mulenge. June, 1959. Slope of terrain, about 30°.

white symbols Ground nests
black symbols Nest elevated two or more feet above ground
□ Silverbacked male
○ Blackbacked male or female
△ Juvenile

to area. Whereas at Kabara 97.1 per cent of the nests were on the ground, the comparable figure for the Kayonza Forest was 53.5 per cent, and for the Utu region 21.8 per cent. The difference is partially attributable to the type of vegetation available for nest construction in the different regions. Successful nest construction in trees requires a type of growth in which a number of small branches grow near a crotch or horizontal branch. To support a gorilla nest, shrubs must either be sturdy enough to bear the weight of the animal or be so entwined with vines as to prevent collapse.

At Kabara the absence of suitable trees, rather than weather conditions (Bingham, 1932), probably accounts for the low frequency of tree nests. The foliage of *Hagenia* clusters in the crown, and the twigs usually snap off when broken, making the construction of a sturdy nest difficult. The branches of *Hypericum* trees are fairly suitable for nest construction, and most of the tree nests found occurred in this species and in *Pygeum*. *Vernonia*, one of the common shrubs in the area, is brittle. Nests above ground were usually located on fallen logs, on old stumps, or in the low crotches of large trees. Bingham (1932) noted nests as high as 60 feet above ground in the forest at the base of Mt. Mikeno.

Bamboo provides material for sturdy and springy nests (Pl. 26). At Kisoro, at Mt. Tshiaberimu, and in other areas many nests were constructed all or in part of this grass. At Mt. Tshiaberimu, where it grew in tall stands, I found most nests 10 to 20 feet above ground; at Kisoro, where bamboo was scrubby and cluttered with vines, most of the nests were located at a height of 10 feet or less (Table 40). In solid stands of bamboo, undergrowth tended to be sparse, and gorillas nested infrequently on the ground. This was probably due to lack of suitable construction material, rather than lack of room as suggested by Donisthorpe (1958).

The dense brush and many saplings which covered the slopes and ridges in the Kayonza Forest furnished abundant nesting materials, and gorillas commonly nested as high as 15 feet above ground. However, they rarely nested higher than 30 feet, even though numerous trees appeared to provide excellent sites.

The Utu region differed from all other areas visited in that gorillas regularly nested high in trees (Pl. 27). Admittedly the dense secondary forest, where I obtained most nest data, was admirably suited for building nests above ground, but the frequency was nearly twice that of the Kayonza Forest and cannot be easily explained by the presence or absence of suitable vegetation alone. Thus, I tentatively suggest that the high frequency of tree nesting in the Utu area represents a regional difference in habit.

The positions of ground nests varied widely. Bingham (1932) presented an elaborate outline, and Donisthorpe (1958) and Bolwig (1959*a*) listed various locations. I could find no consistent correlation with the type of vegetation or topography and conclude that gorillas are quite unselective as long as suitable vegetation is present for nest construction.

Several observers have noted that adult males rarely nest in trees. Donisthorpe (1958), Osborn (1957), and others maintained that the branches cannot support the weight of a large male. But at Kisoro both Bolwig (1959*a*) and I saw a few nests of males as high as 8 feet above ground.

At Kabara dung analysis revealed that the following age and sex classes nested 2 or more feet above ground:

Age and Sex Class	A No. of Nests	B No. of Animals in Population (after Table 18B)	Tree Nesting Index A/B
Silverbacked male	0	25	0
Female or blackbacked male	30	83	.36
Juvenile	20	31	.64
Undetermined	23	..	...

Proportionally, juveniles nested above ground roughly twice as often as females or blackbacked males.

COMPARISON OF CHIMPANZEE VERSUS GORILLA NEST POSITIONS Chimpanzees occur in numerous areas also inhabited by gorillas, and the nests of both species may be found in the same section of the forest. Chimpanzee nests can be differentiated from those of gorillas, not so much in the form or method of construction which is similar, but by the position high in the trees and by the type of dung. I saw no chimpanzee nests on the ground nor did I obtain reliable evidence that they nest there. Bolwig (1959*a*) believed that they nested on the ground in the Kayonza Forest, but these data need further confirmation. Table 41 presents the heights of chimpanzee nests in the Kayonza Forest as compared to those of gorillas from the same area. The figures show that nearly 90 per cent of the gorillas nested within 10 feet of the ground, but that all chimpanzees nested higher than 11 feet above ground. Bolwig (1959*a*) lists the heights of twenty-three chimpanzee nests from the mountain forests of the Ruwenzories and the lowland forest from near Fort Portal, Uganda. The nests in these two areas showed a variation in height from 15 to 140 feet above ground.

Size of Gorilla Nests I measured the outside diameter of seventy-six gorilla nests on or within 5 feet of the ground. Nests are either round or oval. The size varies roughly with the size of the animal, making the nests of juveniles and adult males readily distinguishable from the others, especially if judged in conjunction with dung. Most nests of adult males measure approximately 5 × 5 or 5 × 4 feet, but smaller ones occur and fall into the size range of medium-sized animals, which usually measure about 4 × 4, 4 × 3, 3 × 3, or 3 × 2 feet. Juveniles build nests of an average diameter of 2 feet.

Unused Nests Night nest sites occasionally contained nests which were completed but obviously unused. At Kabara I noted a total of fifty such nests, usually only one in a given site but sometimes several. Such nests were probably more numerous than indicated, for I made no effort to look for them. A direct observation illustrates one basis for the presence of these unused nests:

A silverbacked builds his nest and a juvenile constructs one nearby. When the other group members continue to feed and move past the male, the juvenile abandons its nest and builds another near the main body of the group (Fig. 46B).

Figure 50 shows what probably represented a similar situation involving several animals.

Location of Individual Gorilla Nests within the Group Several investigators have attempted to read organization into the placement of individual nests. Blower (1956) and Lequime (1959) observed that nests were always so situated that the animals had a view of each other and of all approaches. Osborn (1957) felt that each group had evolved its regular nest formation pattern. Sabater Pi (1960) noted that the male in one site observed by him placed his nest in a strategically defensive position. Several authors believed that females and young nested in trees for protection with the male conveniently guarding the approaches below. In examining and mapping four hundred nest sites I found no consistent arrangement to support any of these proposals.

Figures 42 to 52 illustrate some typical nest sites. Data are based either on direct observation of the animals in the nest or on dung and nest size.

The nesting position of the dominant male in relation to the other group members was highly variable as shown in Figures 45, 46, 47, and 51, which all represent groups containing only one silverbacked male. Sometimes the male nests in the center of the group, sometimes at the periphery, and sometimes separated from the nearest animal by 20 or more feet. I cannot see the slightest indication that the male locates his nest with reference to defense of the group members; in fact, since the male usually nests first the others nest in

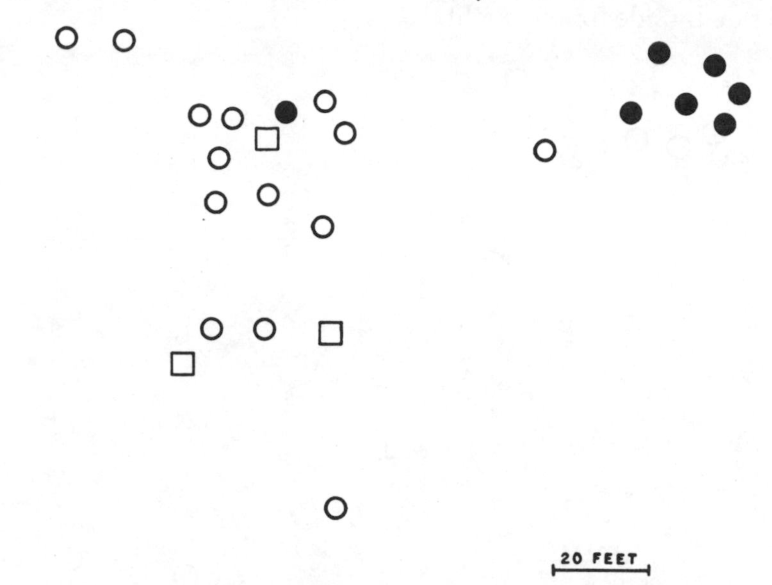

FIG. 50.—A nest site of group IV at Kabara. September 1–2, 1959. Slope of terrain, about 30°. (The nest of one silverbacked male was not found.)

white symbols Nests used throughout the night
black symbols Nests built but not slept in
□ Silverbacked male
○ Blackbacked male, female, or juvenile

relation to his position. On several occasions the male built his nest only to have the other members pass him by and build their nests 50 or so feet beyond (Fig. 46B).

The nesting positions of the extra silverbacked males were quite variable, although there was a tendency for them to nest at the periphery of the group as shown in Figures 42, 44A, and 50. Figure 52 illustrates that females nest near the male lowest in the hierarchy as readily as near the dominant one.

Some blackbacked males tended to nest at the periphery of the group, at times sleeping 60 or more feet from the main site (Fig.

45B). But on other occasions the same animals nested with the group (Figs. 45A, 47A).

Females usually bedded down in the vicinity of the dominant male as shown in Figures 46B and 51, but they seemed to make no particular effort to nest close to him. There was no tendency for females to nest near a certain companion, although at times two medium-sized animals or a large male and a medium-sized animal nested side by side for one night.

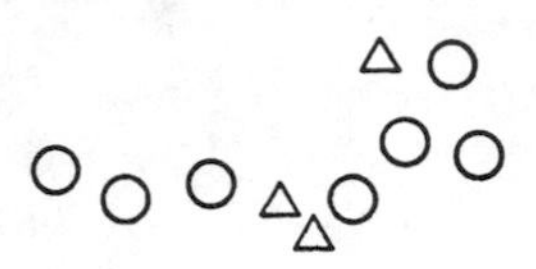

FIG. 51.—A split nest site of group VIII at Kabara. December 6–7, 1959. Slope of terrain, 15°–20°.

□ Silverbacked male
○ Blackbacked male or female
△ Juvenile

The nesting position of individual juveniles varied. In group VII, for example, two juveniles tended to sleep in the same nest or closely adjoining ones (Figs. 44B and 47B), but I could not determine if the same two animals were always involved. Sometimes a particular juvenile nested with a female (Fig. 45), and often by itself (Fig. 47).

Young infants slept in the nest with their mothers (Fig. 44A). Older infants occasionally spent the night in a small nest immediately adjoining that of the female, but not until they were about two and a half or more years old did infants nest independently or with a juvenile.

To obtain some numerical measure of the distances between nests, I noted the number of feet between the nearest nests of the various age and sex classes in 146 sites at Kabara. The data in Table 45 show that the mean distance between nesting silverbacked males was 34.0 feet, a figure which is over twice as high as the nest proximity of silverbacked males and medium-sized animals (females and blackbacked males). The mean distance between silverbacked males and juveniles was 20.7 feet, and between juveniles and juveniles 12.0

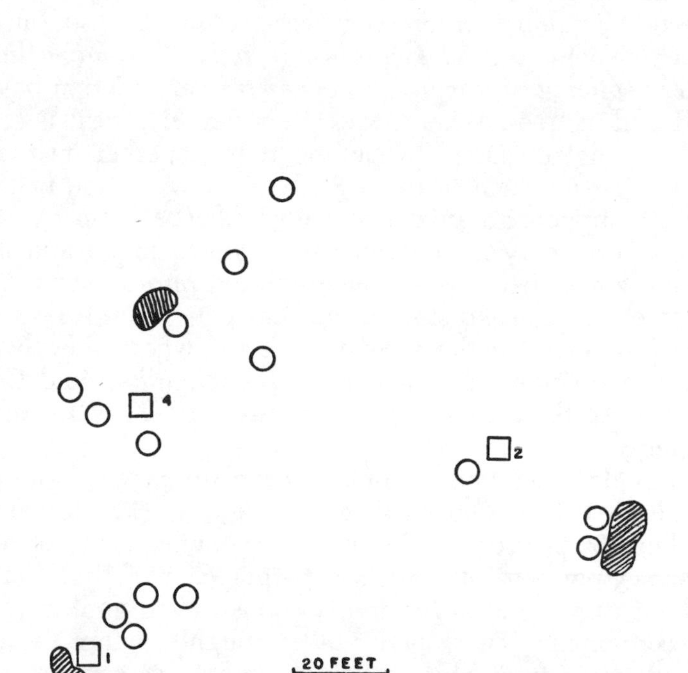

FIG. 52.—A nest site of group IV at Kabara. September 7–8, 1959. Slope of terrain, 25°–30°. The numbers near the nests of silverbacked males indicate the status of the animal in the dominance hierarchy. The data are based on direct observations of the animals in the nest. The hatched areas represent *Hagenia* trees.

□ Silverbacked male
○ Blackbacked male, female, or juvenile

feet, indicating that these animals did not consistently nest very near one another. However, the mean distance between the nearest nests of medium-sized animals was only 5.4 feet, and between medium-sized animals and juveniles only 3.0 feet, suggesting that nests were placed with reference to a certain sex class or particular animal, especially in the case of juveniles.

NEST CONSTRUCTION MATERIALS Gorillas construct nests from nearby branches, vines, and herbs. Only rarely do they carry nesting materials; 15 feet was the longest distance noted. An elaborate classification of nest materials is of little value, for they vary from region to region and merely reflect the composition of the local vegetation.

The main body of the nest At Mt. Tshiaberimu, 71.8 per cent of all nests were of bamboo in an area in which that plant predominated. In the eastern sector of the Kayonza Forest, the stems and foliage of *Mimulopsis arborescens* were frequently used; in the central sector such trees as *Alchornea hirtella, Derplatzia lutea,* and *Galiniera coffeoides* provided branches for construction both on the ground and in trees. Where vines grew densely, gorillas sometimes nested in tangled masses by pulling them together on the ground or by resting on them in trees. Such vines were frequently strong enough to support an animal hammock-like between two trees. At Kisoro, where we worked primarily in the bamboo and mountain woodland zones, bamboo was incorporated into 72 (67.9%) of 106 nests noted. In the Utu region, gorillas constructed nests of those species prominent in the secondary forest where they lived: *Musanga, Aframomum, Marantochloa, Megaphynium,* and *Costus.* In mature forest, the animals frequently used the branches and foliage of saplings.

To provide a detailed example, I have summarized the principal materials used in the construction of 2,439 nests at Kabara (Table 43). About 65 per cent of the nests were built mainly of herbs and lobelias; 15 per cent of branches; 10 per cent of the vine *Galium;* and about 10 per cent of the animals constructed no nest ("soil" and "turf" combined). These percentages roughly match the extent of each vegetation type which provides suitable nest materials.

A more detailed analysis of the vegetation used in nest construction by three gorilla groups is presented in Table 44. Group IV, which ranged primarily along the brushy slopes of Mt. Mikeno where stands of lobelias grew sparsely, constructed 32.4 per cent of its nests of branches and only 3.9 per cent of lobelias. Conversely, group VII, which roamed mostly through the saddle area where shrubs were few in number but lobelias common, built 8.6 per cent of its nests of branches and 37.8 per cent of lobelias. These figures again suggest that for the most part gorillas use whatever vegetation is most readily available and easily obtained for nest construction.

Some nest materials were rarely used. Twice at Kabara an animal placed large pads of moss, which had originally grown on a *Hagenia* branch, along the nest rim. Near Mulenge, in the Mwenga-Fizi region, a gorilla had pulled moss from a boulder and transported it 8 feet to its nest. Long slabs of *Hagenia* bark detached from a nearby tree furnished nest material in three nests at Kabara.

The nest lining Gorillas rarely placed a special lining of vegetation on top of the basic nest structure. A lining of dry *Galium* covered the nest foundation of two nests at Kabara, and grass blades padded the turf bottom of another nest. Gorillas occasionally decapitated lobelias and laid the heads in the nest cup, leaves inward and stems facing outward or along the nest rim.

None of the tree nests examined at Kabara and in the other study areas contained a definite lining of twigs, although two to four small branches broken off near the nest sometimes lay along the rim and in the nest cup.

NEST CONSTRUCTION To a large extent the method of constructing nests can be deduced by analyzing the nest itself. Although the popular literature still contains such uncritical statements as ". . . a gorilla merely gathers leafy branches by armfuls, arranges them in an orderly, comfortable heap, then plunks himself down in the center" (Zahl, 1960), the basic method involved in nest construction has been described correctly for a hundred years (Du Chaillu, 1861), with Donisthorpe (1958) and Bolwig (1959*a*) offering the most recent additions and amendments. The animal stands or sits, and pulls, breaks, or bends in vegetation which it places around and under its body. The precise method employed varies with the particular circumstance—whether the nest is in a tree or on the ground, whether it is on a steep slope or on a flat area.

Ground nests The care with which nests are constructed varies considerably (Pls. 23, 24, and 25). At Kabara about 10 per cent of the animals bedding on the ground built no definite nest. In fact, I was generally impressed by the extreme casualness with which ground nests were constructed. Numerous nests consisted only of two or three handfuls of herbs or lobelias pushed down to form a partial rim (Pls. 23, 24). The center of the nest frequently lacked vegetation and the animal rested on the bare soil or on a few herbs which lay or were pushed there inadvertently. At the other end of the scale were nests with rims up to 2 feet high and with central cups as much as 15 inches deep.

Ground nests frequently incorporated *in situ* logs, stumps, and tree trunks into the edge of the nest. Such vines as *Galium* grew profusely on logs, and the animals merely pulled them down to form a crude but soft nest. The addition of branches or of bamboo to ground nests was no more elaborate than that of herbs. Gorillas simply reached up or around them and pulled in up to a dozen or more convenient branches. Many shrubs and young trees were so leafy and contained so many branchlets that two or three major branches were sufficient to construct an adequate nest. *Vernonia*, a shrub which is easily uprooted, was sometimes pulled over and the animal slept in the leafy crown without attempting to build a nest beyond bending a few twigs inward (Pl. 25).

Figures 53 and 54 illustrate schematically the method of construction of sixteen typical nests on or near the ground as determined by a detailed analysis of each nest. The drawings emphasize five features of construction: (1) the vegetation is bent in from one or more directions; (2) the vegetation is bent toward the animal in such a way that a roughly circular or semicircular rim is formed; (3) there is no particular sequence in the placement of the vegetation; (4) there is no interlacing, weaving, knot-tying, or other involved manipulation; and (5) the major effort is directed toward the nest rim, not the bottom. Some stalks are twisted as the animal pushes them along the rim or under its body.

On steep slopes (25° to 50°) the construction of a nest platform is essential in preventing the animal from sliding. The vegetation on the downhill side was usually broken upward and inward, and stems on either side were pulled across in front of the animal's body. Such nests made of shrubs or young trees were quite sturdy for the limbs remained attached to the main trunk, which also served as an anchor for nest and animal. Platforms of herbs on steep slopes were considerably less firm than those of branches. Occasionally the nest slowly disintegrated during the night with the result that the animal slid farther and farther until by morning it rose some 6 to 10 feet downhill from where it went to bed.

Gorillas rarely scraped a shallow depression in the soil to accommodate their bodies. Frequently turf nests and nests on slopes showed definite depressions, but these were probably formed by the weight and movement of the reclining animal. Red forest duikers (*Cephalophus nigrifrons*) sometimes pawed in the turf, thus forming shallow craters which may have been used by gorillas as nest sites. Only on

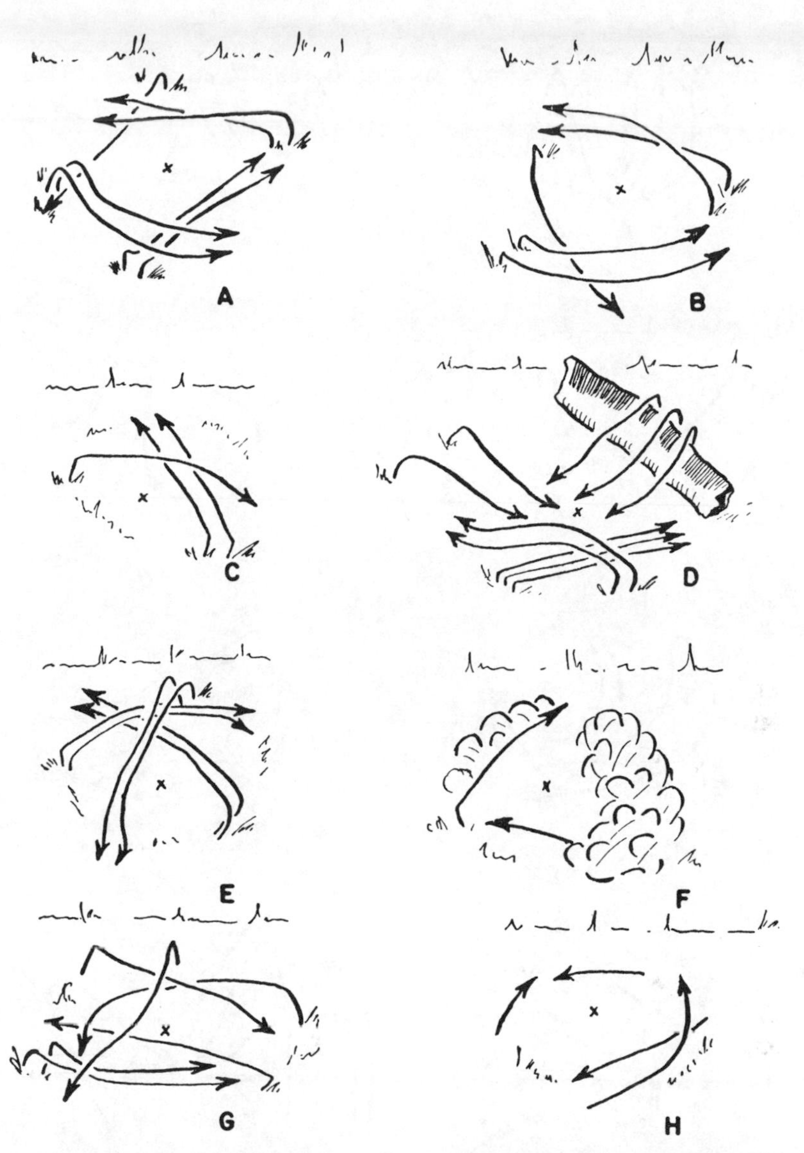

FIG. 53.—Schematic diagrams of gorilla nests on the ground in the Virunga Volcanoes. Each line represents about one handful of herbs, consisting usually of several stems each, or one major branch. The end opposite the arrow point denotes the place where the stem is attached or broken off. The arrow point represents the top of the plant as well as the direction in which it has been bent. The cross marks the approximate center of the nest. *A–E*, Kabara nests constructed of herbs; *F*, Kabara nest constructed of *Galium* vines and a few herbs; *G*, Kisoro nest constructed of shrubby branches; *H*, Kisoro nest constructed of giant senecio branches.

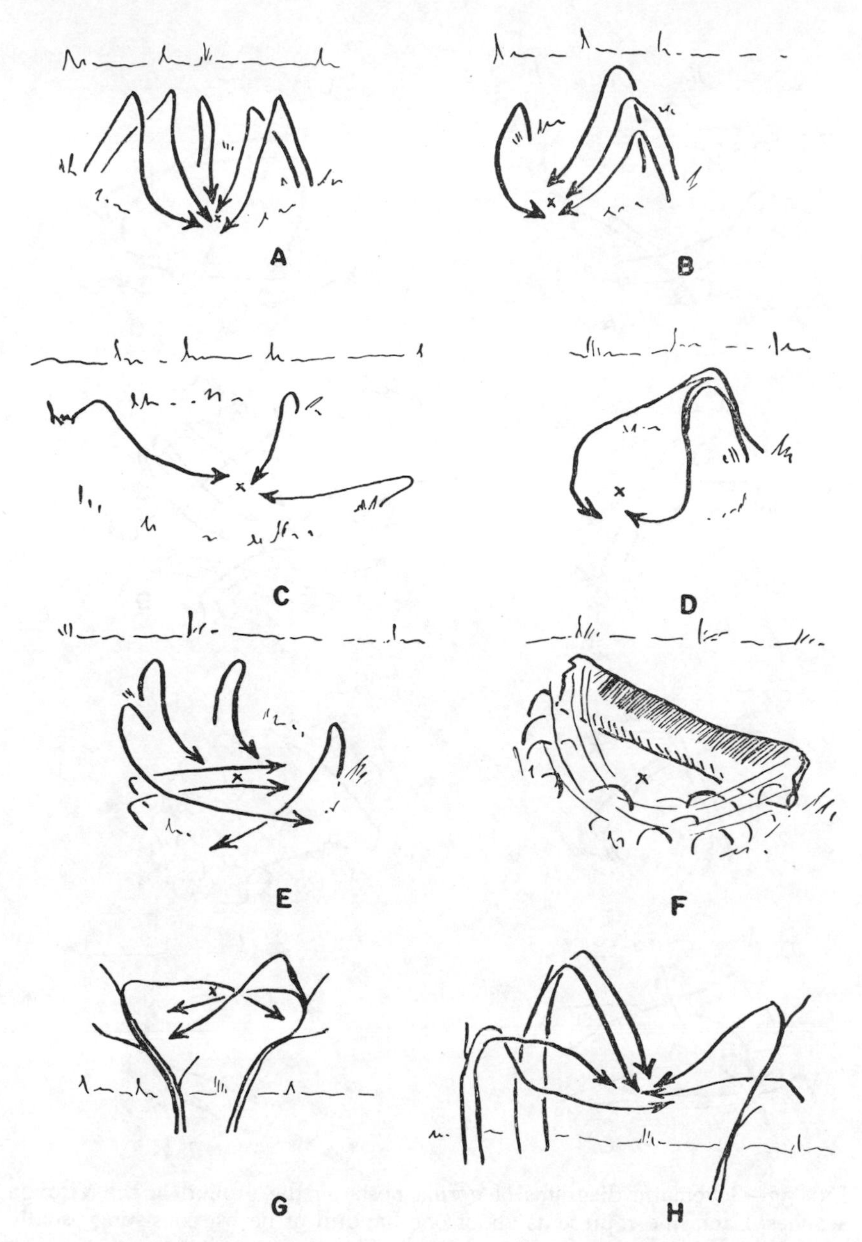

FIG. 54.—Schematic diagrams of gorilla nests on or near the ground in the Kayonza Forest. Each line represents one major branch. The end opposite the arrow point denotes the place where the stem is attached or broken off. The arrow point indicates the top of the branch as well as the direction in which it has been bent. The cross marks the approximate center of the nest. *A–E,* nests constructed of branches and saplings on the ground; *F,* nest constructed of vines by a log; *G–H,* nests constructed in shrubs about 5 feet above ground.

one occasion did a gorilla seemingly excavate a considerable amount of soil at a nest site. The animal had dug into a steep, dry hillside and made a platform of earth on which it nested.

Differences in the care with which ground nests were constructed could not be correlated with age or sex. There was a general tendency for juveniles to build crude nests, but this was highly variable from night to night even among juveniles of the same group. There was no evidence that females with infants build more elaborate nests than females without infants.

Nests above ground Nests in trees require, in addition to branches, a relatively firm substrate. Consequently the animals tended to build in forks, along horizontal branches, in crotches, or merely in a dense tangle of small branches (Pls. 26, 27). Sometimes one large tree furnished a suitable base and an adjoining sapling provided leafy material, which was pulled across and incorporated. Vines frequently entwined the trunks and lower branches of trees, and gorillas used them in the construction of the nest. In building bamboo nests, the animals climbed up the stems until the weight of their bodies bent several stalks, which then served as support for the leafy crowns that the animals pulled and broke in from all sides to form the nest.

The principle of constructing tree nests was the same as that for the ground nests, but more emphasis was placed on the nest bottom than on the rim with the result that tree nests often tended to be relatively flat, lacking a deep nest cup.

The number of main branches required for constructing the frame of tree nests depends on the sturdiness of the substrate. At Kabara from three to about fifteen branches were used. Seven nests in the Utu region utilized two to seven (mean 3.8) branches; of the twenty-seven branches used in construction, only three were detached completely from the tree, the rest being merely bent over into position. Figure 55 presents detailed sketches of the construction of five tree nests in the Utu region.

At Kabara one nest 20 feet up in the crown of a *Hypericum* crashed to the ground when the main trunk of the tree snapped. The nest contained dung and the event probably occurred during the night when the animal was in the nest.

Direct Observations on Nest Building At Kabara I observed nest building behavior in twenty-seven animals, twenty on the ground and seven in trees. Twenty-two of the nests were day nests, five were night nests. In addition I watched at least twenty other nests during part of the construction process. Seven examples condensed from my field notes illustrate the behavior.

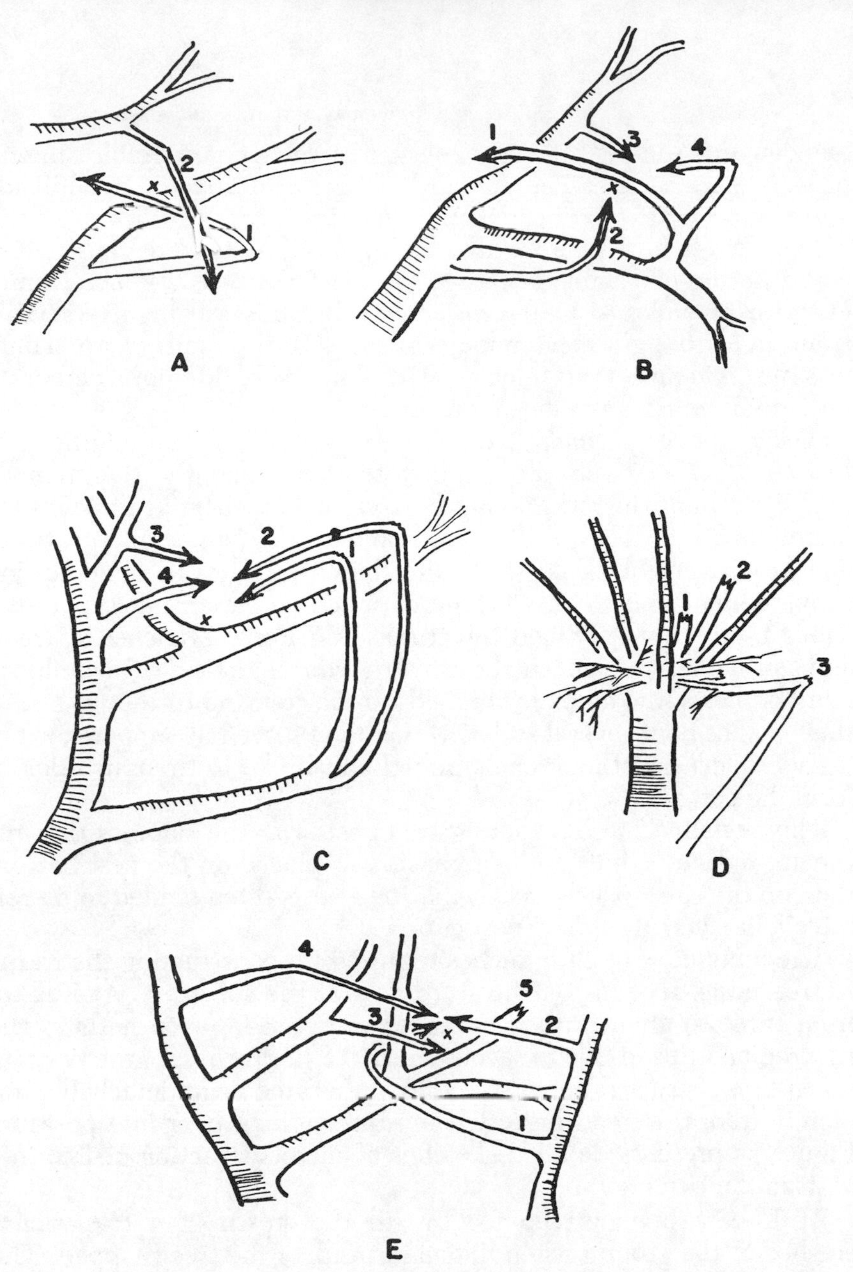

FIG. 55.—Schematic diagrams of tree nests in the Utu region, showing the method of construction. Each arrow represents one major branch as well as the direction in which the limb was bent or broken. The numbers indicate the sequence in which the branches were laid down. The cross marks the approximate center of the nest. *A–C,* nests located 20 to 30 feet above ground in the tops of rather brushy trees; *D,* nest located about 25 feet above ground in the crotch of a *Musanga* tree; *E,* nest located about 40 feet above ground in the crowns of two *Musanga* trees. The nest incorporates branches from both trees.

1. *Day nest. March 23, 1960. Group VII.*

While holding her 6-month-old infant to the chest with her left arm, a female slides the other hand up a nearby lobelia stalk to the base of the leaf cluster; then, with a downward twist of the hand, she snaps the head off and lays it beside her. She repeats this on another stalk before bending in one Peucedanum *with her right hand and pushing it to her left side. Standing on three legs, with the infant clinging to her chest, she breaks off a lobelia in front of her with one hand and places it behind her back. She switches the infant to the other arm, and bends in two handfuls of herbs which she pushes down beside her body. Finally she lies down. The total time elapsed was about 25 seconds.*

2. *Day nest. March 22, 1960. Group VII.*

A juvenile bends in a Senecio trichopterygius *stalk with one hand and with the other breaks the top without detaching it. The animal then places the stalk in front and to one side of its body, a process which it repeats with another stalk. After that the juvenile stops, stands, and presses the two stalks down with both hands. Then it stands on 3 legs and bends in a handful of herbs from farther away, after which it faces in the opposite direction breaking in several more stalks until a total of 7 handfuls has been placed along all sides of its body. Finally it sits and, while rotating its body 360°, pushes the vegetation down along the rim with its hands. The time required for building was about 45 seconds.*

3. *Night nest. April 1, 1960. Group VIII.*

A juvenile sits at the base of a tree and bends 4–5 handfuls of small herbs toward its left side with the right hand. It then stands on two legs, grabs the top of a mass of Senecio trichopterygius, *heavily overgrown with* Galium, *and pulls it in. It sits and breaks or bends the tips of the herbs to fit in a semicircle around its body before pressing the mass down with both hands. Standing on 3 legs, it reaches far out and breaks 2–3 more* Senecio *stalks off at the base and pulls them in. After placing these individually along the edge of the nest, it breaks their protruding tops to fit the rim. It sits, turns around, and sits again. The time required for building was about one minute.*

4. *Day nest. January 9, 1960. Group VI.*

A female sits on a steep slope (about 35°) and faces downhill. She breaks in two handfuls of weeds with her left hand and steps on them with her right foot. Then with her right hand she bends in another stalk to the left side. Sitting hunched over, she rakes in some vines with one hand and breaks off some smaller weeds with both hands, most of which she stuffs beneath her feet. Next she bites off the leafy head of an herb and lays it on the nest rim. Then she turns so that her back faces downhill, and, using both hands at the same time, pulls and breaks in vegetation from above, after which she turns again and relaxes. The time required was about 45 seconds to one minute.

5. Night nest. February 7, 1960. Group VII.

A silverbacked male reaches up with one hand and pulls on a Vernonia, *3 inches in diameter at the base, until the shrub is uprooted. He then sits in the leafy top and breaks a few twigs of it inward along his side. Construction time was about 45 seconds.*

6. Day nest, December 14, 1959. Group VII.

A juvenile ascends a brushy Hypericum *tree to 20 feet. It stands in a crotch and pulls in small pliable branches with one hand and stuffs them down to its feet and steps on them. After about 10 branches have been broken or bent in this manner a crude platform is established. The juvenile then stops building and with hands and feet pushes down on the mass. Next it sits, pulling in all loose ends within reach, one at a time. Each one is treated according to its length—a small one is pushed to one side or along the front, a longer one is bent in, broken about 2½ feet from the base and placed along two sides of the rim. If the branch is still too long it may be broken as many as 4 times until all fits neatly along or around the animal. As a final touch the juvenile snaps off several small branches which stick up from the nest. The construction time was 4½ to 5 minutes.*

7. Day nest. February 9, 1960. Group VII.

A female stands 15 feet up on a main horizontal branch of a Hypericum. *She pulls and breaks in a large green bough so that it rests on the main branch. She squats and pulls and breaks in projecting twigs and other branches, pushing each one under her foot. Standing, she bends in several small limbs from farther away. A final branch is detached completely and laid on the nest rim. The nest is crude, but one side of it is supported by a sturdy semihorizontal branch. The time required for construction was about 1 to 1¼ minutes.*

TIME REQUIRED TO CONSTRUCT A NEST Ground nests which consist of only two or three handfuls of herbs may be completed in fifteen seconds or less. More elaborate nests, made of five to fifteen or more handfuls, may take anywhere from thirty seconds to five minutes to construct, depending primarily on the speed with which the animal works.

Tree nests often require up to five minutes for completion. The shortest time which I recorded was one minute, the longest twenty minutes, but in the latter instance the animal rested between spurts of activity. Osborn (1957) observed the construction of a tree nest which required two minutes to complete.

NEST CONSTRUCTION BY INFANTS The development of nest-building behavior is difficult to trace in free-living gorillas, for the infants usually sleep in their mother's nest for the first two and a half to three years of their life. The earliest age at which I observed traces of nest-building behavior was at eight months: an infant sat propped against the back of its mother. It reached forward with one hand, bent in an herb about one foot long but released it. The action was repeated with another stem. Finally it pulled in a third stalk and pushed it under one foot.

My only other observations of nest building by infants were of two considerably older animals. The youngest animal to build a definite ground nest was one and a half years old. Another infant, two years of age, built a tree nest about 10 feet from its mother, who prepared her own nest at the same time. The infant climbed 7 feet up into the crotch of a *Vernonia* shrub. It reached out with one arm while steadying itself with the other and broke in a branch. It stepped on that branch, and, slowly rotating, repeated the process until every branch within reach had been broken in and packed down. To break in the larger branches, it pulled with both hands and with one leg braced against the stem. Frequently it stopped and pushed the broken branches down with the hands. It then sat on the crude platform for 10 minutes before descending to spend the night with its mother. The time required to build the nest was 10 to 12 minutes. In spite of the effort expended on it, the nest was poorly constructed.

Small nests were occasionally encountered both in trees and on the ground. Most of these nests showed no signs of having been used and apparently represented practice nests built by infants. However, infants slept at times in their own ground nest adjacent to that of their mother. Such infants were at least one and a quarter years old judging by the size of their dung. Thus, infants begin to construct nests at least one and a half years before they use the behavior regularly in their daily life.

The relative roles of the innate and learned components in nest building cannot be determined from observations in the field alone, but I agree with Yerkes (1943) who believed that in chimpanzees the general pattern is innate but that specific adaptations are learned. An observation made by Lang and Schenkel (1961*a*) on a gorilla born in captivity tends to support this assumption. At the age of ten months the animal placed branches around itself at dusk, forming the crude outline of a nest.

Comparisons with chimpanzee and orang-utan Recent direct observations on nest building in chimpanzees have been published by Haddow (1958) and Bolwig (1959*a*). These accounts, as well as my notes on nests in various Uganda forests, indicate that the basic method of tree nest construction in the two African apes is very similar. The main difference between the two is that gorillas usually do not line the nest cup with twigs, whereas chimpanzees appear to do so quite frequently. One chimpanzee nest which I examined in the Kayonza Forest contained no twigs in the nest cup, but one in the Budongo Forest was lined with four twigs. Bolwig (1959*a*) and Goodall (pers. comm.) have both noted definite linings of twigs in the chimpanzee nests observed by them. One chimpanzee required about three minutes to construct its nest (Bolwig, 1959*a*).

Wallace (1869), Hornaday (1885), Schaller (1961), and others describe the nests of orang-utans. The method of nest construction in orang-utans and in the African apes is similar. Schaller (1961) analyzed thirteen orang-utan nests and found that they consisted of three to nine (mean 4.7) main branches each; the number of twigs in the nest lining varied from none to thirty-four (mean 10.0). He observed one agitated female as she built a crude but functional nest in about ten seconds.

Body Position of the Gorilla in the Nest In their day nests, gorillas lie in the various positions of resting animals pictured in Figures 39 and 40. However, at night, when the air and ground temperatures at Kabara fell near the freezing point, gorillas slept either on their sides, with knees drawn up and close to the body, and with one or both arms covering the chest, or curled up on the abdomen with legs and arms tucked under them (Fig. 56C–D). I never saw a gorilla on its back at night. An infant was usually held to the chest of its mother or lay at her side with its chest huddled against her body.

Location of Dung in the Nest The fact that gorillas deposit dung into their night nests has been observed by several investigators. In West Africa, Reichenow (1920) commented on it, and Merfield and Miller (1956) maintained that nest fouling was most prevalent on cold, wet mornings. Sharp (1927), on the other hand, always found nests free from droppings. Donisthorpe (1958), Bolwig (1959*a*), and others have observed that dung commonly occurs in the nests of the mountain gorilla. In a total of 2,451 nests checked during the present study, 99 per cent contained dung (Table 45). This dung was in the nest cup in 73 per cent of the nests, and there was evidence in all of these nests that the animal had lain on its faeces. Dung was on the edge or over the edge of the nest in 26 per cent of the cases. Only 1 per cent of the nests showed no evidence of dung close by.

The percentage of nests in which all or part of the dung has been lain or sat upon by the animal varies considerably from area to area. Kabara and Kisoro show roughly comparable figures of between 73 to 82 per cent, but the Kayonza Forest, the only other area with an adequately large sample, reveals only 8.6 per cent. Similar low percentages were noted at Mt. Tshiaberimu. Cordier told me that in the Utu region gorillas rarely lie on their dung, a statement supported by meager personal observation. It appears, therefore, that the gorillas in the Virunga Volcanoes differ from those in other areas in the frequency with which they lie on dung. I have not been able to explain this difference to my satisfaction. Altitude and temperature are apparently not a responsible correlate, for the Kayonza Forest lies at only a slightly lower altitude than Kisoro, where gorillas lie on their dung at 8,000 feet as well as at 12,000 feet. The amount of dung deposited during the night may be a factor. The average amount of dung eliminated per night appears to be higher in the Virunga Volcanoes than in the other areas. If defecation occurs several times during the night, as is frequently the case in the Virunga Volcanoes, the animal is more liable to lie on the faecal matter than if defecation is delayed until early morning, a time when many gorillas deposit faeces in or near the nest (see section on defecation).

FUNCTIONS OF NESTS Speculations as to the functions of nest building in gorillas raise some interesting problems. Tree nests clearly provide a platform on which the animal can recline without danger of falling. In addition, many tree nests, especially in bamboo, are quite soft and comfortable. Ground nests on slopes prevent the gorilla from sliding downhill when constructed with substantial rims. Nests with padding may also provide comfort and insulate the body from direct contact with damp or cold soil.

On the other hand, nests do not provide protection against rain, wind, and other adverse weather except when placed in a sheltered location. Furthermore, ground nests seem quite inefficient in keeping the animals warm, for they are usually not effectively insulated from the bare soil.

Of the functions listed, only the provision of a secure support in trees is essential to the survival of the animal. Tree nests must be consistently stable and well built, for if they are not the consequence could be fatal to the gorilla. The function of a ground nest is more difficult to visualize, and the occurrence of bedding sites with little or no trace of nest construction in 10 per cent of all records from Kabara clearly indicates that nests are not essential when the animals sleep on the ground. In addition, a large proportion of the other ground nests are so crude that comfort and other possible functions are not evident.

Many ground nests consist solely of a rim. Such a rim is often too sketchy to provide any conceivable function. Thus the animals exhibit and retain vestigial nest-building behavior in an environment where nests seem superfluous—an anachronism which suggests arboreal ancestry.

Interestingly, Köhler (1957) observed nest building on the floor in captive chimpanzees, apes which in the wild nest strictly in trees. On this behavior he comments: "If the material under consideration is anything like stalks or twigs and if there is little of it, then we are confronted with the strange phenomenon that, whatever the circumstances, the first thing is never to make even a scanty support for the body to squat on, but to create a ring round the animal; this is always done first, and if there is not enough material, then the ring is the only thing that is made."

Transport of Materials

Gorillas carry vegetation to eat at a different location, for nest construction, and in play. Transport of materials is infrequent, however, and the distance covered rarely exceeds 25 feet. In an environment which provides an abundance of food and nesting materials there appears to be little need for such transportation.

The animals sometimes carried food to a specific sitting place. About twenty observations were made of which the following are typical examples:

1. *A juvenile ascends a tree to 7 feet, carrying a piece of* Hagenia *root in its right hand. It sits and eats.*
2. *An infant, 2 years old, breaks off a* Carduus, *and on 3 legs walks 20 feet dragging the stalk with one hand.*
3. *A female ascends a* Pygeum *tree to a height of 15 feet, breaks off one branch, and descends, holding it in one hand. She sits at the base of the tree and feeds.*
4. *A blackbacked male leaves his seat, walks about 10 feet to a nettle, breaks it off, and, holding it in one hand, walks on all fours back to his former place and eats.*
5. *A blackbacked male walks quadrupedally for 10 feet, carrying a thistle in one hand and a* Helichrysum *in the other. He sits and eats.*

Nest material was rarely transported as far as 15 feet. A cushion of moss was apparently dislodged from a *Hagenia* branch before being incorporated into a nest rim about 15 feet away. I noted at least four instances in which moss or *Galium* was pulled into a nest from a distance of 6 to 8 feet. Gorillas commonly decapitated lobelias within 5 to 7 feet of the nest and then placed them in the nest cup.

Playing infants occasionally carried an herb or leaf for several feet, a topic more fully discussed in the section on play. Most playful transport was over distances of less than 15 feet, and only once was a stalk carried over 25 feet by an infant while riding on the back of its mother.

All objects were carried by hand except in one instance when a blackbacked male grasped and dragged a stalk for about 15 feet with his hind foot. The animals walked either on three or four legs while transporting vegetation. If they used only three, the material was held in one hand with an inwardly flexed arm. If the hand that grasped the vegetation also functioned in locomotion, the stalk was usually pressed between the fingers and the palm of the hand as the animal walked on its knuckles.

Captive gorillas appear to transport food more readily than wild ones. For example, one male in a large enclosure at the I.R.S.A.C. station carried three bananas as he walked bipedally about 40 feet to the other end of the cage.

Tool-using and Handling of Objects

TOOL-USING

The use of tools, a prominent feature of human cultures, has been demonstrated in captive chimpanzees (Hayes and Hayes, 1954) and gorillas (Yerkes, 1927). It has been suggested that the use of tools may have preceded erect bipedal locomotion in man (Bartholomew and Birdsell, 1954), so any evidence of naturalistic use of tools by free-ranging apes is of great interest.

The literature contains two seemingly reliable observations of tool-using by chimpanzees in nature, and recently Goodall (pers. comm.) noted other instances of it in Tanganyika. Beatty (1951) watched chimpanzees as they cracked dried palm nuts in Liberia: "He then picked up a chunk of rock and pounded the nut which had been placed on the flat-surfaced rock." Merfield and Miller (1956) observed several chimpanzees around a hole which led to the nest of subterranean bees: "Each ape held a long twig, poked it down the hole and withdrew it coated with honey. There was only one hole, and, though for the most part they took turns at using their twigs, quarrels were constantly breaking out, and those who had licked off most of their honey tried to snatch the newly coated twigs. We watched them for over an hour at a range of fifty yards. . . ."

No comparable observations exist for the gorilla. Pitman (1931) noted that ". . . a gorilla had been seen to use a stick for obtaining fruit otherwise out of reach," and Philipps (1950) claimed to have seen a free-living gorilla use a crooked stick to reach food. However, a personal query to the latter authority brought a reply which indicates some uncertainty about the record.

I saw no tool-using in gorillas.

Gorillas spend many hours daily on their haunches either feeding or just sitting with their hands entirely free to manipulate objects. They are in many ways admirably pre-adapted to a life in which tool-using could play a prominent role. However, the need for it is apparently slight. In the lush forest environment, gorillas can easily obtain forage by simply detaching the desired plant with the hand or by climbing into a tree to reach a certain fruit. No preparation of the food is required beyond stripping or shredding which is readily accomplished with the teeth and fingers. In other words, a vegetarian in an environment where forage is plentiful finds little need to use tools.

Handling of Objects

In their daily routine, gorillas handled many objects. They carried food to the mouth almost exclusively by hand, they grasped limbs when climbing, they pushed over vegetation, they broke in herbs for nest construction. Infants sometimes carried leaves and sticks in play. But gorillas rarely handled anything for the sake of manipulation alone. They showed no interest to fondle, pull, and tear objects, traits so prominent in captive chimpanzees.

Gorillas also appeared to lack the inclination to investigate strange inanimate objects manually. On one occasion, my rucksack lay in full view and within 15 feet of a blackbacked male. He glanced at it once and then ignored it. A wad of paper, strikingly white in the surrounding green of the forest, which I had used to mark a trail, was apparently ignored by a passing group. I once placed a shiny tin can on a buffalo trail to mark the site where I had poured plaster of Paris into a gorilla track. The following day a blackbacked male traveled along the trail, and, from his tracks in the soft earth, it was evident that he had passed the can without breaking his stride or investigating it manually. After studying a captive female mountain gorilla, Yerkes (1927) stated: "From my first meeting with Congo I was impressed by the relative infrequency of indications of curiosity about unusual objects or events," a statement which appears to apply to the subspecies as a whole. This hesitancy to explore new situations through manipulation may be related to the infrequency of tool-using in the wild.

Eliminative and Self-care Behavior

Defecation

Gorillas defecate most frequently while walking or while lying in the night nest. A walking gorilla defecates without change in posture or speed, the sections of dung merely dropping on the trail. In trees gorillas characteristically squat while defecating (Fig. 56F, G); in the nest they usually do so while resting on the side. One female wriggled her rump toward the edge of the nest while lying on her side, thus depositing the dung outside the nest cup. Occasionally gorillas defecate while lying on the belly or while resting on elbows and knees. Infants being carried defecate on the arms or on the backs of their mothers.

Gorillas were twice seen to handle faeces. A piece of dung became entangled in the anal opening of one female, and she reached back with her right hand and wiped the faeces to one side. Another female squatted on a branch, urinated, and defecated two single sections. She then cupped her left hand under the anus, caught the next piece of dung, turned her head to glance at it, and dropped it. My wife noted several times that a captive two-year-old mountain gorilla wiped sections of dung from its anus with its hand. The frequent fingering and throwing of dung by zoo animals is well known. I have never seen comparable behavior in the wild, nor have I seen evidence of coprophagy.

Gorillas deposit a prodigious amount of dung in the course of a day. At Kabara the animals eliminated on the average of twenty to thirty-five pieces or sections per day, about half or somewhat more of this being dropped in or near the night nest. The gorillas usually defecated once or twice during the night and once immediately before or after rising. Frequently the morning defecation was dropped along the trail soon after leaving the nest site, although the animal sometimes eliminated while still standing in the nest. Infants frequently did not defecate at night.

I counted the number of dung sections in fifty night nests at four nesting sites (group IV). Infant dung was not included in the count. The extremes are not indicated in this sample count, for twenty-six

No. of Dung Sections in the Nest	No. of Nests
1– 5	5
6–10	10
11–15	23
16–20	10
21–25	2

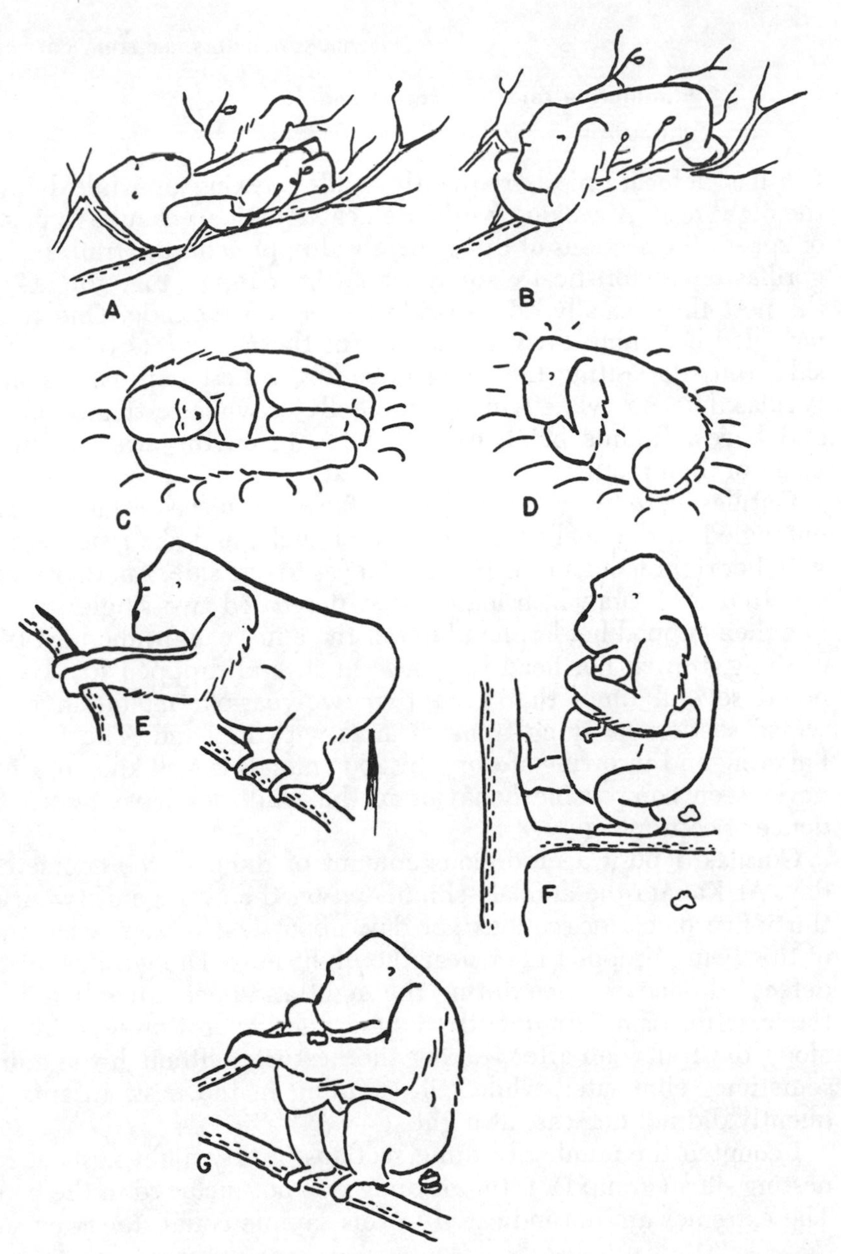

FIG. 56.—Sketches of gorillas at Kabara showing body positions in the nest and during eliminative behavior. *A–B,* two resting positions of the same female in her day nest, which was located about 15 feet above ground in a *Hypericum* tree; *C–D,* sleeping positions of two females in their night nests on the ground; *E,* a female urinating while standing bipedally in a tree; *F–G,* females defecating in trees while holding their infants to the chest.

(1.1%) nests at Kabara contained no dung (see Table 45) and one nest held thirty sections.

The total amount defecated by one animal during the day is difficult to determine. I obtained the best indication from a lone silverbacked male whom I tracked for two complete days, counting all sections of dung. Each of two night nests contained fifteen sections. On the trail a total of twenty-eight sections was found at nine separate places, varying from two to five (mean 3.1) sections each. The total number of sections of dung eliminated during the two days was fifty-eight, or an average of twenty-nine per day. If these figures can be considered typical, each gorilla defecates at least five times per day. This finding differs from that of Sabater Pi and de Lassaletta (1958) who maintained that in West Africa gorillas usually defecate only once a day.

Six oven-dried sections of dung weighed as follows:

Age and Sex Class	No. of Sections Weighed	Average Weight (in gms.)
Silverbacked male.......	2	29.0
Medium-sized animal.....	4	16.3

If an animal defecates on the average of thirty sections per day, a silverbacked male eliminates roughly 1 kilogram of dry matter, and a medium-sized animal about ½ kilogram.

None of the nine fresh nests which were examined in the Utu region contained more than ten sections of dung. Although I made no dung counts in the Kayonza Forest or at Mt. Tshiaberimu, my impression was that gorillas in those areas deposited amounts similar to those found in the Utu region and thus less than in the Virunga Volcanoes.

Urination

Both male and female gorillas usually squat while urinating on the ground. In females, the water tends to splash slightly to the back of the legs; in males between and somewhat forward of the legs. No animal was observed to urinate while walking, although males sometimes urinated while standing quadrupedally with the hind legs slightly spread. Gorillas also urinated while propped on elbows and knees, the stream of urine being ejected caudally with considerable pressure some 8 to 12 inches beyond the rump hairs in both males and females. One female backed to the edge of the nest and urinated over the side. However, the localized moisture in the bottom of some nests indicated that the animals also urinated into the nest cup.

Gorillas usually squat to urinate in trees. Twice females stood on their hind legs, grasped a branch in front of them with their hands, and urinated in a stooped position (Fig. 56E). On one occasion a female in a tree sprayed another female on the ground below her. The latter one glanced up, then dodged rapidly out of the way. An infant, about four months old, wet the hand of its mother, and she shook the urine off.

Sneezing and Coughing

Gorillas sneeze and cough in a manner similar to humans. Explosive sneezes and deep bronchial coughs occurred throughout the year. I noted two spells, one in mid-November, the other in late March, during which several animals of various groups sneezed or coughed; however, only one or two animals in a group were affected at any one time. A clear mucus ran from the noses of several gorillas.

In coughing, the head and upper part of the torso jerks forward and the abdomen moves slightly outward. I never heard more than two consecutive coughs by the same animal. A sneeze is preceded by a sudden inhalation coupled with a raising of the head, and followed by a forward jerk with an open mouth as the air is forcefully expelled. Once a juvenile jerked its hand to its mouth during a sneeze, then three times wiped mucus from its nose with the index finger and licked it. Another juvenile also raised its hand to the mouth at the onset of a sneeze and afterwards licked its palm.

Yawning

Gorillas yawn frequently and with varying intensity. A moderate yawn consists of raising the head and partially opening the mouth during inhalation. A yawn of high intensity exposes the entire cavernous interior of the mouth while the head is tipped back. The upper lip may be curled so that the gums are visible, but this usually does not occur until the animal has ceased to inhale. At the termination of the yawn the head is lowered, but the mouth sometimes remains partially open with both upper and lower lips pulled back to expose gums and teeth. The animal may at this point shake its head back and forth vigorously. The behavior resembles yawning in man.

Gorillas occasionally yawned while appearing uneasy because of my proximity. One silverbacked male yawned over twenty times in a fifteen-minute period as he sat within 60 feet of me. Yawning also occurs in man in mild conflict situations (Tinbergen, 1951), and has been interpreted as a displacement activity.

Hiccuping

Four instances of hiccuping were noted, all in females. On three occasions the animals had ten or fewer spasms, but one female sat for three to four minutes with her torso jerking as she emitted audible hiccups.

Burping and Wind-breaking

The diet of gorillas apparently generates much gas in the alimentary tract. Frequently such gas escapes through the mouth, accompanied by a grumbling or belching sound. If the gas passes through the anus, the sounds produced resemble those of horses or men.

Self-grooming

I recorded a total of eighty-eight instances of self-grooming at Kabara (Table 46). The animals limited their activity to those parts of the body which they could reach with ease and where visual inspection could accompany manipulation. The arms and shoulders, chest, abdomen, and legs were the main areas involved in self-grooming; the rump, back, and head were either scratched or presented to another animal for grooming (see section on mutual grooming).

Although the animals occasionally groom their shoulders and arms with one hand alone (Fig. 57F), they typically use the mouth or chin as well. The gorilla pushes the long hair against the grain with thumb or index finger and tucks it under the chin or presses it back with the mouth, an act which prevents the hair from flipping back and covering the exposed skin. The head is twisted in such a manner that the eyes can observe the actions of the fingers (Fig. 57D). The animal sometimes picks at the skin with the nail of its bent index finger, and any object so detached is removed either between thumb and index finger or with pursed lips; some objects are inspected visually and eaten. Both hands are used for grooming the abdomen, the chest, or the legs. The gorilla holds back the hair with one hand while picking at the skin with the other (Fig. 57A).

Gorillas usually sit when grooming themselves, but occasionally some lie. Legs are sometimes groomed while half sitting, half lying on the back with one leg held in the air.

The animals appear to groom themselves very intently, although they rarely continue the activity for as long as five minutes. Most grooming occurs during the midday rest period, especially on sunny days. The activity appears to be contagious at times in that others

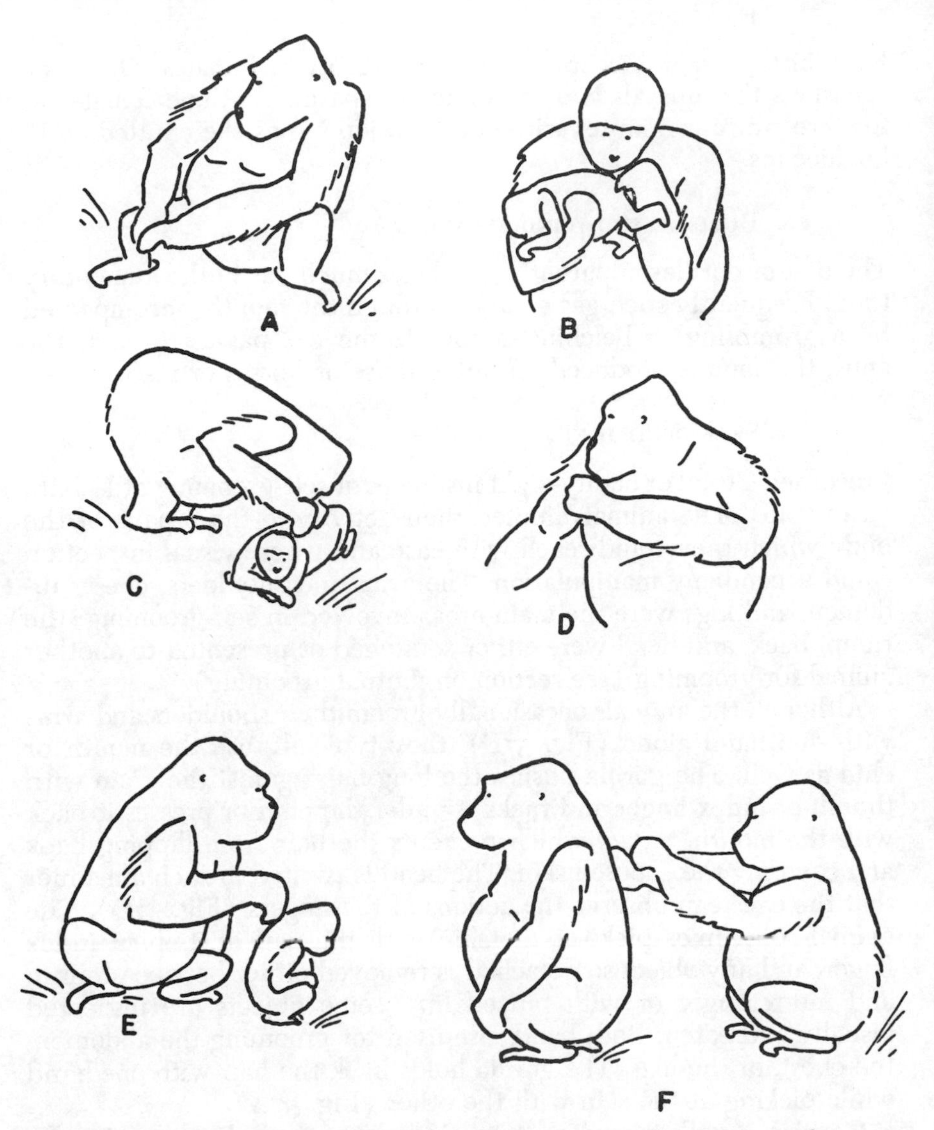

Fig. 57.—Body positions of gorillas at Kabara during self- and mutual grooming. *A*, a juvenile grooms its leg; *B*, a female grooms her small infant; *C*, a female grooms her large infant; *D*, a silverbacked male grooms his upper arm; *E*, a female grooms a juvenile; *F*, a female grooms her own hand while resting it on the back of another female.

ILLUSTRATIONS

FIG. 9.—An aerial view of the Kabara study area in the Mts. Mikeno–Karisimbi saddle, Virunga Volcanoes. (Published with permission of the Institut Géographique, République du Congo.) Geographical names: *M*, Mt. Mikeno; *C*, Mt. Karisimbi; *B*, Bishitsi; *K*, Kabara; *U*, Rukumi grass meadow; *R*, Rweru. Vegetation types: 1, Bamboo; 2, *Hagenia* woodland; 3, *Hypericum* woodland; 4, Tree heath; 5, Giant senecio; 6, Mountain woodland; 7, Herb meadow.

PLATE 1. A typical view in the Utu region, showing a village of mud huts surrounded by stands of bananas, some young secondary forest, and in the distance the mature lowland forest. This picture was taken along the road between Utu and Kasese.

PLATE 2. The Kayonza Forest stretches over the rugged rift mountains. To the south, five of the six dormant Virunga Volcanoes are visible, with Mt. Muhavura on the left and Mt. Mikeno on the right. This photograph illustrates well the short distance which separates the gorilla population in the Virunga Volcanoes from the one in the Kayonza Forest.

PLATE 3. A mountain rain forest covers the chaotic jumble of ridges and valleys in the Mwenga-Fizi region. This picture was taken along the main road near Nzombe.

PLATE 4. My wife sits in the crotch of a tree at Kabara and observes the behavior of a gorilla group. A silverbacked male and three other animals are visible.

PLATE 5. The Lowa River in the Utu region about 20 miles west of Walikale. The river is too broad to be bridged by a fallen forest tree, thus preventing the gorillas from expanding their range to the other bank at this point.

PLATE 6. A field in lowland rain forest near Beni, Congo, which has been cleared by the slash and burn method. Banana trees have been planted among the standing stumps and fallen trunks.

PLATE 7. A view of old secondary forest in the Utu region. This photograph was taken between Utu and Niabembe in an area which has been uninhabited by man for over 40 years. Gorillas typically use fallen trees, as shown in the foreground, to cross streams.

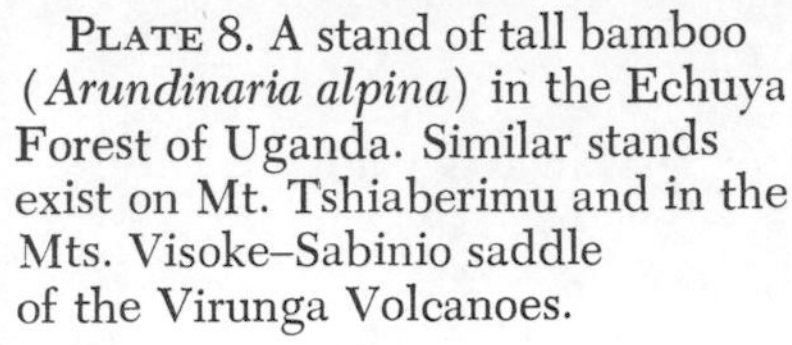

PLATE 8. A stand of tall bamboo (*Arundinaria alpina*) in the Echuya Forest of Uganda. Similar stands exist on Mt. Tshiaberimu and in the Mts. Visoke–Sabinio saddle of the Virunga Volcanoes.

PLATE 9. A typical ridge in the Kayonza Forest, illustrating the sparseness of the groundcover beneath the canopy of trees. Gorillas find little forage here.

PLATE 10. An open valley in the Kayonza Forest. Herbs, vines, and tree ferns are abundant, providing ample food for gorillas. A pygmoid Batwa stands in foreground.

PLATE 11. A view of the Virunga Volcanoes looking east from an altitude of 12,000 feet on the slopes of Mt. Mikeno. The flat-topped peak is Mt. Visoke; the jagged one is Mt. Sabinio; and the one with its summit in the clouds is Mt. Muhavura. Low tree heath and white-flowered *Helichrysum* grow in the foreground.

PLATE 12. *Hypericum* woodland in the Mts. Muhavura–Gahinga saddle, Virunga Volcanoes, at an altitude of 10,000 feet. The tree on the left is a *Hypericum lanceolatum*. The shrubs consist primarily of *Pycnostachys goetzenii*. The dominant herb is *Senecio trichopterygius*. The person in the foreground (the author) is 6 feet tall. The summit of Mt. Gahinga is visible in the distance.

PLATE 13. The zone of giant senecios at an altitude of 12,800 feet on the slopes of Mt. Muhavura, Virunga Volcanoes. The large-leafed *Senecio erici-rosenii* tower above the African in the foreground. Two narrow-leafed *Lobelia* grow on the left. The fern-like plants at the ground level are *Peucedanum kerstenii*. All three species provide gorilla forage.

PLATE 14. Snow-covered Mt. Karisimbi towers above the *Hagenia* woodland in the Mts. Mikeno–Karisimbi saddle area, Virunga Volcanoes. The Kabara meadow lies in the foreground. Carl Akeley's grave overlooks this beautiful scene.

PLATE 15. A typical foggy day at Kabara, altitude 10,200 feet. *Hypericum* trees and lava outcrops break the expanse of the Kabara meadow.

PLATE 16. One female gorilla of group VI squats in the crotch of a tree, her two-year-old infant beside her. Another female beats her chest. A small black-backed male sits on the sloping trunk.

PLATE 17. A female gorilla of group VII with an infant, about 10 months old, on her back.

PLATE 18. The striking individual differences in the facial features of gorillas are evident in these members of group VI. Ten animals are visible. The silverbacked male is yawning, showing his black teeth. The dominant herb in the rank groundcover is *Senecio trichopterygius.*

PLATE 19. A silverbacked male of group V (No. 3 in the dominance hierarchy) in a *Hagenia* tree. His silver-colored saddle is conspicuous.

PLATE 20. A blackbacked male of group IV sits with folded arms

PLATE 21. A hole excavated by a gorilla at Kisoro in order to obtain the root of *Cynoglossum*. The ruler is 6 inches (15 cm.) long.

PLATE 22. A banana stem broken down and torn apart by a gorilla in extracting the tender pith. This photograph was taken at the western edge of the Kayonza Forest.

PLATE 23. A typical ground nest of herbs in the Kabara area. Note that the emphasis of construction is on the nest rim rather than on the nest bottom.

PLATE 24. A crude gorilla nest of lobelia stems and heads at Kabara.

PLATE 25. A ground nest of *Vernonia* branches at Kabara. The animal uprooted the whole shrub and fashioned its nest in the leafy crown.

PLATE 26. A gorilla nest in a tangled mass of bamboo at Kisoro. The nest is 6 feet above ground and sturdy enough to easily support my wife.

Plate 27. A gorilla nest in the crotch of a *Musanga* tree, 35 feet above ground, in the Utu region.

Plate 28. "Symbolic feeding" by the silverbacked male of group VI preceding the chest-beating display. The large leaf between the lips of the male (on the right) is clearly evident. Eight other animals are visible. A blackbacked male sits directly above the silverbacked male. Two females are scratching themselves.

Plate 29. A female of group IV stands bipedally and beats her chest

PLATE 30. A blackbacked male displays the strutting walk

PLATE 31. The silverbacked male of group VII indicates his readiness to leave the rest area by standing motionless and facing in a certain direction with his legs spread. A female with a small infant on her back has joined him. The long hair on his forearms is especially prominent in this picture.

PLATE 32. A juvenile of group VI stands bipedally and rests its arms on the back of a blackbacked male.

PLATE 33. At least five members of group IV have crowded behind the dominant silverbacked male as he watches me

PLATE 34. Two females, a blackbacked male, and two large infants of group VI crowd onto a tree branch, thus obtaining a clear view of me.

PLATE 35. Africans find many uses for bamboo from the forest reserve on the Uganda side of the Virunga Volcanoes. The photograph shows a hut built entirely of bamboo stems, and the African is weaving a basket of bamboo.

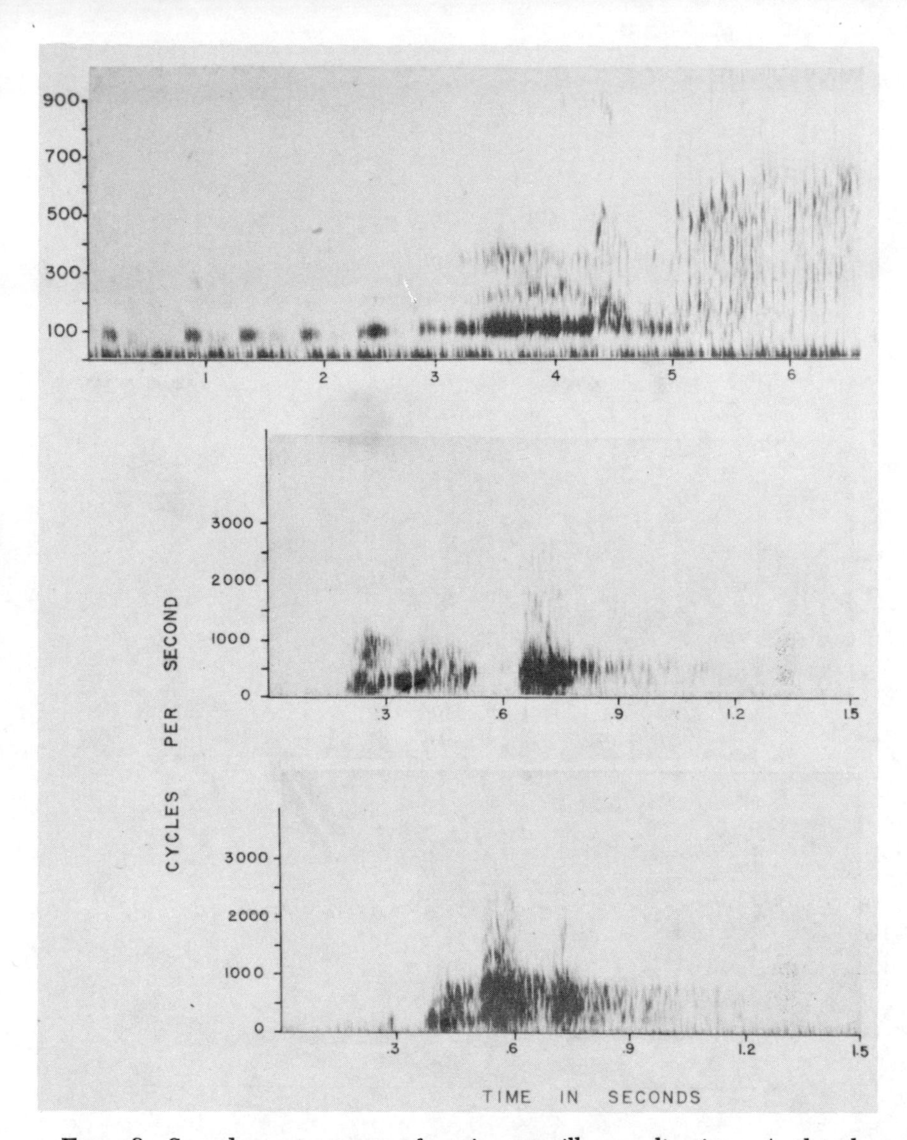

FIG. 58.—Sound spectrograms of various gorilla vocalizations. *A*, the chest-beating display of a captive silverbacked male in the Utu region. The display begins with a series of 5 distinct hoots which are followed by a growling sound, and ends with a chest beat consisting of 17 individual beats. *B*, a series of 3 annoyed grunts by a silverbacked male (group VII) at Kabara. *C*, a roar of low intensity by the silverbacked male of group VII at Kabara.

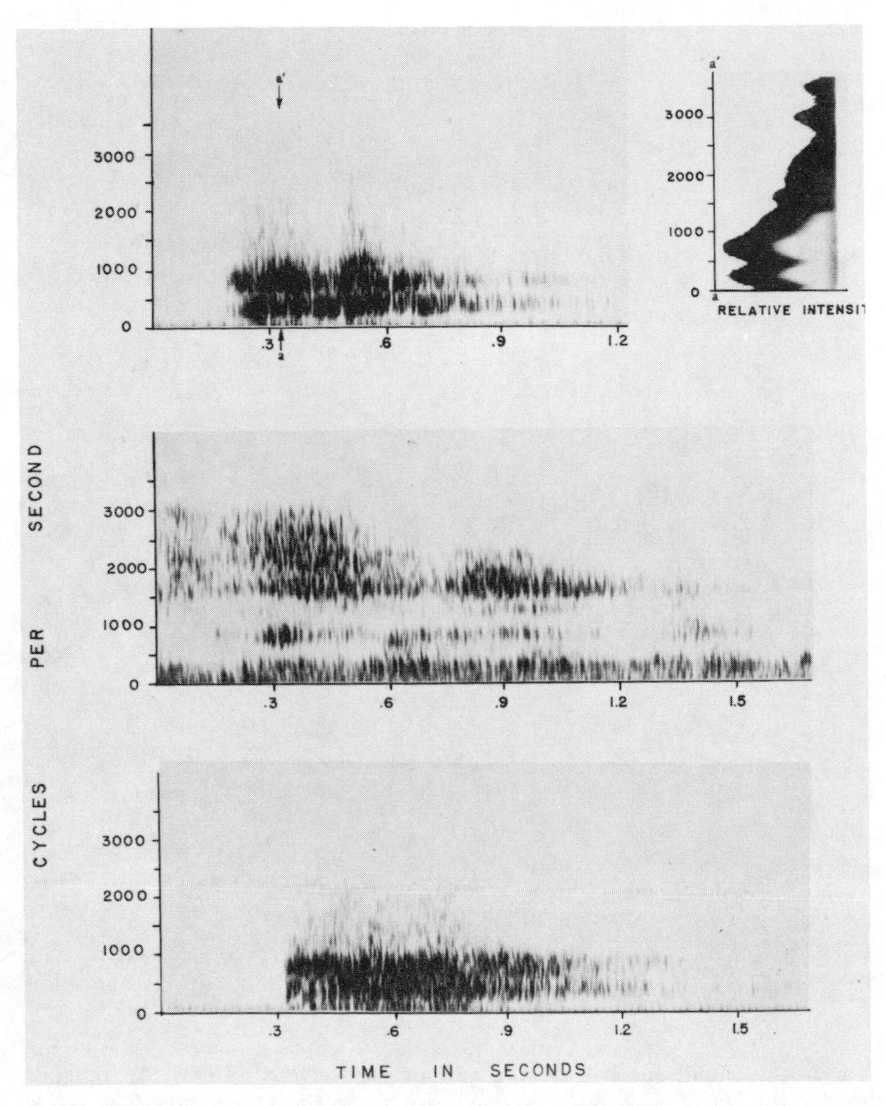

FIG. 59.—Sound spectrograms of various gorilla vocalizations at Kabara. *A*, a roar of low intensity by the silverbacked male of group VII. The section on the right represents the relative amplitude of the sound at the instant of time a-a′. The instant of time is 1/24 second long. *B*, two quarreling females in group VII. As one female screams, the other emits a series of barks. *C*, an intensive roar by the silverbacked male of group VII.

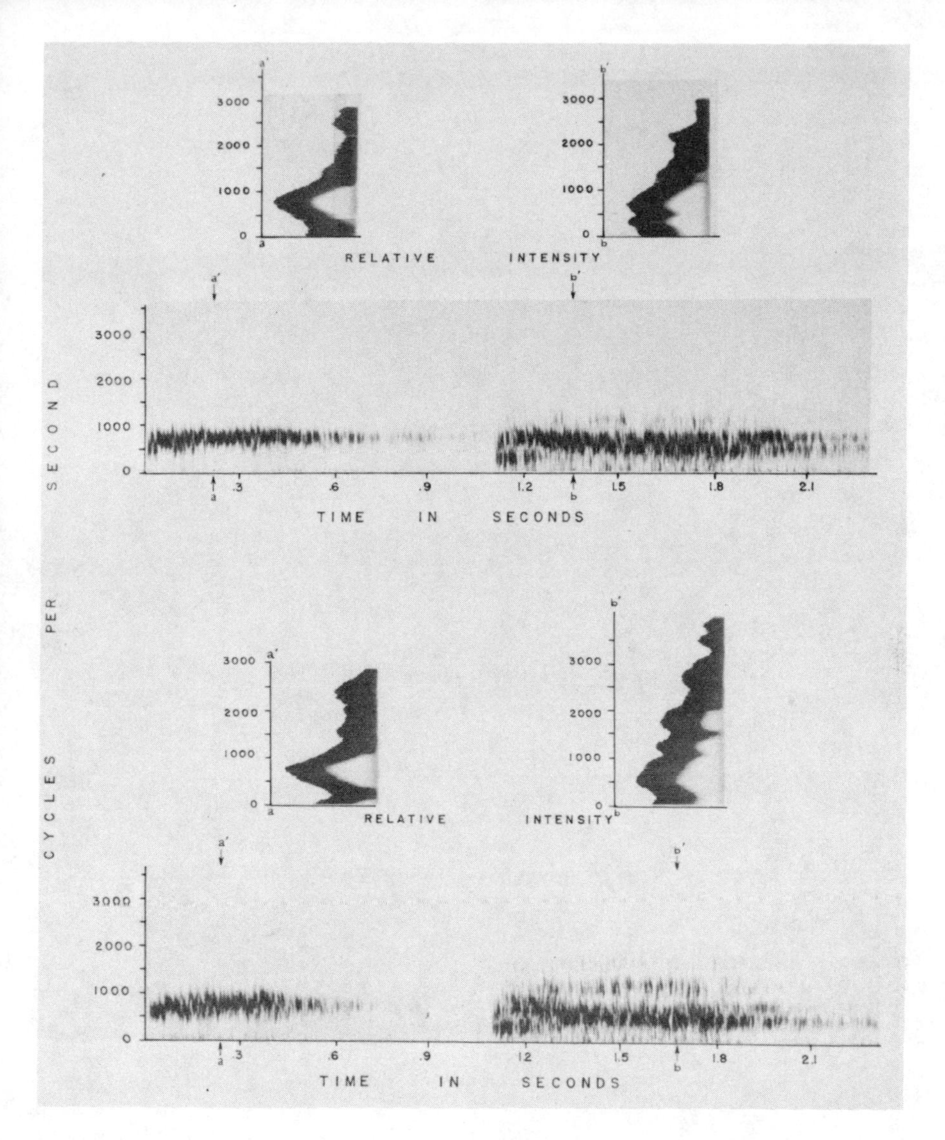

FIG. 60.—Sound spectrograms of intensely roaring male gorillas in group VII at Kabara. *A* and *B*, the first roar is that of a blackbacked male, the second that of a silverbacked male. The section above each roar represents the relative amplitude of the sound at the instant of time a-a′ or b-b′. The instant of time is 1/24 second long. The sections show relative intensity throughout the entire spectrum of the sound, including overtones which did not register on the spectrogram with the settings used on our instrument.

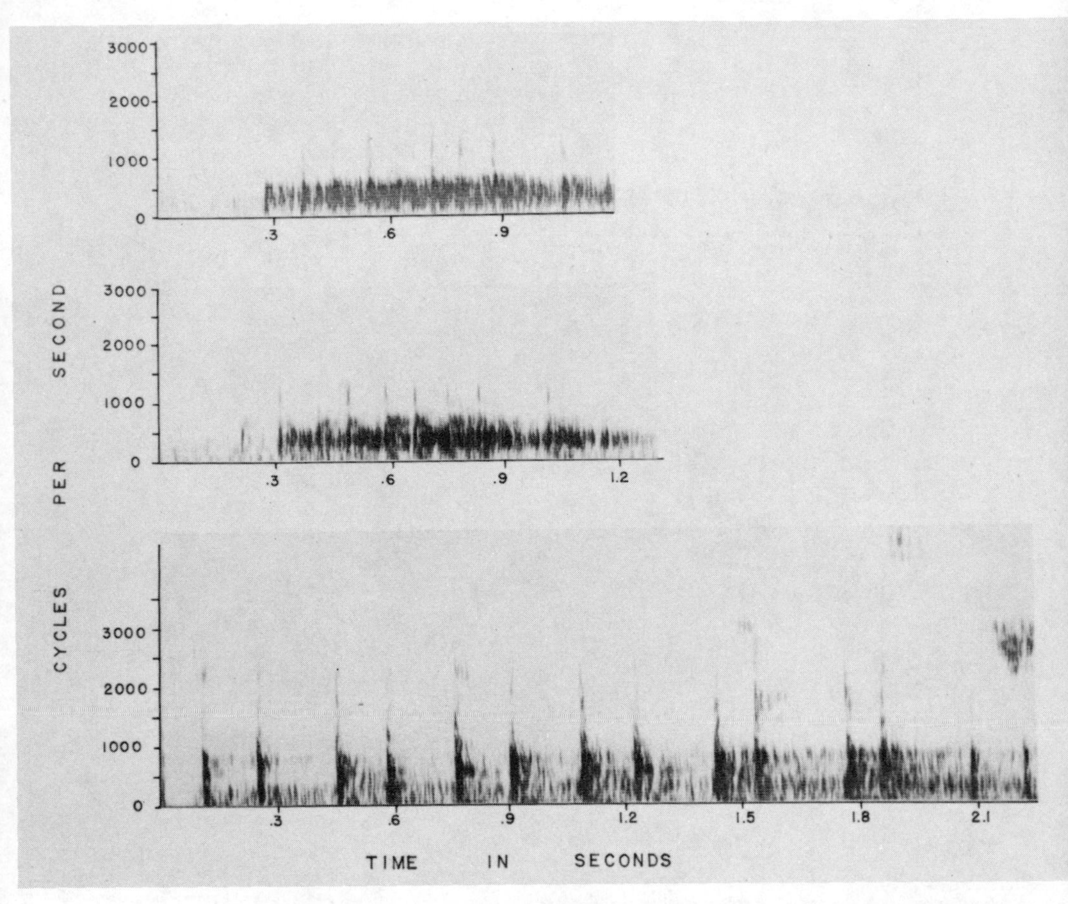

FIG. 61.—Sound spectrograms of chest beats by male gorillas at Kabara. The intermittent vertical lines of high frequency represent the impact of the hand against the chest. The sounds which are shown between each beat apparently consist of resonance and of various background noises. *A*, silverbacked male, group VII, 8 individual beats; *B*, silverbacked male, group VII, 9 individual beats; *C*, blackbacked male, group VII, 15 individual beats. (Two notes from the song of a southern double-collared sunbird, *Cinnyris chalybeus*, are recorded on the right side of the figure.)

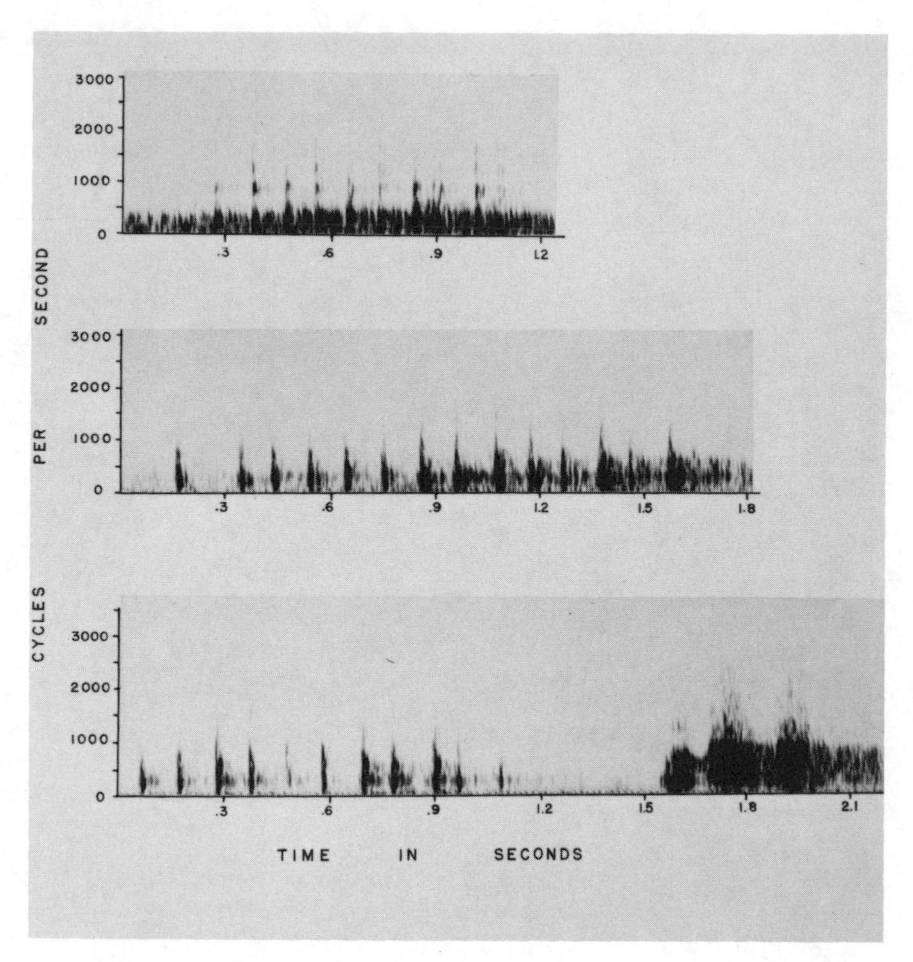

FIG. 62.—Sound spectrograms of chest beats by female and juvenile gorillas at Kabara. The intermittent vertical lines of high frequency represent the impact of the hand against the chest. The sounds which are shown between each beat apparently consist of resonance and of various background noises. *A*, female, group VII, 10 individual beats; *B*, female or juvenile, group VII, 14 individual beats; *C*, female, group VII, 11 beats. A roar of low intensity by a silverbacked male follows her chest beat.

are stimulated to groom when observing the behavior in their companions.

Grooming apparently functions to rid the skin and hair of imperfections, dry skin, and foreign matter. Often the animal merely runs its fingers through the hair, suddenly scratches, and then begins to groom.

Animals of all age and sex classes groom themselves (Table 46), but the differences in the frequency are striking. Proportionally females groom themselves twice as often as silverbacked males, and juveniles groom themselves another half as many times as females. Blackbacked males groom themselves infrequently, and infants rarely. Infants less than two years of age were never observed to groom themselves, although they were occasionally seen to pick obvious pieces of foreign matter from their hair.

Scratching

Gorillas scratch often, especially the arms, legs, abdomen, and sides. If scratching is vigorous, they move the four bent fingers up and down on the selected area with great force, sometimes taking in the whole length of the arm or the entire side from leg to chest. To scratch a delicate area, such as the corner of the eye, gorillas usually use the middle or index finger.

Frequently scratching is of an exploratory nature in that the animal merely moves its fingers lightly and quite slowly against the grain of the hair. When something is noted, it is scratched off and may afterward be picked from the hair, examined closely, and eaten.

Captive chimpanzees sometimes scratch vigorously when in situations that make them feel uncertain, behavior comparable to head-scratching in man under similar circumstances (Tinbergen, 1951). I noted several times that gorillas, which had approached me closely and were obviously uneasy, scratched their arms slowly and with greater intensity than the situation warranted. Such scratching appeared to be a displacement activity.

Gorillas were never observed scratching themselves with their feet. On the other hand, monkeys do so, and I have seen free-living *Cercopithecus ascanius, Colobus abyssinicus, Papio doguera,* and *Macaca fuscata* scratch their sides and head with the hind foot.

Picking Nose and Teeth

Gorillas occasionally pick their nose with the index finger. When they extract something, they first look at the object, then eat it.

Gorillas may pick at their teeth, both at the front and back of the mouth, with thumb and index finger. When an object is dislodged, they sometimes glance at it before eating it.

Facial Expressions

Yerkes (1927) characterized the facial expressions of gorillas when he concluded from a study of a captive female mountain gorilla: "On the one hand, I am certain that Congo's face is capable of varied expression, and on the other, I am equally sure that it seldom clearly reflects her attitudes, or perhaps I might more safely say that it usually reflects a calm, placid, self-dependent state." Several structural features of the gorilla's face such as the ears and forehead, which are often useful in interpreting the emotions underlying the facial expression, are obscured by dense hair. I found that the eyes, the lips, and the mouth furnished the best indicators of emotion. By watching the eyes I could often predict a gorilla's actions. At close range this helped me to plan my own behavior. For example, if their eyes portrayed nervousness by a slight shiftiness, I refrained from focusing the threatening stare of a camera lens on them; if, however, they conveyed curiosity and boldness, I could photograph with impunity.

Many facial expressions of gorillas resemble those of chimpanzees (Foley, 1935; Yerkes and Yerkes, 1929) and man (Darwin, 1873). The brief descriptions below are based on observations of wild gorillas, but most of these expressions were also noted in captive animals, where the behavioral context in which they were given was usually more readily apparent.

In Undisturbed Situations (Quiet State)

The expressions of gorillas when resting and sitting quietly can best be described as placid and peaceful. The mouth is closed, the eyes are bland, and the animal surveys its environment without obvious attention.

In Interest-arousing Situations (Attention)

If something arouses the interest of the animal, either visual or auditory, the eyes become fixed but their soft hue remains. The lips are often slightly parted and the head may be cocked to one side and slightly raised. This gesture is often replaced by one of active interest in which the eyes alertly follow all movements of the object which has aroused attention.

In Situations of Deprivation (Light Distress)

When the animals appear depressed, as when for example the mother of an infant walks away without waiting for it, the lips are pursed but compressed or only slightly parted, and the forehead is raised. Whining is sometimes heard. Captive infants commonly exhibit this expression when deprived of something they want, or when prevented from carrying out a certain action.

In Play (Pleasure)

Playing gorillas partially open their mouths with the corners drawn far back into a smile, but without exposing the gums and teeth.

In Mildly Disturbing Situations (Annoyance)

The eyes grow hard and fixed, the head is often tipped slightly downward, and the lips are frequently pursed and slightly parted. The forehead is contracted into a scowl and the area between the browridges is wrinkled, although this is not easily seen in the dense hair.

In Strongly Disturbing Situations (Anger)

The eyes are hard and usually fixed on the animal causing the anger. The extent to which the mouth is opened depends on the intensity of the emotion, but characteristically the gums and teeth are shown and the lips curled back. The mouth is either half or entirely open, and the head is frequently tipped slightly down. Screams or roars always accompany the alternate opening and closing of the mouth (see section on vocalizations).

In Alarm-producing Situations of Low Intensity (Uneasiness)

The lips are characteristically pulled inward with the mouth remaining tightly compressed (Pl. 30). The eyes are shifty, the head is often slightly tipped up, and the animal refrains from direct staring. This expression was seen only in response to my presence. The biting of the lower lip by man in similar circumstances is a close behavioral parallel.

In Alarm-producing Situations of High Intensity (Fear)

Anger and fear are closely related in their overt expression, and the interplay of these emotions is often such that one cannot distinguish between the two. In fear the mouth appears to be opened wider and to remain open longer, and the corners of the mouth seem to be drawn farther back. The head is frequently tilted slightly back and the forehead raised. The eyes dart nervously back and forth.

Vocalizations

The literature contains numerous brief descriptive references to gorilla sounds, but these are mainly confined to the vocalizations of high intensity in response to the presence of man. Prince Wilhelm of Sweden (1923) noted angry growls, Barns (1922) screaming roars, and Merfield and Miller (1956) screams of rage and short, sharp coughs. Females and young observed by Du Chaillu (1861) screamed when alarmed and moaned when distressed. Gorillas also whined when curious (Barns, 1922), and females clucked to call their infants (Du Chaillu, 1861). Most descriptions in the literature are entirely inadequate and so vague that in many instances I was unable to use them for comparison with my data. Thus, for example, Donisthorpe (1958) mentioned hoarse croaks, a burbling sound, a squeaking noise, and a high-pitched "toot," none of which I could match with sounds heard by me. Kawai and Mizuhara (1959*b*) gave similar inadequate descriptions.

I was able to record only a few vocalizations of free-living gorillas during the one week that I had access to a tape recorder. Consequently my descriptions of vocalizations remain primarily verbal. To aid in understanding these vocalizations and to make them recognizable to future investigators, I have presented (a) as complete a verbal description as possible, (b) sometimes a schematic representation of the sound either as it sounded to me or as recorded by a sound spectrograph (Missile Data-Reduction Spectrograph, Kay Electric Co., Pine Brook, N.J.), (c) the situation in which the sound occurred, (d) its probable function, and (e) my appraisal of the animal's emotion.

I found gorilla calls difficult to classify. When, for example, a male gorilla roars at the approach of a man, the accompanying bluff charge imparts an aggressive function to the roar; the fact that the group tends to cluster behind the male on hearing the roar suggests that to them it functions defensively; and the fact that the group members respond to the roar signifies that it is a warning. The emotions underlying the roar of the male appear to indicate anger but often also fear. In other words, a single sound can be the product of more than one emotion and possess more than one function.

Gorillas are generally quite silent while pursuing their undisturbed daily routine. Most sounds are abrupt and of a low intensity.

The most varied and intense vocalizations are given in response to the presence of man. The function of a particular sound can only be determined if it always accompanies a certain type of behavior, if it immediately follows a certain stimulus, or if other animals repeatedly respond in a particular manner. A few sounds of gorillas are given in no special context and elicit no obvious response. Some of these vocalizations undoubtedly possess communicative value, but the meaning conveyed by them does not incite immediate response. I was, therefore, unable to classify several vocalizations with precision. Table 47 presents twenty-two more or less distinct gorilla vocalizations, twenty-one of which were heard in the wild. For comparison, howler monkeys possess about twenty different calls according to Carpenter (1934), Collias and Southwick (1952), and Altmann (1959); Japanese monkeys emit "more than thirty kinds of vocalizations" according to Imanishi (1957). Carpenter (1940) lists nine different gibbon calls.

I heard four of the twenty-two gorilla vocalizations only once, and seven less than ten times each. Two of the sounds were given only in response to my presence, and one was heard only in captivity. Thus, eight vocalizations occurred fairly frequently in the daily routine of free-living gorillas: a soft grumbling and grunting of contented animals; a series of abrupt grunts, functioning to keep the group together; a hoot preceding the chest-beating display in silverbacked males; a harsh staccato grunt of annoyance; a bark and harsh scream given by quarreling animals; and a screech by infants when in danger of falling behind the group.

Although the number of basic vocalizations emitted by gorillas is fairly small, there is considerable variation in the pitch, intensity, quality, and pattern of each sound. These variations greatly broaden the scope of the vocal repertoire, for the animals respond to the sounds they hear selectively. Their reactions depend not only on the sound, but also on the condition under which it is given and on the member of the group who gives it. For example, sex differences are apparent in that vocalizations by silverbacked males are characteristically deeper and fuller than those of females; age differences are evident in the screams of infants, which are of a higher pitch than those of older group members. In disturbing situations the degree of loudness and harshness of the sound produced by one animal affects the rapidity of response in the others. The same sound may also communicate two different things depending on which gesture accompanies it. For instance, harsh staccato grunts given by a silverbacked male when females quarrel cause them to subside. If, however, the male emits the same sound without apparent reason, all group members first look at him and then face the direction which occupies his attention.

Gorillas have evolved a method of communication comparable to that of many other mammals. Even though all their sounds appear to be no more than expressions of emotions, their communicatory system is entirely adequate for their simple mode of life.

Table 47 lists the vocalizations of each age and sex class as well as the relative frequency with which I noted each sound. Infants were among the least vocal members of the group, with screeches being the only vocalizations which were heard fairly frequently in the wild. The whines and chuckles of infants were of such low intensity that I had to be very close to detect them. I did not hear infants emit calls of annoyance and anger in the wild, although some of their screams contained overtones which seemed to indicate these emotions. Juveniles gave four of their vocalizations primarily in response to my presence. Females emitted the largest variety of sounds, and the panting ho-ho-ho during displays appeared to be peculiar to them alone. Blackbacked males were at an awkward age, vocally speaking, in that their voices appeared to be changing. Thus, young males screamed when angry, and the older ones produced rather squeaky roars. Only silverbacked males were heard to emit the full roar, the clear hooting preceding the chest beat, and the staccato copulation call.

Descriptions of Vocalizations

My information on vocalizations, which I have summarized in Table 47, is described in greater detail below. In attempting to reproduce the calls I have used verbal descriptions of how they sounded to me, and these should be pronounced according to the German vocabulary. For example, the "u" in u-u-u is like the word foot; the "i" in hi is like heehaw; the "u" in hu-hu is like lute; the "o" in wo-wo is like note; the "a" in ha is like sofa; the "ü" in ü-ü is like the German word "Kümmel"; the "ö" like "König."

The number preceding each description of a vocalization corresponds to the one in Table 47.

Vocalizations Given by Quietly Resting and Feeding Animals

Sounds made by groups when peacefully resting or feeding are few in number and their communicatory value, if any, difficult to determine, for the vocalizations elicit no direct response. The sounds are characteristically of low intensity and pitch and do not carry far beyond the animal. Both in captivity and in the wild they appear to signify contentment.

1. *Purr* The purring is so soft that I had to be within about 20–30 feet of the animal to hear the sound, which resembles that of a large cat. It was given only on sunny days when the animals were utterly relaxed.

2. *Soft grumble* Feeding and resting animals commonly emit an irregular series of soft grumbling sounds, resembling the purr, but being somewhat harsher and louder. The vocalization can be imitated by forcing the air through the larynx and vibrating it on the soft palate.

3. *Soft grunt* A series of two to ten soft grunts, at the rate of about one to two per second, is frequently given by leisurely feeding animals, especially in good weather. The sounds resemble the grumble except that they are abrupt rather than drawn out.

4. *Hum* Once, while the group rested, I heard a soft drawn-out mmmmmmmm, rising and falling in pitch.

VOCALIZATIONS GIVEN WHEN THE GROUP IS SPREAD OUT OR MOVING ALONG When the group was scattered in dense vegetation, I occasionally heard a variety of vocalizations, which probably served to keep the members of the group in touch with each other, meaning roughly "Here I am."

5. *Abrupt grunts* This low-pitched vocalization consists of a rather slow series of two to eight abrupt, soft but distinct u-u-u, ü-ü-ü, or wo-wo-wo. It appears to communicate the position of a certain animal, especially that of the dominant silverbacked male, to the other group members. The sound is sharper and louder than the grunts of contented animals, but not as loud and harsh as the grunts of annoyed animals. Several examples from my field notes illustrate the situations in which the vocalization is given:

a) *The group moves slowly along feeding. The male gives an u-u-u, and another animal (a female?) repeats the sound.*

b) *The animals have just spread out after leaving the rest area, and several grunt as they move along.*

c) *All rest near the male. He suddenly rises and, while walking away, grunts softly.*

d) *The male rises, and, followed by several members of the group, disappears over the crest of a ridge. A few gorillas remain on the rest area, until they too move uphill after the others with much grunting and soft u-u-u and ü-ü-ü.*

e) *A juvenile gives a short, soft u-u-u as another juvenile approaches through the vegetation. The second juvenile reaches out and chucks the first one behind the head with his hand. Then both move along together and feed.*

6. *Bo-bo-bo call* I heard this distinctive and highly variable vocalization about fifteen times. It consisted of a series of rapidly repeated notes, loud, clear, high-pitched, often rising and falling, and of about two to five seconds in duration. However, unlike the abrupt grunts, it caused no visible direct reaction in gorillas. The communicatory function of the vocalization, if any, remains obscure, but I think that it probably serves as a location signal when the animals are scattered. Six examples from my field notes illustrate the situations in which the sounds are given as well as their tonal variation.

a) *The group is spread out feeding. One animal emits a series of at least 20 rapid ho-ho-ho-hi-ho-ho* (— — — ‾ — —) *with the hi being of a higher pitch than the ho.*
b) *The group moves through dense vegetation. One gorilla grunts abruptly and another starts a rapid series of bo-bo-bo, rising and falling in pitch,* ,-‾-_-/‾\\.
c) *The animals feed scattered in a nettle field. One animal vocalizes, several series of plaintive, high o-o-o-o . o-o-o-o.*
d) *As the group files into a deep canyon one gives a series of very rapid and soft bu-bu-bu, either* — — — — — *or* ,-‾-_-/‾\\.
e) *A group moves out from its rest area. One animal emits an abrupt u-u-u and a female begins a high-pitched ho-ho-ho-ha-ha-ho.*
f) *We surprise a feeding group in dense bamboo and the animals scatter. As we continue on along the trail, several gorillas call for about 45 seconds: a high-pitched rapid ha-hu, ha-hu-u-u, a "laughing" sound resembling the gibbering of chimpanzees.*

7. *"Neighing horse" call* Once, in April, 1959, I heard a gorilla in the Kayonza Forest emit a sound which I compared to the neighing of a horse in my notes. I was unaware of the situation in which it was given, but the group was spread out and feeding in a tangled valley. Garner (1896) and Burbridge (1928) mention a similar sound. Yerkes (1927) heard a captive female mountain gorilla give a "shrill tremulous and also tenuous cry similar to the distant neighing of a horse or call of a screech owl." Both Yerkes and Yerkes (1929) and Garner (1896) assumed that it is a sex call.

VOCALIZATIONS GIVEN DURING COPULATION Three different vocalizations were given by copulating animals, but only one of these appeared to be characteristic of this activity alone.

8. *Staccato copulating call* I noted this vocalization only in free-living silverbacked males when copulating. In its initial phases the

sound appears as a rapid series of distinct, fairly low-pitched but loud ö—ö-ö-ö-ö, with the first vowel forceful, abrupt, and separated from the others by a distinct pause of half a second or less, and with the remaining ones in the series following very rapidly and less emphatically. The male produces this vocalization almost continuously when copulating intensively, and less than a second often separates each series of sounds. He inhales visibly at the end of each series, after which he forces the air out through the glottis with slightly pursed and parted lips. As excitement in the male mounts only the first ö remains distinguishable while the others are forcefully expelled in one long, harsh, trembling note.

Short scream In the middle of one copulation, a free-living female emitted a single brief scream. I could not determine if this was a scream of pain or one with a sexual connotation.

9. *Loud growling and grunting* During copulation and immediately preceding it, the male and female at the Columbus zoo growled and grunted loudly. The sounds were much louder and harsher than those of peacefully resting animals.

VOCALIZATIONS GIVEN BY DISPLAYING ANIMALS Males, females, and juveniles occasionally produce vocalizations preceding or during the chest-beating sequence. Silverbacked males vocalize at such times quite frequently, but the others rarely do so. A detailed discussion of the functions and other aspects of these vocalizations is found in the section on displays.

10. *Hoot* Silverbacked males often emit a series of two to twenty or more fairly loud, low-pitched but clear and distinct hu-hu-hu, at first slowly, sometimes with one second elapsing between successive hoots, then more rapidly, until near and at the climax of the display the sound turns into a continuous harsh growl.

The sound spectrogram (Fig. 58A) represents hooting followed by a chest beat of a silverbacked male who was trapped as an adult and held in captivity by Cordier in the Utu region. The display sequence can be broken into three distinct parts.

a) A series of five hoots are given over a period of 2.75 seconds. Each hoot is roughly .1 to .25 seconds in duration. The greatest intensity of the hoot lies at about 100 cycles per second.

b) The hoots become slurred, blending into a growling sound. This period lasts 2.5 seconds. The main intensity of the sound remains at or slightly above the level of the hoots, but overtones of the growl reach 400 and more cycles per second.

c) In the final part of the display, the animal slaps its chest seventeen times in about 1.5 seconds. The hooting portion of the display is not evident now, but the growling continues, reaching a frequency of about 700 cycles per second.

11. *Yip* While watching me, a silverbacked male once emitted a loud, squeaky yip before beating its chest. (A male chimpanzee in the Budongo Forest gave a similar sound in a similar situation.)

12. *Panting ho-ho* Females gave a brief series of low-pitched rapid, panting ho-ho-ho or hö-hö-hö three times while beating their chests.

13. *Squeak* One juvenile emitted a short, high-pitched, squeaky "a" and beat its chest. The sound resembles the yip of the male.

VOCALIZATIONS GIVEN BY PLAYING YOUNGSTERS One sound appeared to be characteristic of this activity alone.

14. *Chuckle* When infants played, I could occasionally hear a very soft panting chuckle (a-a-a-a), which was expelled through an open mouth. The same sound can be elicited in captive infants by tickling them.

VOCALIZATIONS GIVEN IN DISTURBING SITUATIONS (ANNOYANCE AND ANGER) The differences between vocalizations of annoyance and anger lie primarily in the intensity of the sound, and the animals readily switch from one to the other depending on the behavior of the stimulus object.

ANNOYANCE Gorillas emitted two distinct vocalizations which seemed to signify annoyance.

15. *Harsh staccato grunt* Gorillas, both in the wild and in captivity, give a series of about two to five short, harsh grunts in rapid succession. The vocalization may sound like o-o-o, u-u-u, uh-uh-uh, oh-oh-oh, or ö-ö-ö. The sound is best imitated by forcefully contracting the muscles of the abdomen and abruptly expelling the air through the glottis.

The sound spectrogram (Fig. 58B) shows an annoyed u-u-u by the silverbacked male of group VII. The vocalization begins abruptly and the three notes follow each other rapidly, the first two almost blending into one another. The main part of the vocalization is only about .6 seconds long, but the last note is slurred, extending the duration for another .3 seconds. The greatest intensity of the sound lies at a frequency of 200 to 500 cycles per second.

Gorillas emitted the sound in two types of situations: (a) Females used the vocalization when quarreling among themselves, and the silverbacked male gave it when the strife between females was prolonged. The vocalization seems to contain elements of threat as well as annoyance, for females tend to stop quarreling upon hearing it given by the male. (b) When I arrived at the periphery of the group, the male, but sometimes also the females, gave the call, whereupon the other animals looked up. Several examples from my field notes illustrate typical circumstances in which the vocalization is given.

a) *The whole group rests. A female sees me and grunts harshly once. Another female joins in. The male looks up and roars.*
b) *A female tilts her head back, purses her lips, and gives a series of harsh u-u-u while facing me. She follows this with a loud scream and ends it with a harsh grunt.*
c) *At the end of a quarrel, in which several females screamed, one female faces another and grunts harshly, o-o-o-o.*
d) *Quarreling females are screaming loudly. The silverbacked male rises, walks toward his group, and looks at the screaming animals as he emits an u-u-u. The quarrel subsides within 3 seconds.*
e) *A silverbacked male lies on his belly as I approach. He raises his head, grunts harshly, then turns his back to me. The others look, but soon pay no further attention.*

16. *Bark* Blackbacked males, females, and juveniles occasionally emit one or more loud, often fairly high-pitched barks in situations similar to those which elicit the harsh grunts. The barks resemble the barking of a dog: waw, waw-a-a, wa-wa, or wo-wo. If several successive barks are given, the latter ones tend to rise in pitch, become slurred, and end in screams.

The sound spectrogram (Fig. 59B) illustrates the end of a quarrel between two females in group VII. Two different vocalizations are shown: as one female screams harshly (see vocalization No. 18), the other emits a series of barking wo-wo-wo. (a) The first of the two screams is about .7 seconds in duration, and its main intensity ranges between 1,500 and 3,000 cycles per second. The second scream is about .5 seconds long, and has a main frequency range of 1,500 to 2,000 cycles per second. (b) Although the barks are somewhat obscured by the screams, at least four of them can be distinguished at a frequency of 500 to 1,000 cycles per second. One bark is roughly .3 seconds long.

The barks may serve as a warning of approaching danger, and they also appear to contain elements of threat, especially if directed at another gorilla. Four examples illustrate the situations which elicit vocalization:

a) *A juvenile barks at me for 15 minutes—a rather begging wa-wa. After looking up initially, the others continue to feed. I move slightly. The barks rise in pitch and loudness, become less abrupt, and are finally drawn into a scream.*

b) *A female looks at me. She grabs her baby tightly to her chest and then gives a loud "waw." The male looks up, roars at me, and the others stand up in the vegetation to look around. Then all continue to feed.*

c) *A male runs down a path on which a female sits. She sees him and barks two times, wo-wo. As he advances further she screams loudly.*

d) *A juvenile sits in a tree and watches me in the rain. He emits a high, squeaky waw, waw-a-a several times before descending. The other animals pay no heed.*

ANGER Males roar and the other members of the group scream when angry. I have heard males roar only in response to the presence of man, and when being swooped at by ravens. Females often scream when quarreling with each other.

17. *Roar* Only silverbacked males and large blackbacked males emit the loud and intense roars so often described in the literature. The roars of small blackbacked males lack the fullness of tone, being somewhat squeaky. Only once did a female roar—a squeaky sound similar to those of small blackbacked males.

Roars vary greatly in intensity. Those of low intensity consist usually of an abrupt, single, but two- to three-toned sound:

```
                                    a                 a
              u                  u |               u |
            u |                u   |             u   |
          u   |              u     |           u     |
uu–a    u     |uu          u       |u        u       |uu
```

The animal contracts the abdominal muscles visibly, and, as the sound is ejected, lifts its head. The roar often rises in pitch before dropping to a final somewhat slurred tone as the head is jerked back down. The lips remain slightly parted, except for the "a" when the mouth is opened.

Figures 58C, 59A, and 62C show some typical sound spectrograms of low-intensity roars, all emitted by the silverbacked male of group VII. The three roars are similar in that they start abruptly and end in a slurred note. The main intensity of the sound appears to have two peaks, one at about 270 cycles per second, the other at 750. This is most evident in the section of the roar (a–a′) in Figure 59A. The duration of the roars is about .5 seconds if the slurred tone at the end is omitted. The three roars vary somewhat. Those in Figures 58C and 62C consist of three tones each, but in the former the third tone is of briefer duration and of lower intensity than the second one, whereas in the latter the two are roughly equal. In Figure 59A the roar consists of only two tones.

Roars of high intensity are probably among the most explosive sounds in nature, consisting of but a single violent tone forced out through an open mouth. Such roars are illustrated in the sound spectrograms of Figures 59C, 60A, and 60B, all taken from the same blackbacked and silverbacked males in group VII. These roars, like those of lower intensity, begin abruptly, and trail away rather gradually. The most intense part of the roar is about .5 to 1.0 seconds in duration, but the slurred ending may continue on for another .5 seconds. The sections of the roars (a and a′, b and b′) show that the peak of intensity in the blackbacked male is at about 750 cycles per second; however, the silverbacked male has a peak once at 500 cycles per second (Fig. 60B), and the other time he has two peaks, at 300 and 700 cycles per second (Fig. 60A).

The roar apparently serves as a warning of danger, for the other members of the group tend to respond by moving in and clustering around the male. It also intimidates human intruders; at least, the response of most Africans to a sudden roar is flight, and I too lost all inclination to advance.

18. *Harsh scream* Small blackbacked males, females, and juveniles emit a series of harsh, fairly short screams through open mouths (see Fig. 59B and the spectrogram description under annoyance). Such screams are usually given by quarreling gorillas, and those animals not directly involved tend to ignore the strife. Two examples illustrate circumstances in which screams are given.

a) *Several animals sit beneath the leaning bole of a tree during a rainstorm. A female and juvenile approach and try to crowd in. Suddenly several scream and the female backs away, her teeth bared, followed by the juvenile.*

b) *Several females scream, and others, including the male, give harsh grunts. The screaming continues for 30 seconds, as animals crash around in the vegetation; I briefly glimpse a female as she sits, her teeth bared, facing another female.*

VOCALIZATIONS GIVEN IN LIGHTLY DISTURBING SITUATIONS BY YOUNG ANIMALS As observed at the Columbus zoo and in the Cordiers' home, captive infants whine readily when uncomfortable, when left alone against their will, when they want something which is not immediately given them, and when something is taken from them which they want. The whine often turns into a high-pitched screech which appears to be part anger, part distress, and part frustration: it resembles the temper tantrums of a child. Free-living infants emit similar vocalizations when they are in danger of being left behind or of being hurt. Wild infants whine and screech far less than captive ones.

19. *Whine* The whine of infants is soft and tremulous, sometimes rising and falling in pitch, and resembles that of a young dog

in distress. I heard it only three times in the wild. In all instances the sound apparently communicated distress, for the adult nearest the infant reacted to the vocalization.

a) *An infant, about one year old, stands on a trail as an excited male dashes in its direction. The infant gives a soft whine and the male stops before contact.*
b) *A female moves very slowly up a slope, her 3-month-old infant behind her on the ground barely able to crawl. She sits down out of sight behind a screen of herbs. The infant whimpers softly. The female rises, peers down the slope at her young, and then reclines without further action.*

Only once did I hear a free-living juvenile whine—a soft, plaintive uuuu. In this instance the whine rose in pitch and ended in a scream of fear. One female whined as she lost her balance and fell out of the low branches of a tree.

20. *Screech* Infants often emit one or two high-pitched, short screeches when their mothers move away without them. Only the female reacted to the sound and she did so selectively. Sometimes she stopped and waited, sometimes she slowed her pace, and at other times she ignored her infant. The screeches frequently carried quarrelsome overtones as if the infant were annoyed as well as distressed. Two examples illustrate the behavior:

a) *A female turns suddenly to leave her resting place. Her 1½-year-old infant, which sits a few feet from her, emits a loud, brief, high-pitched scream as it dashes after her, grabs her rump hairs, and climbs onto her back.*
b) *A female rapidly descends a tree. A 1¼-year-old infant hurries down after her, slips and slides down a sloping branch, while the female continues. Finally the infant lets out one short screech as the female disappears into the vegetation, but does not stop.*

VOCALIZATIONS GIVEN IN ALARM-PRODUCING SITUATIONS (FEAR) I heard vocalizations of fear primarily in response to my sudden appearance at the periphery of the group. If one animal screamed and fled, the behavior sometimes elicited a panic reaction in the group as a whole, resembling a human stampede at the cry of "Fire" in a crowded building.

21. *Screaming roar* The screaming roar of silverbacked and large blackbacked males in alarm-producing situations always appeared to contain elements of anger as well as fear. In general, the higher the element of fear, the more the roar resembled a scream. Screaming roars are similar to roars of anger in their explosive beginning and in their single long tone of high intensity.

22. *High-pitched screams* In females and young, screams of fear are of longer duration, of higher pitch, and of clearer tone than those of anger. Other members of the group always reacted to such screams, in sharp contrast to ones denoting anger. Four examples from my field notes illustrate the situations eliciting screams of fear:

a) *A juvenile sits alone in a tree, hesitating to descend because of my proximity. The rest of the group is some 80 feet away. The juvenile purses its lips and several times emits a soft, plaintive uuu, starting low and rising in pitch, and ending in a loud scream with wide-open mouth. At the sound of the scream the males of the group roar.*

b) *A juvenile spots me in a tree, wheels, and runs away, screaming loudly. All animals except one rush after the juvenile. Only a black-backed male remains to look at me.*

c) *When I suddenly appear from the vegetation, a female lets out one high, long scream. The dominant male immediately runs up and looks around.*

d) *Several animals begin to scream—loud, high, piercing notes—while running excitedly about in a dense stand of lobelias and crowding around a particular spot. The male walks up and gives several harsh grunts. The screaming subsides. Investigating the site later, I found that one animal had apparently fallen through a hidden opening some 5 feet into a cave. (At least 3 animals from the same group entered this cave through another entrance a few days later and walked the whole 100-foot length of it.)*

Displays

Two displays were recognized in the gorilla: the chest-beating sequence and the strutting walk.

The Chest-beating Sequence

One hundred years ago Du Chaillu (1861) first described a male gorilla "beating his chest in rage." Almost every hunter, traveler, and scientist who since that time has encountered gorillas in the wild mentions this striking display, in which the anmal rises on its hind legs and beats a rapid tattoo on the chest with its hands. But it is a curious fact that none of these observers noted that the chest beat is merely the climax of a complex series of actions. Although other apes and man share some of the basic movements of the display, several of its distinct manifestations appear to be species-specific and lend themselves well to an ethological analysis. Before this can be attempted, however, each part of the display and the numerous variations, as well as the stimuli which release them, the functions, and the ontogeny will be discussed.

INDIVIDUAL ACTS IN THE CHEST-BEATING SEQUENCE The chest-beating display consists of nine more or less distinct acts. These are given individually or in several combinations of two or more, although there is a definite tendency for some to precede others and for several to be united in a series. The whole sequence is given infrequently and then only by silverbacked males. The complete series of events occurs typically as follows:

1. Hooting. The display begins as the animal emits a series of soft, clear hoots, which start slowly but grow faster and faster until the individual sounds merge into one another at or near the climax of the display.
2. "Symbolic feeding." The hooting is sometimes interrupted at one point as the animal plucks a leaf from nearby vegetation and places it between the lips.
3. Rising. Just before the climax the gorilla rises onto its hind legs and stands bipedally.
4. Throwing. As the animal rises, it often grabs a handful of vegetation and throws it into the air.
5. Chest beating. The chest beat occurs at the climax of the display, and usually consists of a rapid, alternate slapping of the chest.
6. Leg kicking. While beating its chest, the gorilla may kick one leg into the air.
7. Running. Immediately after the climax and sometimes during the latter part of it, the gorilla runs sideways for several feet.
8. Slapping and tearing. While running or immediately afterwards, the animal slaps the vegetation and tears off branches by hand.
9. Ground thumping. The final gesture in the sequence is a single loud thump of the ground with the palm of the hand.

This impressive display may require as long as thirty seconds to complete, although all but the first two acts follow each other in one continuous, violent motion which is usually finished in five seconds or less. A separate account of each act as it occurs in the sequence or alone serves to illustrate the existing variation.

1. *Hooting* At the start of the display the gorilla may be sitting, lying, or standing, either on the ground or in a tree. While hooting, the head of the animal is characteristically tipped up slightly, or, sometimes when sitting, far up like a howling dog; the lips are pursed, and the hu-hu emitted through parted lips. The number of hoots before the climax varies. Only about 10 per cent of the displays by silverbacked males which reached the climax were preceded by hoots. Occasionally a single hoot is given; sometimes about five calls

follow each other about one second apart. However, a complete display includes a series of some ten to forty distinct hoots gradually fusing into a slurred growling sound at the climax of the display (Fig. 58A). One silverbacked male took twenty-two seconds to emit thirty-five distinct hoots.

Sometimes the male begins to hoot only to stop, look around, and begin again. Such false starts may be repeated four or five times. On a few occasions the pitch of the "hu" apparently rose too high at the beginning of the display, for the animal either lowered the tone perceptibly or ceased vocalizing.

Apparently males find it difficult to reach the climax if distracted or if the rhythm of their hooting is broken. One silverbacked male, for example, was interrupted by the chest beat of a blackbacked male immediately behind him. The silverbacked male stopped vocalizing, turned and looked, then began to hoot again.

In group IV two or more silverbacked males occasionally vocalized together. One began hooting only to trail away to nothing before trying again. Then another joined in, and a third one. Their clear hu-hu rose and fell as each stopped and started independently. But when one reached the climax and beat his chest the others followed. Then they usually settled down for a few minutes before repeating the whole procedure.

Hooting is usually a preliminary to several vigorous actions of the display, and it seems to be recognized as such by the other members of the group. At the first hu-hu the animals near the performer move back. Infants apparently have to learn to avoid a displaying male: the youngest animal which fled the vicinity of a male was about ten months old.

2. *"Symbolic feeding"* About 5 to 10 per cent of the chest-beating displays by silverbacked males which reach the climax are preceded by "symbolic feeding." The gorilla sits or stands, reaches out, very gently plucks a leaf or herb between thumb and index finger, and places it between his lips (Pl. 28). Sometimes it pulls a leaf from the plant with a quick backward jerk of the head, using the lips directly. The leaf, herb, or branch is often held between the lips until or immediately after the chest-beating climax, although occasionally the animal drops the vegetation right after plucking. Once a blackbacked male placed crumbly lichens between his lips with the result that he spit and wiped the material from his mouth for over one minute after the display.

The amount of vegetation which an animal stuffs into its mouth during the display is highly variable, ranging from a single leaf or twig to several consecutive handfuls of leaves and vines, not necessarily belonging to a food species. When such "symbolic feeding" occurs, members of the group close to the displaying male seem to recognize the gesture and generally move away from him, thus escaping the subsequent violent parts of the display.

3. *Rising* Most gorillas rise up on their hind legs just preceding the chest beat and remain bipedal for several seconds if the intensity of the display is sufficiently high. Although the position of the body is upright, it is not as erect as is usually portrayed in the stuffed specimens of museums. The legs remain quite bent at the knees and the head and torso are hunched forward (Pl. 29).

If no other acts in the display sequence precede the rising, the gorillas often indicate by a sudden restlessness that they are about to rear up and subsequently beat their chests. They stuff food into the mouth, they lower the head only to jerk it up or sideways, and they tentatively reach out to grab an herb only to release it.

4. *Throwing* Gorillas commonly throw vegetation during the display. The animal may throw while sitting or standing; it may throw while on the ground or in trees. Throwing takes several forms: (a) an underhand motion in which the gorilla tears off the vegetation and propels it upward with one continuous swing; (b) a sideways motion in which the animal first tears off a branch or herb and then throws it forward or sideways with the arm straight or flexed at the elbow; (c) an outward flick of the wrist or lower arm in which the vegetation is first snapped off and then thrown; (d) a simple dropping of an object in which the animal detaches and releases a branch, usually without propulsion.

Slight modifications of these throwing patterns occur. A black-backed male, for example, picked up a dead branch while standing quadrupedally, and threw it backwards and upwards so that it landed on his back. Sometimes an animal retained vegetation in its hand while beating its chest, with the result that leaves flew in all directions during the rapid hand motions.

Occasionally gorillas throw with seeming deliberation. First the animal looks around for an herb or branch, detaches it, usually with only a twist of the wrist but sometimes using considerable effort when a branch is involved, and flings it away, following the flight of the object with its eyes. One male looked intently to one side, then at a branch above him as if debating whether or not to break it off, but finally rose and beat his chest. Another male lightly pulled at a large branch with one hand as if to test its strength, and then slapped his chest instead.

Sometimes an animal, quite aside from other parts of the chest-beating display, leans over and bites off a twig or herb with a jerk of the head. The vegetation may then be either dropped or thrown. Simultaneous biting into a branch and breaking it by hand also occurs.

On occasion gorillas go through sham throwing motions such as an upward jerk of the arm or a flick of the wrist without propelling anything.

Thrown vegetation rarely flies farther than about 20 feet. The direction of flight depends primarily on the motion of the arm. A sideways flick of the wrist may, for example, propel a branch straight up, or to one side. Although vegetation frequently flew in my general direction, this appeared to happen inadvertently.

5. *Chest beating* The climax of the sequence is the chest beat, which is the most frequently heard and seen part of the gorilla's display. The animal characteristically stands on its hind legs, raises the bent arms laterally, and rapidly beats the chest over the lower edge of the *pectoralis major* muscle with open, slightly cupped hands (Pl. 29).

Chest beating can occur in almost any body position and the only requisite is that at least one hand is free to do the slapping. The display is seen in animals standing bipedally or on three legs, in sitting ones, and in those lying on their backs or sides. And one juvenile hung from a branch by one arm while beating its chest with the other.

The hands are almost invariably held flat while beating the chest; that is, with fingers extended and the palm often slightly cupped. The animal tends to hold its hands within six inches or less of the chest, and the alternate beats are rapid and direct. The fingers are often spread in females and young animals, but those of adult males tend to touch each other.

The sound produced when a silverbacked or large blackbacked male beats his chest may be described as a hollow "pok-pok-pok" somewhat resembling the noise produced by rapping an empty gourd with sticks. Under favorable conditions the sound carries for as much as a mile. In silverbacked males the prominent air sacs presumably act as resonators, for their sudden inflation on each side of the throat is sometimes readily apparent before the chest beat. Small blackbacked males, females, and youngsters produce a mere slapping sound when beating the chest.

The six sound spectrograms of chest beats by males and females or juveniles in Figures 61 and 62 show that the individual slaps in each chest-beating display follow each other in rapid succession about .1 seconds apart. The series of beats tends to be evenly spaced, although considerable variation is at times evident as in Figure 61C, in which from .1 to .2 seconds elapse between successive beats. The greatest intensity of the sound falls into a frequency range of less than 1,000 cycles per second, but overtones of some slaps reach 1,800 cycles per second.

Between March and September, 1960, I attempted to count the number of times an animal slapped itself during the chest-beating display by watching its moving hands and at the same time listening to the number of sounds produced. The number of slaps in the majority of chest beats did not exceed eight, making counts relatively easy. However, in displays where the chest was rapidly slapped more than 8 times, I usually was not able to record the number of beats precisely and hence such data are not shown in Table 48. This table lists the number of slaps in 255 displays. Chest beats in which individual beats exceeded eight were taken from sound spectrograms. A captive adult male belonging to Cordier slapped his chest seventeen times during one display, and I have several times seen youngsters beat themselves twenty or more times in seemingly playful situations. Thus, it appears that most chest beats consist of ten or fewer slaps but that as many as twenty may be given. There was no apparent difference in the number of beats in the various age and sex classes, nor was there any consistent difference between animals which were habituated to the observer and those which were not.

At the climax of the display, gorillas may beat not only their chests but also other parts of the body, and even other objects, in a variety of ways. The animals frequently slap their abdomen. One blackbacked male beat the outsides of his thighs, a method also commonly employed by a male in the Columbus zoo. Females beat the backs of their infants while holding them to the chest or while they are sitting in front of them. Branches and logs are often slapped either with one or both hands.

Animals frequently slap themselves with one hand only. Some infants and one blackbacked male beat their chests with greatly exaggerated gestures, extending their arms considerably after each slap. Two females exhibited a circular motion of the arms instead of direct alternate slaps. They rapidly moved the lower arms and hands close to the chest in a clockwise or counterclockwise direction and struck their breasts in passing. Burbridge (1928) observed a gorilla "play a tattoo upon his chin, causing his teeth to collide in a rattle." No comparable behavior was seen during this study. Several observers (Barns, 1922; Akeley, 1923) maintained that gorillas beat their chests with fists. This was exceedingly rare in the gorillas at Kabara. I saw it only two times in roughly 3,000 observations on chest beating, once in a female and once in a juvenile.

Burbridge (1928) wrote that gorillas "slap each drawn cheek with open palms"; Zenker (in Yerkes and Yerkes, 1929), Baumgartel (1960) and others also assert that they do this. I have never seen such cheek slapping, nor have Cordier or Rahm (pers. comms.) after considerable gorilla experience. This suggests that the behavior is rare at best. It is possible, in fact, that the similarity of the sound of chest beating to the slapping of hollow cheeks may have led to erroneous interpretations of the method of sound production.

In two instances gorillas beat unusual parts of the body, seemingly in play. One juvenile patted the top of its head alternately about thirty times; and a blackbacked male lay on his back with legs stretched toward the sky, slapping the soles of his feet.

Quite frequently an animal moves its hands in the chest-beating pattern without actually striking anything. Although the hands are sometimes raised to the chest, they often lie in the lap or hang at the sides with only an alternate twitching of the fingers indicating the pattern.

Gorillas sometimes begin to beat their chests only to cease suddenly. I have several times observed an animal which raised one or both hands as if to strike the chest, look around, and then lower them. This occurred also after the chest had been slapped once or twice.

Like hooting, chest beats frequently elicit the same display in other gorillas. Sometimes one or two animals wait as the male goes through the preliminary displays, then rise, and all beat their chests in unison at the climax. Chest beats by a male in one group occasionally stimulated a male in another group some distance away to behave similarly.

At the completion of the chest beat, gorillas frequently remained standing for ten or more seconds to watch me as though waiting for a response on my part. Sometimes a gorilla popped from the vegetation, displayed, and ducked back behind the screen of brush, after which it peered at me intently. Blackbacked males and juveniles occasionally displayed with great abandon, then sat quietly, and looked all around as if to judge the effectiveness of the behavior.

6. *Leg kicking* Gorillas often throw one leg into the air while beating their chests. Usually the animal centers the weight of its body on one hind limb and kicks the other one sideways with the knee bent. Only rarely is the leg brought forward unless it is sharply

bent at the knee. Sometimes the toes grab some herbs, which are thrown into the air. The leg is usually lowered gently to the ground, but occasionally the animal stamps it down forcefully, producing a dull thud. The tendency to kick the leg, which appears in approximately one-third of the standing chest-beating displays, is also evident in sitting animals on occasion. While beating the chest, they push one leg to the side, or merely flex it slightly. In a few instances an animal in a standing display flexed both knees and actually jumped several inches into the air.

7. *Running* Immediately after and sometimes during the chest beat the animal tends to run bipedally sideways anywhere from 5 to about 20 feet before dropping to a quadrupedal position and terminating its dash by checking momentum and by slapping at the vegetation.

If the gorilla displays on a slight elevation such as a stump or log, it may dive forward, with arms extended over the head or in front of the body, landing with a crash on the ground below. On a steep slope the fall may be 5 feet or more, and, judging by the forcefully expelled air through mouth and anus, the animal hits with a considerable jar to its body. It usually lands with the hind legs slightly before the front ones, and then continues its run forward for several steps in a quadrupedal position.

8. *Slapping and tearing* Gorillas tend to slap, break, and pull at any vegetation that grows within reach while in the running stage of their display. This may take several forms:

a) The animal may run and sweep its arm through the vegetation, or swat at the weeds with an overhand motion. Vegetation is sometimes slapped or batted sideways by sitting animals entirely apart from the sequence of a full display.

The overhand swipe with extended arm is one of the most characteristic gestures during the run. In adult males it may be dangerous, not only because of its violence, but also because of its unselectiveness. Any member of the group standing near the male is likely to be hit. One juvenile was picked up by a male and bowled down the slope. Another male reached over with both hands, pulled in a female by the hips, and swatted her on the back.

Man in the path of a running male gorilla apparently is also treated to the overhand swipe. Two quotes from Merfield and Miller (1956) illustrate this: "He opened his mouth, showing long yellow, canine teeth, and let out the most terrible scream I had ever heard as he brought his arms crashing downward, sweeping the canes with them, and striking me heavily on my left side . . . ;" and, "N'Denge was holding his gun loosely pointing downward, and was looking toward me, when a big male gorilla suddenly crashed out of the bush and swept him aside with a terrible blow full in the face."

b) A running gorilla frequently grabs a sapling or log near its path. The momentum may carry the animal in an arc around the tree.

c) Running or sitting animals may break branches growing above them. A gorilla sometimes reaches above its head and with a violent downward pull rips a branch from the tree. Once a large silver-backed male jumped up and grabbed a branch which, however, failed to break. The male hung by his arms and looked around—a somewhat incongruous anticlimax to the preceding forceful display.

Animals in trees frequently pull at branches or vigorously shake them in a manner suggestive of display.

9. *Ground thumping* The final gesture in the complete display, and one fairly often heard or seen by itself, is a single, vigorous ground thump. The gorilla stands bipedally or on three legs with one or both hands raised to face height or above the head before slamming them down onto the ground with a dull thud. Once a silver-backed male clasped both hands over one shoulder in the manner of a winning prize fighter and then slapped them down. The animal usually slaps the ground with flat palms or rarely the back of the hand.

The overhand motion may be carried into a sweeping movement which causes vegetation and dirt to fly. A loud panting sound sometimes accompanies the ground thumping in males.

Combinations of Individual Acts in the Sequence With the exception of the leg kick, all acts in the full chest-beating sequence occurred also by themselves, although many of them, such as the bipedal posture, running, and slapping could not always be classed as displays under such conditions. By far the most frequently heard and seen display is the chest beat followed by slapping, throwing, running, ground thumping, hooting, and "symbolic feeding" in approximately that order.

Numerous combinations of the nine acts exist, with some tending to precede or follow one another more frequently than others. The bipedal posture and the chest beat are seen during almost every encounter with gorillas, preceded sometimes either by "symbolic feeding" or by hooting. The latter two are readily interchanged as the first act in the sequence. Chest beating is often followed by slapping or ground thumping. Other frequent combinations are: hooting followed by ground thumping; rising and chest beating followed by running; "symbolic feeding" and running followed by chest beating and slapping; throwing followed by ground thumping; hooting and thumping followed by rising and chest beating. Hooting and "symbolic feeding" never occurred after the climax of the display.

The frequency with which gorillas exhibit the various combinations of the display depends on their sex, on the intensity of the display, and on individual variation. I have only seen silverbacked males give the complete sequence of nine acts, and even with them I have observed it only about a dozen times. Silverbacked males also tend to average more individual acts in each display than the other members of the group.

Individual Variation in the Chest-beating Display Some individual variation occurs in the form and frequency with which the various displays of the sequence are given. Thus, the silverbacked male in group VII tended to climb trees and shake the branches. Only two females, one in group IV, the other in group VIII, repeatedly gave the rotating chest beat. One of the silverbacked males in group IV often leaped wildly down slopes, slapping the ground as he landed. A young blackbacked male in group VI kicked his leg higher and more often than other males; another blackbacked male, in group VII, frequently exhibited the exaggerated chest beat usually seen in youngsters.

The number of slaps in each chest-beating display also appeared to show some variation among individual males (Table 49). The general level of excitability and the extent to which the animal had been habituated to my presence appeared to have had little effect on the average number of slaps per chest beat. For example, the male in group VI was generally more excitable than the male in group VII, yet his average number of beats is similar. The male in group IV, who showed the lowest mean number, was new to the group and not habituated to me.

Responses to Thrown Objects When animals display, thrown or slapped vegetation sometimes flies inadvertently in the direction of another group member. The response of gorillas to such flying objects is similar to that of man: they flinch, duck, or cover the head or face with hand or arm. When a female flipped a branch toward the face of a juvenile, it held one hand with spread fingers in front of its nose. Once while eating lunch, I flicked a piece of paper at a blackbacked male who sat within 15 feet of me and he jerked his arm above his head. Similar responses can also be elicited by making incipient throwing motions. When, for example, I suddenly raised my camera to photograph a blackbacked male, the animal lifted its bent arm so that both the fore and upper arm covered the side of the head. Geddes (1955) observed that a captured adult male jerked his arm across his face when a revolver was raised.

ONTOGENY OF THE CHEST-BEATING DISPLAY SEQUENCE Infants displayed various acts of the sequence at an early age. When only four to four and a half months old they rose shakily and very briefly onto their hind legs and beat their chests. They also swatted at vegetation. Small infants never reacted to my presence by displaying; the earliest age at which an infant reacted to me by beating its chest was one and a quarter years. Infants were first seen to place a leaf between the lips prior to chest beating at the age of one and a half years, and I also first noted throwing at that age. Juveniles exhibited all acts in the display sequence except the hooting and running (Table 50).

I heard hooting in its full form only in adult males, never in females. Although the latter occasionally emitted a panting sound while beating the chest, this bore little resemblance to the clear hu-hu of males. Small blackbacked males never hooted. However, they exhibited the first rudiments of the sound when before a chest beat they sometimes gave a single squeaky note. The sideways run likewise reached its fullest expression in adult males; I saw it only twice in females.

Although all parts of the display sequence, except the hooting and possibly the kick, were given by females as well as males, displays in females were less frequent, less intense, and of shorter duration than those of males. One female placed a leaf between her lips, rose, stepped sideways while beating her chest, and finally slapped the vegetation—an unusually complete performance for a female. The threshold of excitation also appeared to be higher in females than in males, which is perhaps related functionally to the male's being the protector of the group.

Field observation alone cannot determine the innate, maturational, and learned components of each display. Chest beating and attempts to stand bipedally appear at an age when co-ordinated movements and bodily balance are still rudimentary. The infant lowland gorilla in the Columbus zoo beat its chest without the opportunity of having learned the act from other gorillas. Goma, an infant born at the Basel zoo, beat objects when she was only four months of age. Lang and Schenkel (1961*a*) suggested that in gorillas the motor pattern for chest beating is inborn, a view with which I concur. Hooting seems to be under maturational control, appearing in males at about puberty. The placing of a leaf between the lips in small infants occurs so spontaneously and in such characteristic form that the learned components of the behavior, if any, are not discernible. On this limited evidence I suggest that chest beating, "symbolic feeding," and perhaps other acts in the display sequence are basically innate.

SITUATIONS ELICITING CHEST-BEATING DISPLAYS Six situations were observed to elicit all or part of the chest-beating display.

1. *The presence of man* The most intense, prolonged, and diverse displays are exhibited by gorillas unhabituated to the presence of man. Detection of man is usually visual, but at times also auditory or olfactory. All members of the group, except the younger infants, display, but the silverbacked males do so with greater frequency and intensity than the others.

2. *The presence of another gorilla group or a lone male* If one group hears the distant hooting, ground thumping, or chest beating of another group, it either ignores the sounds or responds by displaying. Frequently the silverbacked males are the only members of the group to react at such times. Individual differences in the readiness to respond are readily observable. One peripheral silverbacked male in group IV, for example, beat his chest several times during the night whenever a lone male displayed uphill, but the other three silverbacked males in the group did not react.

When seeing each other, two groups commonly display by beating the chest, thumping the ground, and slapping the vegetation. Most displays are given by the males, although some chest beating occurs also in females and juveniles.

3. *In response to an undetermined disturbance* Gorillas which note fleeting glimpses of moving objects such as a partially hidden observer, become uneasy until they have determined the nature of the disturbance. The whole group sometimes approaches, with several animals beating their chests, throwing herbs, thumping the ground, and slapping the vegetation.

4. *Displays by another member of the group* Chest beating and other displays are contagious in that such behavior in one animal sometimes induces another to act likewise. This is especially prominent among males. Females, blackbacked males, and juveniles frequently display simultaneously and in unison with the silverbacked male.

5. *Play* Playing infants rise to beat their chests, they slap at vegetation, and they sometimes place a leaf between the lips.

6. *Without apparent outside stimulus* Gorillas occasionally start seemingly spontaneous movements of the fingers and hands in the pattern of the chest-beating display while sitting or lying quietly.

FUNCTIONS AND CAUSATION OF THE CHEST-BEATING DISPLAY The hooting, the bipedal stance, the vegetation held in the mouth and thrown about, the kicking, the running and breaking of vegetation—all are actions which serve to make the animal conspicuous. It seems to be advertising its presence or "showing off." Human response tendencies and those of other gorillas suggest that the display serves to repel intruders by intimidation.

The displays also appear to have secondary communicatory value both within the group and between groups. Group members become alert if the dominant male beats his chest without obvious reason, and they become aware that another group or a lone male is in the vicinity upon hearing displays in the distance.

Yet intimidation and communication alone do not explain the causes of display in several other situations. Why, for example, does it occur prominently in play, and why do gorillas beat their chests when there is nothing obvious to intimidate or communicate? Yerkes (1927) "gathered the impression that chest-beating indicates impatience or other mild dissatisfaction, sometimes lonesomeness or slight irritation, and that it may be done to attract attention or to startle or intimidate the observer."

The most general emotional term which encompasses all the diverse manifestations of the display is excitement. Gorillas are excited in the presence of man, at the visual or auditory proximity of another group, and during play. When displaying, the animals find release for the tension which has accumulated in their system in an excitable situation. Thus, the primary causation of the chest-beating sequence appears to be the build-up of tension (excitement) above a certain threshold. After the display, the level of excitement temporarily drops below the threshold, and the animals behave calmly until a new accumulation of tension erupts in display.

ETHOLOGICAL INTERPRETATION OF THE DISPLAY SEQUENCE Following the concept of causation as developed by Moynihan (1955) and Hinde and Tinbergen (1958), a potentially dangerous situation, such as a meeting with a human being, arouses two conflicting impulses in the gorilla. On the one hand, the animal exhibits a tendency toward flight, on the other aggressiveness, with the result that it neither flees nor approaches. The conflicting tendencies of "flight or fight" generate tension in the animal. This tension finds release in some functionally inappropriate but fairly stationary act like throwing, beating something, and jumping, or in a more restrained act like "symbolic feeding." Thus, in a situation which arouses conflicting tendencies, the animal expresses itself by some behavior which is not actually relevant to the situation at hand. Such behavior is termed a "displacement activity."

Several of these displacement activities have apparently become quite stereotyped in the gorilla; they have become incorporated into a definite display, and in this development have achieved secondary functions such as intimidation. In the ethological terminology they have become ritualized.

With the gorilla I have found it difficult to distinguish between behavior which is simply displacement and that which has become ritualized. However, two of the acts in the display sequence retain displacement and ritualized elements that are readily discernible. From an early age there is a noticeable tendency for gorillas to beat something with the hands, and animals often slap themselves or an object in situations where there is no apparent occasion for intimidation or communication. At such times chest beating appears to be a displacement activity. However, in response to the presence of man at least, the chest beating acquires intimidating functions and may be considered a ritualized display.

In the "symbolic feeding" portion of the full display, displacement and ritualized elements can also be seen. At their first sight of me, males occasionally began to feed very intensively for a minute or so. This appears to be displacement feeding. Sometimes a male stuffs three or four handfuls of vegetation into his mouth before rising to beat his chest. At other times, the vines or herbs are not pushed into the mouth, but gently placed between the lips, the amount varying from a big handful to a single leaf. These latter acts appear to be almost entirely ritualized.

Although gorillas show ritualization of several displacement activities, the displays are not as stereotyped and inflexible as those often found in birds. However, the gorilla's chest-beating sequence is one of the most complex ritualized displays among mammals.

Comparison with Displays in Other Apes and Man Various aspects of the chest-beating display sequence are present in the gibbon, orang-utan, chimpanzee, and man, although the specificity is sometimes lacking. For example, wild gibbons in Sarawak hooted several times before the climax, which was marked by sounds of very high pitch, bipedal running, and a final rapid swinging through the trees. The similarity to hooting, rising, and running in gorillas is obvious. Orang-utans broke off and threw branches when excited (Schaller, 1961). Chimpanzees in Uganda shook branches and beat the edge of the nest with both hands in response to my presence. A large aggregation of chimpanzees in the Budongo Forest in Uganda hooted at first slowly, then increased the tempo and loudness of the vocalization until at the climax they screamed, slapped branches, and beat the hollow buttresses of ironwood trees (*Cynometra*) with their hands to produce a loud drumming sound. In zoological gardens I have observed chimpanzees throw objects, slap floors, walls, and themselves, and stamp their feet. Emlen saw a chimpanzee beat its chest in the zoo at Elisabethville, Congo. Chimpanzees, therefore, exhibit most of the displays noted in gorillas.

Man behaves remarkably like a chimpanzee or a gorilla in conflicting situations. Sporting events are ideal locations for watching the behavior of man when he is generally excited and emotionally off guard. A spectator at a sporting event perceives actions which excite him. Yet he cannot participate in them directly, nor does he want to cease observing them. The tension thus produced finds release in chanting, clapping of hands, stamping of feet, jumping up and down, and the throwing of objects. This behavior is sometimes guided into a pattern by the efforts of cheerleaders who, by repeating similar sounds over and over again, channel the displays into a violent but synchronized climax. The intermittent nature of such behavior, the transfer of excitement from one individual to the next, and other similarities with the displays of gorillas are readily apparent.

Other situations in the daily life of man likewise show similarities to the behavior of displaying gorillas. Dictators know well the usefulness of repeated slogans and cheers to excite a crowd, and native peoples chant and beat the same simple tune on a drum over and over again to produce the desired climax. Marital squabbles and similar human interactions, where neither person cares to attack or retreat, may end with objects being thrown, doors being slammed, furniture being kicked—all means of reducing tension.

The Strutting Walk

Gorillas both in the wild and in captivity occasionally exhibit a very rigid walk. In the wild, the animal usually chose an open area or a fallen log within 80 feet of me. It then walked at right angles to me, displaying the side of its body. The arms were bent outward at the elbow, giving them a curious curved appearance and making the long hair on the forearm look impressive (Pl. 30). The body itself was held very stiff and erect, the steps were short and abrupt, and, except for brief glances, the head was turned slightly away from me. However, the animal watched me out of the corner of its eye. The display never exceeded fifteen seconds in time and the basic behavior from infant through silverbacked male was similar. If the display was directed toward another gorilla, the two animals were closer together, within 10 feet of each other, but the display remained the same. The gorillas uttered no vocalizations and they were not obviously excited.

I saw the strutting walk infrequently in the wild. Infants, one and a half to two and a half years old, strutted several times alone on a log. Twice blackbacked males exhibited the walk in front of a female, seemingly in a somewhat playful manner. About ten other displays were directed at me, twice by a juvenile, once by a female in very rudimentary fashion, about five times by blackbacked males, and twice by a silverbacked male.

In captivity the strutting walk is commonly seen. A two-year-old male and a one-and-a-half-year-old female, which belonged to Cordier, both exhibited the display in response to my presence. The adult male in the Columbus zoo strutted in front of the female before copulation (see section on copulation). Benchley (n.d.) observed the display in two adult male mountain gorillas in the San Diego zoo. In the London zoo, the male "Guy" strutted just before he sloshed water at his audience and dashed about his pen. Sambo, a male at the Milwaukee zoo, displayed immediately before running rapidly across his cage. Although the strutting walk followed by a run is frequently seen in captive animals, I never observed this behavioral sequence in the wild.

The strutting walk is a display which makes the animal appear big and powerful; it is apparently a form of "showing off."

CHAPTER 6

SOCIAL BEHAVIOR

Leadership

The focal point of each group is the leader, who is, without exception, the dominant silverbacked male. The entire daily routine—the time of rising, the direction and distance of travel, the location and duration of rest periods, and finally the time of nest building—is largely determined by the leader. Every independent animal in the group, except occasionally subordinate males, appears to be constantly aware of the location and activity of the leader either directly or through the behavior of animals in his vicinity. Cues reflecting a changed pattern of activity are rapidly transmitted through the group and the subsequent behavior of the members is patterned after that of the leader. This response insures cohesiveness and co-ordination of action. The complete adherence of gorillas to one leader must be considered in all discussions of groups, for variations in behavior between groups tend to reflect individual idiosyncrasies of the leader.

The leader usually communicates his readiness to move on by a simple but characteristic gesture. He rises and without hesitation walks rather stiff-legged and rapidly in a certain direction. This gait differs markedly from the leisurely pace employed when merely ambling a few steps. By watching the leader I was able to predict a group movement as rapidly as the gorillas. The response of the animals to the leader is most clearly shown in the rest area. Whenever the leader rises to change his position, at least one animal turns its head to watch. If the leader rises and walks slowly several feet, the others usually do not respond. But if he indicates his readiness to leave, at least one animal rises to join him within half a minute, and others do likewise until the whole group has become alerted. In general, the speed of response to the leader depends on the abruptness of his actions. When he moves slowly out to feed, some gorillas remain occasionally in the rest area for five or more minutes, but when he departs suddenly, the animals generally run to his side.

Sometimes the leader employs a characteristic posture which apparently serves as a signal to the other members of the group, indicating his imminent departure. He faces in a certain direction and stands motionless for as long as ten seconds with front and hind legs spread farther from each other than usual (Pl. 31).

On occasion, the male emits a series of two to eight rather short but forceful grunts, either as he stands or within 200 feet after having moved from a site. This vocalization seems to signify "Here I am, follow me," for the members of the group respond by moving in his direction. Once, the leader gave this sound while the rest of the group watched me from the branches of trees. Although he was out of sight behind a curtain of vines, all except one animal descended and joined him, only to return into the trees when the leader did not move away.

When the animals move rapidly in single or double file, the dominant male is usually in the lead position as the following procession in group VII illustrates: silverbacked male, female and three-month-old infant, female and nine-month-old infant, juvenile, juvenile, blackbacked male, female and seven-month-old infant, two-and-a-half-year-old infant, female and one-and-a-quarter-year-old infant, two females and two juveniles together, female and nine-month-old infant, blackbacked male. However, when the general movement is slow, the leader may travel in the center or near the end of the group. Often when the leader has started to move in a certain direction, the other members surge past him and occupy the head of the procession as illustrated in the following incident:

The silverbacked male in group VIII feeds slowly as the other members of the group move by him and continue on ahead. The leader then rejoins the group at a rapid walk. As he passes through the middle of the group, several gorillas anticipate his direction and hurry ahead with the result that he remains in the middle of the group.

In groups containing several silverbacked males, the females and subadults usually react only to the leader. Subordinate males tend to stay with the group, but they follow the leader less directly as they occasionally wander in the vicinity of the group or remain behind only to rejoin the others a few minutes later.

Subordinate silverbacked males sometimes assume leadership of a few animals for brief periods. Two juveniles apparently joined two silverbacked males in leaving group V for several days. Once, the number two male in the hierarchy of group V moved about 300 feet from the main body of the group followed by a blackbacked male and two juveniles, but they all rejoined the others within half an hour. When group IV split into two subgroups for several hours, the number two male in the hierarchy assumed leadership of one subgroup. However, when the two subgroups rejoined, those females which had previously followed the subordinate male did not respond to him in the presence of the leader.

In times of danger any animal may assume temporary leadership. A juvenile, on seeing me, ran screaming down the slope followed by the whole group which was unaware of the nature of the danger. One female in group VI tended to run at the first sight of me and by her action elicited similar behavior several times from all or part of the group, including the leader. A female shot by Akeley (1923) was followed by several others as she rolled down into a ravine.

We observed several times that at Kisoro the rest of the group fled while the leader remained behind to display at the human intruders. We could not determine if a certain individual assumed leadership of the retreating animals in the absence of the displaying male.

Internal Organization of Gorilla Groups

Several investigators have noted that certain primate groups possess a definite organization and spatial distribution of age and sex classes. Thus, in *Macaca fuscata* the core of the group consists of certain males and females surrounded first by a ring of peripheral females and finally by a ring of peripheral males (Imanishi, 1960). In moving baboon groups the less dominant males travel in front, followed by females with juveniles, dominant males and females with infants, and finally the most subordinate males (Washburn and DeVore, 1961).

In gorilla groups which contain one or more blackbacked males or two or more silverbacked males there is a tendency for the extra males to assume peripheral positions. Frequently one of these males travels last in line when the group moves. Thus the spatial organization of some gorilla groups consists of a central core composed of all females and young as well as the dominant male, and surrounded by a varying number of peripheral males.

Zuckerman (1932) and Haddow (1952) suggested that the monkey groups studied by them split into small family parties for the night. Nothing comparable was observed in gorillas, whose spatial organization at night tended to resemble the one found during the day.

Field studies on several primate species have shown that certain individuals within a group may associate more closely than others. Clusters comprising a male, a female, and a young are frequent in baboons (Washburn and DeVore, 1961) and in rhesus monkeys (Carpenter, 1942). The same authors also mention more or less permanent play groups among the juveniles, and close association between a female with infant and a juvenile. Consort relationships of males and females have been described for howler monkeys (Altmann, 1959) and rhesus monkeys (Carpenter, 1942).

In gorillas persistent aggregations of specific individuals within the group are infrequent. Males and females never consorted very closely for more than the duration of a rest period, and play groups lasted only for a similar length of time. Once two females remained near each other for several days, the only such instance noted. A close association between a female with infant and a juvenile was usually intermittent, rarely lasting for more than two hours but occasionally as long as one night. A detailed discussion of the interactions is found in the following section on intragroup relations.

Interactions of Individuals within the Group

Each member of a group belongs to a definite age and sex class which tends to have a certain status and function in the group as a whole, and which to a large extent determines the interactions between individuals.

The small number of direct social interactions was to me the most striking aspect of intragroup behavior. Except for mother-infant relations, the most frequently noted interactions were dominance, mutual grooming, and play, each treated separately on the following pages. But even these were relatively uncommon. In 466 hours of direct observation, most of it during rest periods when such activities were most prominent, I saw dominance only .23 times per hour of observation, mutual grooming .28 times, and social play .11 times.

Another prominent feature of intragroup relations was the marked individuality in the behavior of the animals. An observer soon realizes that he is not only studying gorillas as a group or as a specific age and sex class, but also as individuals within that group or class.

Dominance

Definite dominance interactions were observed 110 times at Kabara (Table 51). These occurred in five different situations.

1. Dominance was most frequently asserted along narrow trails when one animal claimed the right of way. The subordinate gorilla usually stepped aside, or, where this was not possible, crouched down. One infant suspended itself beneath a branch while a female walked by above. Quotes from my field notes describe the behavior:

a) *A female, carrying a one-month-old infant against her chest, walks down a path. A young blackbacked male, who has been approaching, rapidly turns, walks in the opposite direction, and then leaves the path.*

b) *A female moves down a trail. A juvenile, who sits in her way, moves to one side and lets her pass. The female then sits down on the trail, but moves on when one of the peripheral silverbacked males approaches.*

c) *A blackbacked male walks along a horizontal branch. A juvenile on the branch ducks down, and the male steps over it.*

d) *A silverbacked male sits on a log, a juvenile beside him. He leans over and gives the juvenile a light push with his forearm. The juvenile moves over one foot. Five minutes later, the male rises and faces the juvenile, who ignores the male even when touched lightly with the forearm. The male then suddenly pushes the juvenile sharply, and the juvenile rapidly clambers to one side while the male descends from the log.*

2. A dominant animal frequently asserted its rank by supplanting another animal at a sitting place.

a) *A female with infant climbs toward the crotch of a tree which is hastily vacated by a blackbacked male, who climbs higher while his seat is being taken by the female.*

b) *A juvenile sits under the dry canopy of a leaning tree trunk during a heavy rainstorm. A female walks toward the juvenile who rapidly vacates its seat, while she appropriates the dry spot. Shortly thereafter the silverbacked male arrives and pushes the female with the back of his hand on the lower part of her back until she is out in the rain and he under cover.*

c) *A silverbacked male sits and watches me. A juvenile moves in front of him and squats on a knoll, which obscures his view. First the male looks to one side, then to the other, and finally reaches over to slap the juvenile with the back of his hand on the upper arm. The juvenile descends the knoll and departs.*

d) *A female walks up to a sitting juvenile and touches it lightly with the back of the fingers of her left hand on its upper arm. It moves and she sits.*

e) *The silverbacked male lowest in the rank order of group IV sits within 15 feet of the No. 2 male in the hierarchy. The latter jerks his head toward the former, who then moves 10 feet uphill.*

3. Captive gorillas commonly show dominance behavior in re-

sponse to food. However, in the wild where forage is plentiful, I saw it only once, seemingly in play:

A juvenile appears restless: it lies on a sloping branch and waves its arms, kicks moss off the trunk with one foot, scratches its abdomen vigorously, and rolls its head back and forth. Behind it, an infant, 1½ years old, sits and feeds on Galium. *The juvenile suddenly turns and grabs at the* Galium *in the hands of the infant. The infant jerks back, but on the second try the juvenile pulls the vine away and eats it.*

4. I noted dominance only once in a sexual situation when the dominant male in group IV walked toward a copulating peripheral male, who ceased his activity, but later resumed it (see section on copulation).

5. Twice females behaved dominantly toward other females when protecting their young.

A female sits with her one-month-old infant in her arms. A female without infant ambles up, stops, and, with her face two feet from the infant, looks at it. She reaches out as if to touch the infant, but the mother jerks up one hand and bats her arm away.

From these observations it is apparent that, in general, dominance is asserted with a minimum of actions. The most frequently noted gesture involving bodily contact was a light tap with the back of the fingers or hand against the back or arm of the subordinate animal. Slightly more forceful means involved pushing with both hands, or walking into the animal and pushing it over with the chest.

The relative infrequency of direct dominance interactions is probably attributable to four main causes. (1) Competition for food and mates, two factors which commonly initiate dominance behavior in some mammalian species, provides little basis for strife since forage is abundant and sexual behavior not prominent. (2) The animals are frequently alert to possible encounters, and subordinate members tend to circumvent possible issues by avoiding close contact with animals which assert their rank. (3) Gorillas are not aggressive. (4) The dominant animal frequently does not assert its rank, as one example illustrates:

Two females sit as the male walks up, grunting softly. When the male stops, one female drapes an arm over his rump and a juvenile suddenly takes a swipe at his leg but misses. One of the females, with a one-month-old infant, moves ahead of the male and he follows slowly. She requires about 10 seconds to climb up on a fallen log, but the male just waits until she has moved before he also ascends the log.

All members appeared to occupy certain positions in the dominance hierarchy of the group. Silverbacked males were dominant over all blackbacked males, females, and young. Therefore, in groups with only one silverbacked male this animal was dominant over all group members. If more than one silverbacked male was present in a group, these males appeared to have a linear dominance hierarchy. Although groups IV and V contained two or more silverbacked males, the dominant animal was easily recognizable, being the focal point of all females and young. Peripheral males appeared to be subordinate to males which remained in the center of the group. Two lone males, which joined with group VI at separate times, retreated from the advance of the dominant male. Table 52 illustrates the dominance gradient in group IV for the silverbacked males present in August–September, 1959 (see also Fig. 17). Splitnose was an extremely peripheral silverbacked male who twice moved to another part of the slope when The Outsider walked in his general direction. Since mere avoidance of close proximity to a specific animal was frequently difficult to determine with certainty, I have not included such observations in Table 51. The second and third males in the hierarchy subsequently left the group. A new male, who entered and remained in the group for the duration of the study, assumed the number two position in the rank order, with the two peripheral ones remaining in their lower positions.

The dominance hierarchy in silverbacked males appeared to be based in part on age. The dominant animal was characteristically a fully adult male and in the prime of life. All very young and very old silverbacked males occupied subordinate positions in groups. The rank order and approximate age of the males in two groups was as follows:

Group V (Nov., 1959) Rank	Age
1	middle-aged
2	old
3	young middle-aged
4	young

Group IV (Sept., 1959) Rank	Age
1	middle-aged
2	middle-aged
3	middle-aged
4	young middle-aged
5	young

This suggests that males in the prime of life sometimes replace aging males as the dominants before the latter die.*

Blackbacked males were dominant over all juveniles and infants which had wandered away from their mothers. The relationship between blackbacked males and females varied, being perhaps dependent on size and on personal differences. Females were dominant over blackbacked males four times and the reverse occurred twice.

Females were dominant over all juveniles and infants, either their own or those away from their mothers. I tabulated dominance inter-

* See p. 405, n. 2.

actions between females and infants only if the infant was not sitting or riding directly on its mother at the time. Although females were involved in dominance interactions fairly often, they appeared to lack a stable hierarchy among themselves. Situations which elicited dominance behavior between other age and sex classes frequently brought forth no response among females. Of the twelve female-female interactions noted, four females with young infants were dominant over females with older infants, and six over females with no infants. In one instance the infants were of about equal age. One old female without infant was dominant over a young female without infant. These limited data suggest that females have a changing hierarchy in which females with young infants are dominant over those with older infants or with none. Individual differences in temperament of females perhaps influence their status also.

Dominance among juveniles and infants appeared to be correlated with size. In every interaction between juveniles the dominant animal was somewhat larger than the subordinate one. On eight occasions juveniles were dominant over infants, all of which were one or more years old. A tentative schematic outline of a dominance hierarchy in a typical gorilla group is presented below:

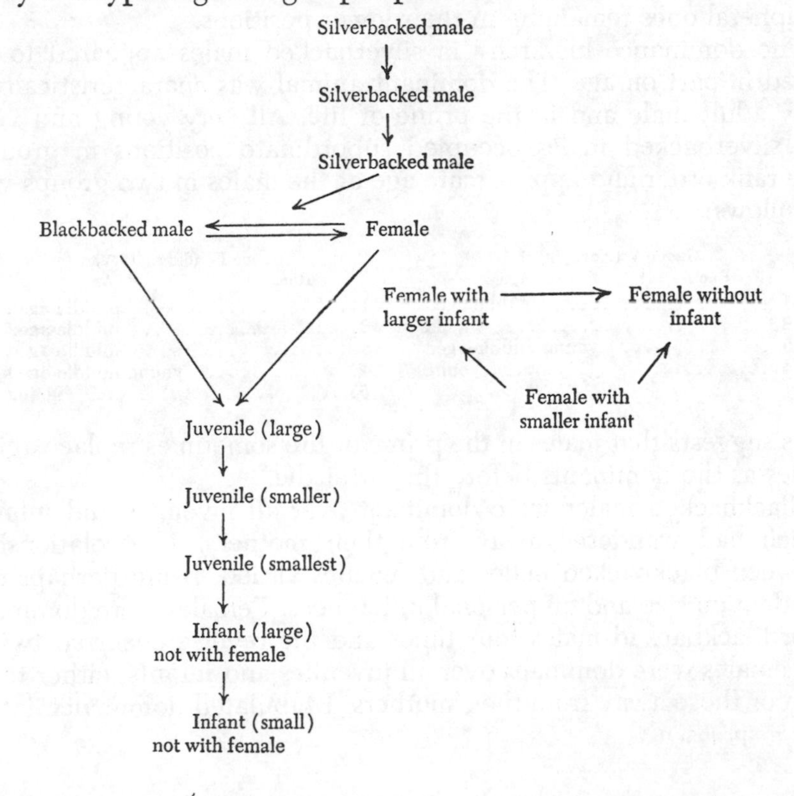

MUTUAL GROOMING

I recorded all instances in which one animal groomed the pelage of another, and the 134 observations are summarized in Table 53. Mutual grooming was not a prominent activity among gorillas, and the behavior was never reciprocal. The duration of mutual grooming was similar to that of self-grooming (see section on self-grooming), usually one minute or less, but sometimes five to ten minutes, and once about fifteen minutes.

GROOMING BETWEEN ADULTS Grooming between adults was infrequent. Gyldenstolpe (in Prince Wilhelm of Sweden, 1923) observed two males help each other with their "morning toilet." I watched one female as she briefly groomed the back of a small blackbacked male. Females groomed each other a total of five times. Although silverbacked males were the leaders and dominant animals in the groups, I never observed females groom them. Thus, the frequency of mutual grooming between adult gorillas differs strikingly from baboons, in which females commonly groom the male (Table 54), but resembles the red-tailed monkeys where grooming among adults is infrequent (Haddow, 1952).

ADULTS GROOMING YOUNG Silverbacked males groomed infants three times, but blackbacked males never did so. Over half of all the observations on mutual grooming were made on females grooming infants, the activity sometimes being prolonged and intent. Young infants were characteristically held in one arm, or in the lap, with the body draped in such a way that the female retained free use of at least one hand. She usually held her face within six inches of the part she was grooming, and, with pursed lips, folded back the hair or removed small objects. Every hairy portion of the infant was groomed, with emphasis on the rectal region (Fig. 57B). Young infants never invited grooming from their mothers; in fact, they occasionally resisted the process as when the female held them upside down. Three observations from my field notes illustrate the method which females employed in grooming infants one year old or less.

1. *A one-year-old sits by its mother. She reaches over and with two hands carefully parts the hair on its arm. After about 10 seconds she suddenly switches to her own arm and grooms that. Then she lifts her infant, turns it upside down, and holds it so that the rump is nearly in her face. She picks around the anal region with one hand, using primarily the index finger.*
2. *A female turns her 3½-month-old infant upside down. Its legs are spread and the anal region is rather bare. With pursed lips the female manipulates the genital area until the ¾-inch flesh-colored penis of*

the infant stands erect which she then places between her lips several times. This was my only observation of this type.

3. *A sitting female holds her 3-month-old close to her face as it lies on its belly supported by one of her arms. With index finger of the free hand she parts the hair against the grain, her pursed mouth nearly touching the skin. In this manner she checks the head, arms, and rump.*

Females groomed infants older than one year similarly but less completely, concentrating much of their activity on the arms, shoulders, back, and rump instead of the anal region (Fig. 57C, E). One example illustrates the behavior:

A two-year-old sits beside its mother. She reaches over, takes her offspring into her arms, and nuzzles its shoulder with her face before grooming its arm briefly. The infant suddenly reaches up and puts both arms around her neck and they sit holding each other closely.

On three occasions females groomed infants other than their own. Twice, the same female inspected the same large infant even though she carried a small one of her own.

Females also groomed juveniles readily, although they were not picked up and inspected in the same manner as the younger infants. Except in two instances, attention was usually given the head, back, and rump. Either the juvenile or the female initiated the grooming activity. The three examples below are typical:

1. *A large juvenile stands beside a female with infant. She reaches over and gives a cursory pick at the hair on the juvenile's back. But something attracts her and she grooms very intently for some 30 seconds, her pursed lips only 6 inches from the hair.*
2. *A juvenile walks by a sitting female with infant. She reaches far over with both hands, pulls in the juvenile, and grooms its back.*
3. *A lone female sits. A juvenile ambles up, thrusts its head into the bend of her elbow, and retains that position while the female grooms its head for about 15 seconds. The juvenile then wanders off.*

JUVENILES AND INFANTS GROOMING ADULTS One juvenile inspected the anal region of a silverbacked male briefly, and another groomed the rump of a blackbacked male. All other observations involved juveniles or infants and females. I saw infants groom their mothers only twice. A one-and-a-quarter-year-old infant groomed the back of its mother; it was the youngest animal in which grooming activity, self or mutual, was observed. Juveniles groomed females nine times, occasionally when the latter carried small infants in their arms. One juvenile very intently parted the hairs on the arm of a female with both hands, then advanced upward until it gently inspected her brow ridge. Another female held her four-month-old infant to her chest when a juvenile approached and sat facing them. It groomed the arm of the female and then the back of the infant.

YOUNG GROOMING YOUNG I observed only one infant, two and a half years old, groom another infant, one and a quarter years old. Juveniles were not groomed by infants, although the reverse occurred quite commonly. One juvenile, for example, held an eight-month-old infant in its lap and groomed its back. The infant toddled off but returned, and the juvenile reached out and pulled it in and continued to groom. On another occasion a juvenile groomed a two-year-old and then both sat holding each other chest to chest. Juveniles also readily groomed each other. Juveniles concentrated their grooming on the head, back, and rump of the other animal; I noted only two instances in which a juvenile groomed the arm of an infant.

INVITATION TO GROOM Infants seemed to make no overt gestures to indicate that they wished to be groomed. Juveniles, however, invited grooming from other juveniles and from females. Females presented themselves only to other females to be groomed except in one instance when a female invited a juvenile to do so.

The desire to be groomed was communicated by simple but obvious gestures, never by any detectable vocalizations. Usually one animal walked up to another and presented a certain part of the body for inspection. I have already given examples illustrating the behavior, but two more are of interest.

1. *Two females sit about 5 feet apart. One reaches far over, taps the other on the upper arm with the back of the hand, rises, and walks over. She turns and backs her rump toward the face of the sitting female, whereupon she is groomed.*
2. *Two juveniles sit 10 feet apart. One rises, walks over and faces the other, then swivels around and backs its rump close to the face of the sitting one. It stands still for several seconds. Nothing happens. But when it looks over its shoulder seemingly to determine the cause for the delay, the other juvenile begins to groom the rump.*

FUNCTIONS OF GROOMING The function of mutual grooming among gorillas seems to vary with the age and sex classes. Adult gorillas, including blackbacked males, rarely groom other adults, and the grooming that does occur is concentrated on those parts of the body which the animal cannot itself reach with ease. This, together with the fact that females do not attempt to groom the dominant male or rarely groom each other, suggests that mutual grooming in gorillas has little or no social function. Mutual grooming in adults appears to be primarily utilitarian, with the animals usually grooming their own bodies (see Table 46).

Infants and juveniles undoubtedly derive benefit from being groomed by females, for the activity rids the skin of foreign matter, imperfections, and in some areas perhaps ectoparasites. However, females sometimes groom old infants and juveniles in an extremely cursory fashion, with the whole behavior on the part of the female suggesting more a desire for physical contact than a concern for the condition of the skin.

Juveniles employ grooming seemingly as a means of initiating social contact with females, especially if the latter carry a small infant. When juveniles groom other juveniles and infants, the activity appears to be in part utilitarian, for juveniles sometimes invite grooming of certain parts of the body from other juveniles. However, much of the grooming which is directed toward infants appears to possess more of a social than a utilitarian function.

Infants rarely groom others or themselves (see Table 46). This is probably correlated with the fact that they are readily groomed by and have continuous social contact with females.

Thus, the functions of mutual grooming in gorillas differ to some extent from other primates such as chimpanzees and baboons. Zuckerman (1932) felt that, in general, grooming was "involved in the maintenance of a social group of sub-human primates." In captive chimpanzees, Yerkes (1943) noted that grooming aids in the ". . . development of mutual interest, sympathy, and cooperation." In baboons "grooming is pleasurable to the individual, it is the most important expression of close social bonds. . . ." (Washburn and DeVore, 1961.) Mutual grooming in gorillas lacks such all-encompassing functions, the activity being primarily utilitarian between adults and only to some extent social between females and young, and young and young.

On at least three occasions mutual grooming appeared to be a displacement activity. Twice a silverbacked male groomed an infant in cursory fashion while watching me. Another time a juvenile inspected a blackbacked male briefly while disturbed by my close proximity.

Play

After the period of intensive morning feeding is over and the animals have settled down to rest, the younger ones sometimes wander away from the mothers to play alone or perhaps to initiate play with others. However, on the whole, gorillas are not playful. Frequently several days went by without my observing a single instance of this behavior, especially when the clouds hung low in the saddle and the vegetation was wet.

I recorded as play any relatively unstereotyped behavior in which an animal was involved in vigorous actions seemingly without definite purpose. Thus, for example, an infant climbing into a tree was not considered to be playing, but one swinging on a vine or hanging upside down on a branch was listed as doing so. I observed ninety-one separate instances of play, spread throughout the day from rising to bedding time, in a total of 156 animals (Table 55). All but five of these animals were juveniles or infants, suggesting that playfulness in free-living gorillas ceases almost entirely by the age of about six years. For convenience I have divided the activity into lone play, involving only one animal, and social play, involving two or more.

About half (43.4%) of all my observations were of animals playing alone. Infants played alone about twice as much as juveniles; in social play the proportion of infants to juveniles was roughly equal. Although a few infants only three and a half to four months old grappled playfully with older infants or juveniles while being held in the arms of their mothers, vigorous play activity began at the age of about four and a half to five months, or as soon as the infants were capable of walking away from their mothers with a fair degree of co-ordination.

Most play involved some form of running, climbing, and wrestling. Lone animals sometimes shredded leaves and slapped the vegetation, but inanimate objects were only twice the focal point of social play. Adults, even though some took the brunt of the youngster's exuberance, rarely entered actively into play, and I noted only one instance in which a female reciprocated with an infant. Nevertheless, adults, including the silverbacked males, were extremely tolerant of playing young, sometimes under conditions which could not but have been uncomfortable.

LONE PLAY Most lone play involves climbing and swinging (Fig. 63C, D, E), jumping and sliding, waving of arms and legs, batting of vegetation, somersaults, and running back and forth with exaggerated gestures. My notes describe some typical play situations:

1. *An 8 month old bumbles around by the reclining dominant silverbacked male. With a wide overhand motion it swats the male on the nose, but he merely turns his head. The infant then runs downhill and turns a somersault over one shoulder and ends up on its back, kicking its legs in bicycle fashion and waving its arms above the head with*

great abandon. A ten-month-old infant watches these proceedings while propped against the rump of the male. Suddenly the ten-month infant rises, hurries to a sitting juvenile, and pulls the hair on its crown with one hand. When this brings no response the infant yanks at the hair with both hands, but the juvenile remains oblivious. The infant desists, sits briefly, suddenly rolls forward over one shoulder, and with arms and legs flailing like a windmill rolls over and over down hill and disappears in the vegetation.

2. *An infant, 1¼ years old, sits and handles a leafy lobelia head. It bangs the leaves on the ground several times and swings them around in an arc. Finally it places them upside down on its head and remains motionless for several seconds under the green hat (Fig. 63F).*
3. *A juvenile lies on its back, holding its arms straight toward the sky. It shreds a handful of leaves and lets the pieces fall over its chest and face.*
4. *A one-year-old infant sits on a large horizontal* Hagenia *branch, which is covered with cushions of dark green moss. It rips off a piece of moss about 8 inches in diameter, and places the mat on its head like a cap. Then it walks back and forth along the branch, carefully keeping its head steady. But the moss slides, first to the nape and finally off. The infant lunges, retrieves it, and replaces it on the head. When the moss falls again, the youngster slips it under its arm like a briefcase and continues walking. The moss, however, crumbles and the infant sits and stuffs the pieces into the bend of its arm. Then it hurries away.*
5. *A blackbacked male lies on his back, one leg pointing upward. He rolls his head from side to side, with mouth wide open; suddenly he grabs his leg with both hands and wrestles with it.*
6. *A 7-month-old infant runs back and forth on a fallen tree, rather stiff-legged and cantering slightly. Its mouth is open and pulled into a smile. Very shakily it rises on its hind legs and moves its arms as if to beat the chest before losing its balance. It runs again, then sits, and with very exaggerated gestures beats its chest.*
7. *A one-year-old infant stands on a steep slope. It leans away from the slope while holding on to the vegetation with one hand. As it swings its body from side to side, it lashes the ground with a weed held in the other hand.*
8. *A 7-month-old infant climbs up a lobelia stalk and slides down; up again and down. Again and again.*
9. *A juvenile runs down a trail. Suddenly it throws itself on its side, rolls over, jumps up and without pause continues on, repeating the actions like a bouncing furry ball.*

SOCIAL PLAY In mutual play infants have for the first time in their lives an opportunity to interact closely with other youngsters in the group. In spite of the fact that a juvenile sometimes weighed at least five times as much as the infant with which it was vigorously

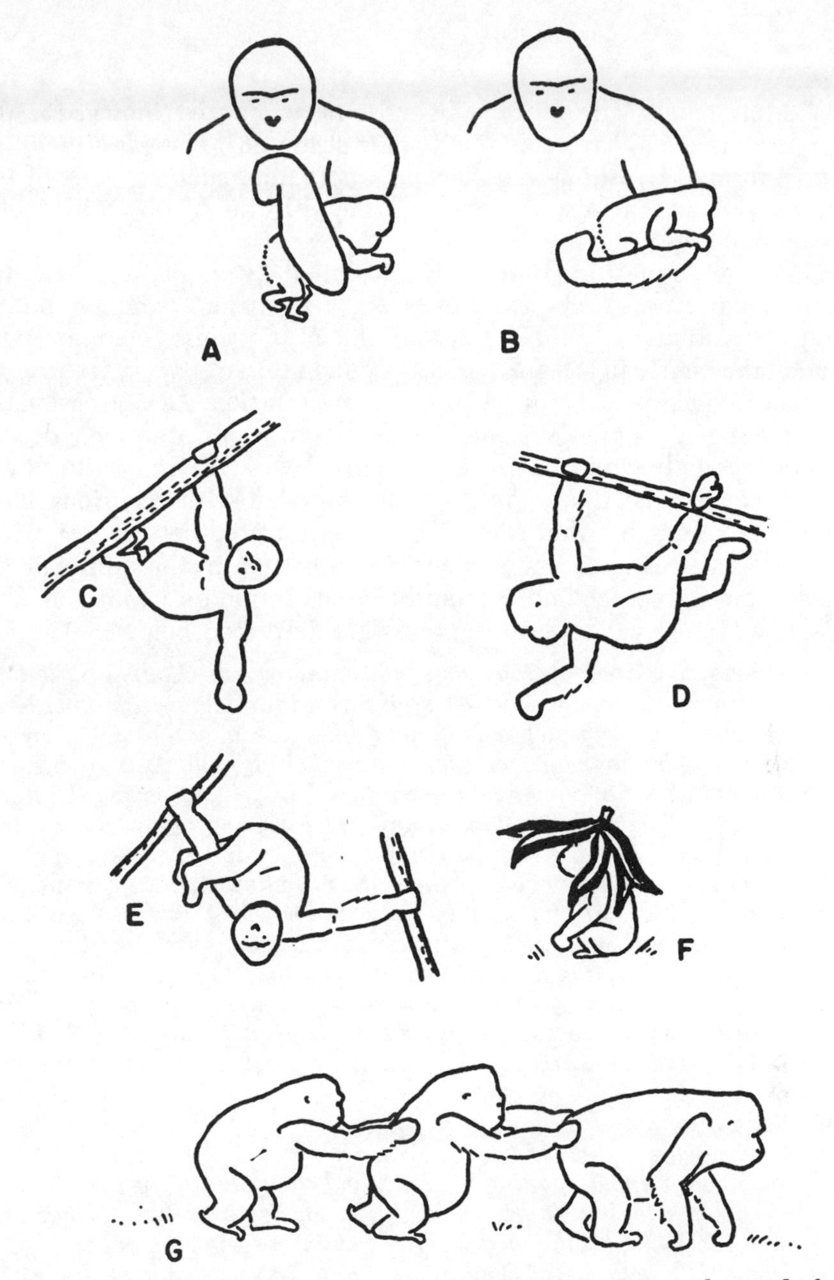

Fig. 63.—Body positions of gorilla youngsters at Kabara during play, and of a female carrying her wounded infant. *A–B,* position of a wounded infant in the arm of its mother. The dotted line on the rump of the infant represents the wound. *C–E,* body positions of infants during lone play in trees; *F,* an infant with a lobelia which it placed playfully on its head; G, three juveniles "snake dance" in play.

wrestling, such a disparity in weight never caused injury, for the larger animals always contained their strength. Most play included some form of wrestling and chasing, and games such as "king of the mountain" and "follow the leader." These forms of play have their close parallels in human society.

Playing youngsters frequently exhibited two characteristic behavior patterns; adults sometimes showed similar behavior but in different situations. (1) Playing infants and juveniles commonly bit each other softly in the angle made by shoulder and neck. Quarreling females mock bit each other in the same location. (2) Occasionally two youngsters either sat and faced each other or approached each other bipedally while waving their arms alternately and quite slowly over their heads. Upon contact they grappled with a curious slow-motion effect, somewhat resembling the ritualized behavior of oriental human wrestlers. Similar behavior was seen in the adults at the Columbus zoo preceding copulation (see section on copulation).

Nine typical social play situations are described below.

1. *A 6½-month-old infant, Max, leaves his mother and climbs up the back of a reclining female, who has no infant. Max advances onto her head, climbs down, rips off a leaf, and attempts another ascent of the female, who has now rolled onto her back. Finally Max reaches the summit of her abdomen and lies down. The female covers the infant with her hand, holding him down. Then Max struggles and squirms and tries to free himself. One hand emerges and then the other; his mouth is partially open and the corners pulled back in a smile. The female loosens her hand and Max grabs it and gnaws at her fingers. Several times she repeats the whole process of holding the infant down until he struggles free. The female then toys with Max, touching him here and there, and in turn he attempts to catch her elusive hand; then he waves his arms and legs with wild abandon. Suddenly he sits, and with arms thrown over his head, dives backwards into the weeds.*

Moritz, a 7-month-old infant, arrives on the scene. He walks up to the lone female and pulls her leg. Max rushes from the weeds, tumbles over Moritz while pulling his hair, and then dashes up the abdomen of the female and jumps in her face. She gathers Max in her arms and rocks him back and forth, her mouth open in a smile.

Nearby, a dense curtain of Galium *has formed a small cave at the base of a tree. Moritz sits in the entrance to the cave, Max advances. Moritz rises on his hind legs, beats his chest, and with arms held above the head and mouth wide open, jumps at and falls over Max. They wrestle, arms and legs flying, they mock bite each other's shoulders, and finally Moritz throws a headlock on Max. Max pushes and pulls and rolls over and they break apart. Moritz rushes back to the cave entrance, Max closely behind. Max bulls his way through this time and they wrestle inside. Max emerges and leaves. But as he ambles along, Moritz rushes from the cave, and rams Max so hard with his shoulder that both sprawl. Moritz regains the cave, but soon emerges carrying a dry piece of log under one arm. The other infant pursues and grabs Moritz by the rump, who then drops the log, turns, and they wrestle again.*

A juvenile walks through the entrance of the cave and tears the structure down. Max and Moritz arrive and both tumble all over the juvenile, who wrestles lightly with them.

2. *It is late afternoon and growing dusky; the group has spread over its nest site, and some have already built their nests. Four infants, from one to two years of age, play on the sloping, moss-covered trunk of a* Hagenia. *They run in single file up and down, they descend the trunk sliding on their seats or on their bellies, and they grapple with each other.*

A female comes up and with folded arms leans against the trunk and looks up at the infants above her. The first infant pushes its way between her arms and emerges farther down; the second climbs onto her head, canters down her back, and in a sitting position slides down her rump; the third runs along the branch and takes a flying leap forward with arms and legs extended in front, and lands with such force on her back that she slides two feet down; the fourth follows the female from the tree. All return to their mothers.

3. *A juvenile and a 1¾-year-old infant sit about 4 feet apart. Suddenly the juvenile twists around and grabs for the infant, who rushes away hotly pursued by the juvenile. The juvenile catches the infant and covers it with its body, propped on elbows and knees. Twisting and turning, struggling and kicking, more and more of the infant emerges from beneath the juvenile. Freedom gained, the infant grabs an herb stalk at one end and the juvenile snatches the other end. They pull in opposite directions; the juvenile yanks hard and the infant is jerked forward. They then sit facing each other, mouths open, and swing their arms at each other and grapple slowly. Another juvenile comes dashing up and in passing swipes at the juvenile, and all three disappear running into the undergrowth.*

4. *A 7-month-old infant sits in the branches of a brushy* Hypericum *and watches the ascent of a 1¼-year-old into the same tree. When the latter grabs the branch on which the other sits, it gets its fingers stepped on. Undismayed, the large one repeats the attempt, but the small one reaches over and jerks the wrist of the other so that it loses its grip and nearly falls from the tree. After repeated attempts the 1¼-year-old grabs a leg of the one above, but is kicked in the face for its effort. However, the end is at hand when the smaller infant retreats to a higher branch closely pressed by the other young. Firmly*

entwined in each other's grip the two youngsters half-descend and half-fall into the weeds below.

5. *Four infants, one to two years old, play in a vine-covered* Hypericum *as high as 20 feet above ground. By themselves they swing back and forth on lianas. Then, when one races along a branch, all others follow. In a row they climb up some vines and slide down others, grabbing a mouthful to eat in passing or pulling and pushing each other. One stops on a branch, and, holding on to a vine, leans far out. The vine snaps and the infant tumbles head first through the tangle of vines which are so dense that they deposit the infant gently into the vegetation 12 feet below. The others immediately jump and join the other, tumbling and sliding carelessly in a shower of dead leaves.*
6. *An 8-month-old infant sits. A one-year-old reaches over and touches it lightly with the back of its hand. The small one rises on its legs and throws itself at the larger one, whereupon they wrestle.*
7. *A juvenile sits, apparently oblivious to two 1½-year-old infants creeping up silently from behind. Suddenly the juvenile throws itself on its back and snatches wildly at the infants beyond its head. The infants, however, have scampered off, their mouths open.*
8. *A group moves out from the rest area with three juveniles walking together. One grabs the rump hairs of the first one with both hands, the third animal does the same to the second one, and, thus in snake-dance fashion, all three careen wildly down the slope (Fig. 63G).*
9. *A female sits, her young infant held to her chest. A blackbacked male ambles past her. As he passes, she slaps his rump. He takes a step, turns partially and flicks her on the shoulder with the back of his hand, and rapidly runs a few steps. Two minutes later he returns down the same path. But when he approaches the female, he breaks into a run and in passing slaps her on the shoulder. She just sits.*

Interactions between Individuals

The interactions between members of a group vary considerably in kind, frequency, and duration. Some of the specific types of interactions such as dominance, mutual grooming, social play, and sexual behavior have been or will be discussed in other sections of this report. Here I am primarily concerned with summarizing the nature of interactions between the various age and sex classes.

Silverbacked Male–Silverbacked Male Interactions Many silverbacked males had little or no opportunity to interact with others of their kind except at widely spaced intervals when lone males attempted to join a group or when two groups met. Only groups IV, V, and IX contained more than one silverbacked male throughout the period of study.

The primary impression which I received from watching the interactions of males was that they generally ignored each other, or, at least, were highly tolerant of each other. Although they possessed a dominance hierarchy, with some of the subordinate males tending to remain at the outer edge of the group, I never observed the slightest indication that the dominant male chased others to the periphery, kept them from coming into the group, or prevented their having social intercourse with members of the group. I once observed three silverbacked males in group IV as they rested within a diameter of 15 feet; another time four of them lay in a diameter of 30 feet. The number one and number two males in the hierarchy of group IV frequently sat within 5 feet of each other. Three males in group V usually remained quite close to each other, and only one tended to roam around the periphery.

SILVERBACKED MALE–BLACKBACKED MALE INTERACTIONS Although most large groups contained one or more blackbacked males, I never observed an interaction involving physical contact with a silverbacked male. Blackbacked males were fully tolerated and usually ignored. They often remained in the center of a resting or moving group, thus being near the silverbacked male, but dominance interactions were rare since the blackbacked males circumvented direct meetings.

SILVERBACKED MALE–FEMALE INTERACTIONS Even though females watched the actions of the dominant male closely, and the male spent much of his time surrounded by females, the number of interactions involving bodily contact was not large. Females sought the vicinity of the male on nine occasions and rested their heads on his saddle or leaned their bodies heavily against his side (Fig. 65A, C). One female rested her head on the rump of the male; two females lay by the male's side touching his flanks; once a female and a male sat back to back. A female with a newborn infant, still wet, approached the male and rested against him (Fig. 64E). When he rose, she stood up on her legs and propped herself up on his back only to place her head on his saddle as he lay down again. Finally she moved to his other side and slightly below him. The male reached down with one hand and very intently fondled either the infant or the female. Philipps (1923) observed that ". . . at the edge [of a cliff on Mt. Muhavura] the male appeared to assist the female with his right arm with an absurdly solicitous air. . . ." I saw nothing comparable.

The females usually tolerated the somewhat harsh behavior of the males following their chest-beating displays: one female merely ducked when a male slapped her on the back with both hands; when

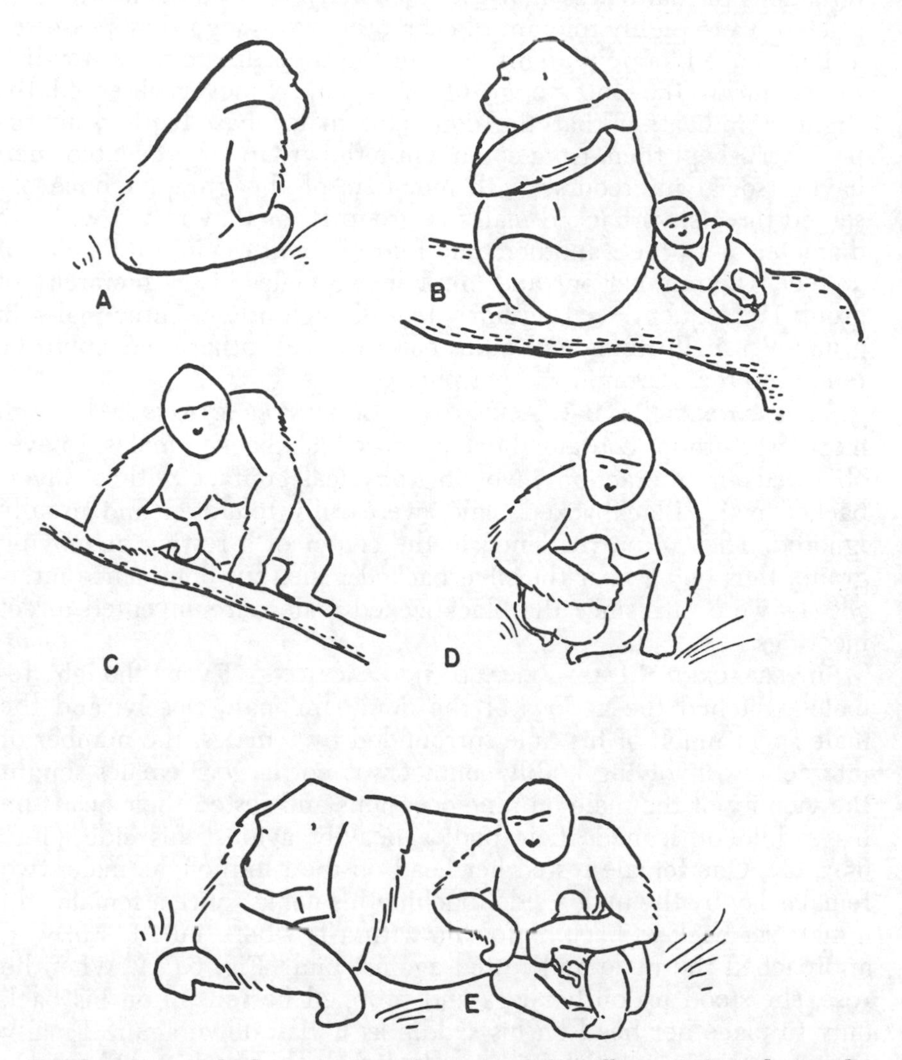

FIG. 64.—Various body positions and activities of gorillas at Kabara during the midday rest period. *A*, a sitting female; *B*, a sitting female with her infant resting against her; *C*, a female watching me intently; *D*, a 1½-year-old infant suckling on the breast of its mother; *E*, a female with a newborn infant resting against the side of the dominant silverbacked male (group IV, March 12, 1959).

another male cuffed a female in the side, she simply ambled away. One silverbacked male ran up to a sitting female for unknown reasons, grabbed her leg (arm, according to Emlen) and jerked her two to three feet down the slope. The female showed no response as the male ran off after this seemingly playful gesture.

Sometimes either the male or the female discouraged contact. One male ran past a female without infant and slapped her. She, however, bared her teeth while attempting to grab his hand. On another occasion, a female stood on her hind legs and rested her hands on the rump of a male for support whereupon he turned his head and stared at her until she moved away.

The interactions of subordinate silverbacked males and females were again peaceful. Females never reclined against subordinate males, but they readily sat or nested close to them.

SILVERBACKED MALE–JUVENILE INTERACTIONS Males rarely responded to the presence of juveniles. Once, after beating his chest, a male ran at a juvenile, gathered it in with his left arm, and bowled it down the slope with a sideways motion. During a sudden meeting with group IV, the dominant male grabbed a juvenile around the waist with one arm, and, holding it against his hip, raced downhill for 30 to 40 feet before sitting and releasing it.

Juveniles, like females, sometimes sought the vicinity of silverbacked males. I observed juveniles as they rested against the dominant male on five occasions. Once a juvenile placed its head on the outstretched hand of a reclining subordinate male (Fig. 65B). In group V, juveniles readily tagged behind the subordinate silverbacked males and apparently even left the group with them.

SILVERBACKED MALE–INFANT INTERACTIONS Infants were definitely attracted to the dominant male. When the group rested, one or more infants often left their mothers and sat by the male or played on him (Fig. 65D). Once three infants, two juveniles, and one female were crowded around the dominant male. Another time four infants played around and on the male. Frequently the behavior of the infants could only be classed as exuberant. When slapped in the face by an infant the male merely turned his head. Twice within the space of an hour a large infant grabbed the rump hairs of the dominant male in group IV and hitched a ride for 30 to 50 feet without eliciting a response. On two separate occasions one infant sat on the back of the reclining male, and another time one infant climbed into the lap of the male and sat for several minutes. If, however, the play of the

FIG. 65.—Body positions of females and youngsters while resting in physical contact with a silverbacked male. *A*, a female and her large offspring sit by the reclining dominant silverbacked male, who rests his hand on the shoulder of the infant (group IV, September 23, 1959); *B*, a juvenile rests its head on the outstretched hand of a subordinate silverbacked male (group V, September 19, 1959); *C*, a female crowds close to the dominant silverbacked male (group IV, August 31, 1959); *D*, two juveniles rest against the legs of the silverbacked male (group II, November 7, 1959).

infants grew too uninhibited or bothersome the male reacted—as two examples illustrate:

1. *Two infants approach the male. One climbs onto the rump of the male and the other tries to ascend but is pushed off. Finally both tumble down and continue to wrestle over and around his legs. When the male turns and calmly stares at them, one infant leaves, and the other sits quietly.*
2. *A two-year-old sits on the back of the reclining male and reaches out to touch his head. But the male jerks his head forward out of reach. Another infant stands on two legs by the male and pulls the hair on his side, whereupon the male swivels his head around with mouth partially open. When the male moves uphill a few minutes later, 4 infants follow him in single file.*

BLACKBACKED MALE–BLACKBACKED MALE INTERACTIONS Interactions between blackbacked males were infrequent. Although these males often came close to each other, they showed no obvious tendency to associate with or to avoid each other.

BLACKBACKED MALE–FEMALE INTERACTIONS Blackbacked males spent considerable time near females, who tolerated them readily. In spite of this frequent close proximity, definite interactions were few and these were largely based on dominance or, rarely, on seeming playfulness. One small blackbacked male in group VI interacted with four different females over a period of several months.

1. *He stands on his hind legs and supports his hands on the rump of a female. She shows no response.*
2. *A female runs down a slope. As she passes him, he slaps at her with one hand but misses.*
3. *He runs up to a female and stands beside her, but she pushes him away with one hand.*
4. *A female climbs on a log and slaps at the male sitting there; he simply moves over two feet.*

Once a blackbacked male was briefly involved in a quarrel with two females; another time a male was groomed by a female.

BLACKBACKED MALE–JUVENILE INTERACTIONS Most blackbacked males and juveniles neither sought nor avoided contact with each other. However, some juveniles showed a seeming preference for the vicinity of certain blackbacked males. In group IV one juvenile consistently tagged behind the sole blackbacked male of the group for over one month; in group V two juveniles sometimes followed a blackbacked male up to 100 feet from the main body of the group. One juvenile rested its arms on the back of a male while standing bipedally (Pl. 32). When a blackbacked male suddenly stopped on a trail, the juvenile behind him pushed his rump with one hand, and the male moved on.

BLACKBACKED MALE–INFANT INTERACTIONS No interactions between infants and blackbacked males were seen. There was no tendency for infants to seek them out and sit with them.

FEMALE-FEMALE INTERACTIONS Females, with or without infants, were closely and continuously associated with each other, although interactions involving bodily contact were not frequent. They usually rested near each other, occasionally with bodies touching. Two females, both with small infants, once consorted closely for several days. Sometimes quarrels, accompanied by grappling, brought females into brief contact; when excited, they sometimes slapped at each other (see section on aggressive behavior). Mutual grooming was infrequent, and a dominance hierarchy, if any, was weak.

FEMALE-JUVENILE INTERACTIONS Although most females and juveniles interacted consistently only in dominance and grooming (see Tables 51 and 53), some juveniles associated closely with a particular female, possibly their mother. The closeness and duration of such associations varied between individuals and apparently depended in part on whether or not the female carried an infant.

In group V, a juvenile about three to three and a half years old remained constantly near a female who had no infant. One year later the female still carried no infant but the juvenile associated with her only intermittently. Another juvenile, in group IV, judged to be three and a half to four years old, stayed with an infant-less female most of the time. The juvenile was with this female as late as January 14, 1960, but, when the group was contacted on April 25, 1960, the female carried a newborn and the juvenile was not near her; nor did it rejoin her when the infant died two days after birth. In group II, a youngster barely three years old was still closely attached to its mother, and a juvenile roughly three and a half years old slept several times in the nest of a female even though their association during the day was very intermittent. A four-year-old juvenile in Group VII approached a certain infant-less female infrequently during the day, but it apparently slept with her during some nights.

A close association between a specific female and a specific juvenile sometimes persisted even after a new infant was born. Juveniles occasionally approached and groomed a female with an infant in her arms. A juvenile in group VII, about 4 years old, sat beside a female with a four-month-old infant several times. Once, when the infant was eight months old, the female reached over, put her arm around the back of the juvenile and pulled it to her chest. All three sat quietly together for a minute before parting. A close examination of dung in the nests revealed that juveniles occasionally slept in the same nest with a female and infant.

Although the bonds between females and their offspring were sometimes severed during the juvenile period, females rarely chased juveniles forcefully away from them. One female swatted at a juvenile, and another snapped at one. Most females used such mild rebuffs as detaching the hands of juveniles which held onto their rump hairs while traveling. The social break between females and juveniles was further expedited by the fact that the latter actively sought contact with other group members as they grew. Thus rebuff by the female and the seeking of social contact by the juvenile were complementary phenomena. All obvious mother-offspring ties appeared to be severed by the time the juveniles were about four and a half to five years old. Interactions between old juveniles and females were primarily those of utilitarian grooming and dominance.

FEMALE-INFANT INTERACTIONS The relationship between mother and offspring changes from the infant's complete dependence during the first few months of its life, through a period of gradual lessening of the physical and emotional bonds, to a stage when the youngster becomes integrated into the group. During this period the functions of the female change. She is at first the complete source of all such necessities of life as food and transport, but later the infant merely utilizes the social comfort which it appears to derive from her proximity. Concomitant changes occur in the female's behavior also. Constant watchfulness and great care when the infant is helpless and very young change to less concern and an occasional light rebuff.

Feeding I found it difficult to observe suckling in small infants, for the young were held so closely to the mother's chest, and the forearm of the female was so large and hairy, that I could rarely determine if the infant was actually suckling or merely resting its mouth in the vicinity of the nipple. Consequently I recorded suckling only eleven times. The ages of the infants in these observations were: 1, 3, 5, 5, 8, and 10 months; 1, 1¼, 1¼, 1½, and 1½ years. In the older infants suckling was more easily observed and the fact that it was noted only five times in animals of over one year suggests that the infants were then partially weaned.

Infants ingested some solid foods by the age of two and a half months as indicated by direct observation and remains in the dung.

By the age of five months they ate considerable amounts of vegetation, and by seven to eight months the bulk of their nourishment was probably derived from forage.

Infants appeared to suckle infrequently. I have, for example, spent as long as seven consecutive hours near a group without observing any suckling. It is possible that nursing occurred more commonly during the night or in the early morning while the animals were still on the nest site. Jambo, an infant lowland gorilla born at the Basel zoo, suckled on its mother for roughly three minutes every two to three hours when about six days old (Lang, 1961*c*). The infant Goma at the Basel zoo was satisfied with three daily bottle feedings at the age of three months (Lang, 1959).

No infant was ever noted to suck its thumb, a common practice in captive gorillas.

Females held small suckling infants to the chest with one or both arms while sitting or lying (Fig. 64D). A five-month-old, for example, was supported by one arm. It suckled on the right nipple while grasping the left one with one hand. It then switched and suckled on the left one, for a total time of about one minute.

Large infants suckled rather casually, appearing to seek social contact more than nourishment:

A 1¼-year-old suckles about 30 seconds while sitting in the lap of the female. Something catches its attention and it leans far back and turns its head with the nipple still in its mouth.

Twice females rebuffed their infants gently when they attempted to suckle. The one infant was eight months old, the other one and a quarter years old.

A 1¼-year-old squats on the ground in front of a sitting female and suckles. She lifts her arm after about 15 seconds, bends it, and with the forearm pushes the infant away.

Once a five-month-old infant suckled on a female which was not its mother. It tried one nipple, then the other; ten seconds later it switched back to the first one. The breasts of the female probably contained no milk for she lacked an infant of her own. Finally the infant raised its arms above its head, stood up, and fell playfully backwards into the lap of a juvenile.

Females exerted little direct control over the foraging of their infants. The young appeared to learn what to eat and what not to eat largely by observing their mothers and other group members, and by trial and error (see section on food habits). One infant pulled down the lip of a female, extracted a piece of *Galium*, and ate it.

Once a female removed an unpalatable *Hagenia* leaf from the hands of an infant who had been chewing on it. On another occasion a female aided an infant in obtaining food: a five-month-old pulled with both hands at a one-foot section of broken lobelia stem rooted in the ground. It jerked and pulled but could not break it. The female watched. Finally she reached over with one hand, snapped the stem off, and laid it on the ground. The infant picked it up and gnawed on it.

Transport Only mothers carried their own infants over long distances; males and juveniles were never observed to transport infants farther than 50 feet. The method of transporting the infant varied with its age. Newborn infants appeared to lack the strength to clasp hair securely, and the mothers had to continuously support them with at least one arm. When the female sat, she usually held the infant against her chest with her whole forearm lending support to its body. Often both arms were used, the newborn being cradled either high on the chest or low in the lap. The body of the infant was usually perpendicular or oblique to the long axis of the female's body. When walking, the female supported her infant with one arm against her chest as she herself proceeded on three limbs. The infants grasped any hair available, but their hold was insecure and frequently one leg or arm slipped off and waved in the air. Lang (1960*a*) observed that the grasping reflex in one newborn gorilla was weaker than that of chimpanzees and orang-utans of comparable age.

During the first month of life, infants were so rarely without support that I found it difficult to determine at what age they were first able to cling to their mothers without assistance. At the age of one month they were able to do so long enough to permit the females to climb into trees, using both hands to grasp branches. At such times the infants lay sprawled parallel to the long axis of the female's body, grasping the hairs at her side, usually in the vicinity of the armpits, and similarly clasping the hairs with the feet on each side of her abdomen. However, the physical structure of the female is such that infants of all ages apparently find it difficult to hang with ease from her chest while she walks. Her chest is broad and sparsely haired and the abdomen is very rounded. Her back, on the other hand, is ideally suited for carrying infants. The hair is long, the back is very broad and relatively flat, and the infants can sit, lie, and even walk around without much danger of slipping off.

One infant, only one and a half months old, appeared once on the back of a female. It was shaky and barely able to hold on. It is perhaps significant that this infant disappeared shortly thereafter and probably died in some mishap. The earliest age at which infants appeared more or less regularly for short periods on the back of their mothers was about three months. However, these infants were pulled to the chest whenever the females sat or moved rapidly. At the age of four months, the infants appeared on the back of the females more and more frequently, and by one year of age they were carried against the chest only infrequently and under special circumstances, as when suddenly disturbed and put to flight.

While climbing in trees, large infants were sometimes held to the chest, apparently for convenience and safety. One infant, one and a half years old, clung to the back of the female, but several times it lost its hold as its feet slipped whenever she stood upright. The female reached one arm back to support her offspring at least three times when it slid as low as the rump, and finally she hauled it to her chest where the infant fastened one of its arms around her neck. Most infants over one and a quarter to one and a half years of age followed their mothers under their own power while climbing.

The characteristic position of infants on the back of moving females was prostrate, with the head in the vicinity of the shoulder region, the hands grasping the hair on each side or sometimes around the neck, and with the legs spread (Pl. 17). The clasping power of the toes was relatively poor, and frequently the only support seemed to be derived from pressing the legs against the body of the female. Occasionally an animal lay sideways across the back. When the females walked slowly, the infants sometimes walked up and down on the back, sat and picked up vegetation in passing, or rested on their elbows and knees. When the females sat, the older infants slid off and the younger ones were pulled to the chest and held there.

Although the youngster is thus transported by the female throughout the period of infancy, the female encourages it to travel on its own at an early age. One female placed a three-month-old infant, barely able to crawl, on the ground, and walked ahead very slowly. This was the earliest such instance noted. By four and five months, however, youngsters sought social contacts with others and crawled away from their mothers, until by six and seven months they were climbing by themselves and occasionally toddling behind their mothers. Development continued rapidly, and infants over one to one and a quarter years of age spent much time in sitting by the female rather than on her. They often were out of contact with her for considerable time during the rest period. By the time the infants were one and a half years old and weighed some 25 pounds, the females occasionally appeared unwilling to carry them. Large infants often walked behind slow-moving females, either by themselves or with one or both hands grasping her rump hairs. Infants over two and a half years old sometimes ran behind their mothers even when travel was rapid. Once, when I surprised a group, a large infant climbed onto the back of a female and another jumped on her rump and both were carried about 30 feet.

When an infant ventured away from its mother it usually remained aware of her location and behavior, for at the first sign of movement it rushed back to her and climbed aboard. Small infants less than about seven to eight months of age, did not show such responses and had to be gathered in, apparently not having learned to recognize the slight forward jerk or sideways twist of the female preceding her rise from a sitting or lying position. A common sight was a female sitting quietly with an infant resting by her side but slightly behind her. When she moved, the infant merely reached out, held onto the hairs on her rump or back, and was automatically lifted aboard. When infants "missed the boat," they sometimes screamed, or they simply chased after her. If movement was leisurely females showed little concern for the infants and most caught up to their mothers within 30 feet. Three examples from my field notes illustrate the behavior.

1. *An excited male runs. Infants in his way scatter and rush to their mothers. However, one female moves off before her offspring, about 1 year old, can reach her. It runs and screams once; she slows down without turning her head until the youngster climbs up on her rump.*
2. *A female leaves. Her 1½-year-old infant screams once as she moves away, but she neither turns nor slows down. But it catches up to her, grabs her rump hairs and climbs up.*
3. *As the female leaves, a 1½-year-old stands up on two legs with hands above head and falls forward, just managing to grasp the hairs on her rump. It runs on two legs a few steps and swings up—like catching a moving train.*

Protection Guarding the infant from harm during the average day-to-day existence was entirely the responsibility of the mother. When less than four to five months old the infant was usually not permitted to venture more than about 10 feet from her without being pulled back. One female, for example, lay on her back in the sun. She pushed her three-and-a-half-month-old infant off her chest with the back of one hand. The infant returned to her chest, but was again pushed away. Reversing its direction it crawled away from the female. When it was 4 to 5 feet from her she reached far over, picked it up by one arm, and dropped it by her side. This was repeated twice more until finally the infant lay still. Young infants spent most of their time in contact with their mothers, and were often enveloped by her large arms, which protected them not only from the vagaries of the weather and precipitous topography, but also from inquisitive members of the group.

As the infants developed they were permitted to travel farther afield. By eight months of age some were occasionally seen 20 feet or more from their mothers, and by one year they climbed trees and wandered freely throughout the resting group, frequently out of their mothers' sight. Such behavior occurred only when the group was resting or moving very slowly.

Some of the common situations which elicited protective behavior by the female are best illustrated by examples:

1. *A 1¼-year-old infant ascends a vine about 5 feet, thus obtaining a better view of me. Its mother reaches over and pulls the infant by the waist with both hands until it releases its hold, after which it is placed beside her on the ground.*
2. *An 8-month-old infant clambers on a horizontal dead branch of a fallen tree while its mother sits about 7 feet below. The infant slips and hangs by two hands. The female looks up, stands on two legs, and barely reaches the foot of her infant. She pulls it to her chest but it wriggles free, repeats the climb only to slip at the same place again, to be rescued once more by its mother.*
3. *A female spots me and looks intently. She hurries to her 4-month-old infant on a log about 8 feet away and pulls it to her chest. Then she just sits and watches me.*
4. *A group rapidly moves away because of my presence. A female descends the vines of a tree and her 1¾-year-old infant follows over the same route. She waits at the base of the tree, and when her infant is within reach she grabs it under her arm and hurries after the group, while the infant climbs onto her back.*
5. *A silverbacked male rests on his belly with a female and 10-month-old infant beside him. The infant climbs onto the back of the male, but the female grabs its arm and pulls it off.*
6. *A female climbs through the crown of a* Hypericum *slowly followed by her 1½-year-old offspring. The infant reaches a gap between two vertical branches across which its mother has just passed. It places one hand and one foot on the opposite branch, but then can neither advance nor retreat. The female stops, retraces her steps about 8 feet, grabs the infant by the arm, and pulls it across.*
7. *Two females lie near each other. The 3-month-old infant of one, barely able to crawl, moves 6 to 7 feet and rests by the back of the other*

female. When this female makes a motion as if to turn and thus roll on it, the mother rises quickly and pulls the infant to her chest.

8. *A 3½-month-old infant lies next to a reclining female on a smooth log. It moves, slides, and falls three feet into the vegetation. The female immediately turns over, reaches down, and pulls it to her chest.*

Social The comfort and security which the infant derives from close contact with the female, even after she has ceased to provide food, is probably essential to its bodily and mental well-being. This continuous social contact undoubtedly contributes to the infant's later successful integration into the group as a whole. From the age of about two and a half months the infant exhibits a growing awareness of the environment and with it a desire to explore and seek social contact with other group members. Through the behavior of its mother it probably learns to recognize the meaning of various gestures and the proper response to certain vocalizations, essential knowledge in a closely knit society.

The mother is also the only object in the environment of the infant to which it can turn at all times. Consequently, the actions of the infant are centered primarily on its mother, as one typical example illustrates:

A 5-month-old infant totters about, 5 feet from its mother. Other animals mill around and the infant appears to grow uneasy. It attempts to return to its mother rapidly but stumbles over a lobelia stem. Its mother extends one hand, the infant falls into it, and is pulled to the chest. Soon, however, it ambles away again, across the chest of a juvenile and over the arm of a female before suddenly hurrying back to its mother. It stands bipedally at her side, raises both arms above its head, and she gathers it in with one arm to her chest.

Females too sought contact with their offspring, seemingly for social reasons alone, as shown by two typical instances:

1. *A female and infant, 1½ years old, sit side by side. She reaches over, pulls the infant into her lap, and envelopes it with her arms.*
2. *A female and her one-year-old infant sit side by side. Every few minutes she briefly places one of her arms over the shoulder of her offspring.*

Sitting females frequently pushed, pulled, or bodily lifted their infants from one part of the body to another, usually from the chest to the back and vice versa. By the time the infant was about six to eight months old, it often clambered from the chest to the back of the female under its own power when given a light tap or push. Several typical examples from my field notes illustrate the behavior:

1. *A female walks up with her 1½-year-old on her back. She sits and raises her left arm, reaches under it with her right one, and pulls her baby to her chest. When she moves on, the infant swings back up to its former position.*
2. *A female sits, reaches over her left shoulder with her right hand, and pulls her 3-month infant to the chest. She feeds a little, then grabs one arm of the infant with her right hand and pulls it upward over her left shoulder and to her back before ambling off.*
3. *A female sits, her one-year-old by her leg. The group begins to move out. When the female taps the infant on its back with her fingers, it climbs up to her chest, over the shoulder, and onto her back.*

Gorilla infants, like their human counterparts, apparently find it difficult to remain quiet for long periods of time, and when the group rests they frequently make determined efforts to escape the hold of their mother.

1. *A female lies on her back, her 2½-month-old infant on the chest. The infant is alert. Shakily it reaches one arm toward another female nearby. Then it crawls several times in a row toward its mother's shoulder, but each time she pulls it down by a leg. Finally it pushes with one hand against that of its mother and tries to wriggle loose. No luck.*
2. *An 8-month-old infant sits at the feet of its mother. She grabs the hair on its head and gives it a pull, and the infant climbs to her chest. But it descends again. So she lifts it by one arm. The youngster struggles, it arches the back, stiffens the legs, squirms, and wriggles until the female finally gives up and releases it.*
3. *A 4-month-old infant is unruly in its mother's arms. It pokes at her eyes and she turns her head away, it twists its body and wiggles downward. Finally the female lays it on the vegetation next to her. It clambers back to her chest. She pushes it down. Then it totters off along a log and she reaches over, picks it up by one arm, and, holding it to her chest, walks off.*

Mothers began to sever the social ties with their offspring by the age of one year. Rebuffs were always gentle. One female pushed an eight-month-old infant from her nipple. On another occasion, when a one-year-old youngster ran up to its mother and held on to her rump hairs as she walked slowly, she brushed its hands off with a backward swipe of her arm. Again, when a one-and-three-quarter-year-old infant ran up to its sitting mother, she gave it a light push with the back of her hand whereupon the infant rolled onto its back, threw its feet into the air, and waved its arms in an exaggerated, playful gesture before ambling off.

Response to wounded infant That females may show a strong

concern for wounded infants is illustrated by the reaction of a mother to her eight-month-old offspring with a serious wound on its rump. The infant never rode on her back for it was probably too weak to hold on securely. She always cradled it gently in her arm in such a way that no part of the wound touched her body: it was held continuously belly downward with one of her arms around its waist or with its body draped over her elbow (Fig. 63A, B). Once the female looked at the wound intently and picked at it briefly with one hand. On one occasion, a female approached who had lost her own infant about two months previously. She bent over and with pursed lips touched the face of the injured infant. When she attempted to repeat this ten minutes later, the mother of the infant shifted her weight and pushed the other female away with one foot.

Behavior of a mother with a dead infant I observed the reactions of a female to her infant from its day of birth to its death two days later of unknown causes, and for three additional days until the female discarded the body. When I saw the infant alive on April 25 and 26, the female carried it gently but firmly pressed against the chest, her forearm and hand supporting the whole of its body. She cradled it in both arms when she sat. I did not see the group on April 27, but on April 28 the female rested with the infant, not held to the chest, but loosely in the lap. She supported it with a hand on its back, but she made no attempt to restrain the lifeless limbs from dangling and the head from rolling. The eyes of the infant were closed and its mouth open. While the group rested for over one and a half hours the female neither cuddled the corpse close to her nor gazed down at it. When the dominant male sat next to her briefly, he ignored the infant. When walking, the female held the body low on her chest, but made no attempt to support the head.

On the following day, I watched the female for two hours as she sat in her day nest with the dead infant lying on the ground before her. Her arms were folded and occasionally she snacked. The corpse was never handled during that period. Just before leaving the nest she briefly lifted it once by an arm, then dropped it. She carried it held loosely against her abdomen, and once she took several steps with it dragging on the ground. Similar behavior was observed on April 30. On May 1, four days after the death of the infant, she remained hidden from view. The group was last seen at 1310. Around 1500 to 1600 she apparently dropped the corpse on the trail as the group moved down a slope. I found it there on May 3, face down, arms and legs askew, quite decomposed, and smelling strongly.

Infants with females not their mothers Infants occasionally left their mothers and joined other females either fleetingly or for an hour or more. Even though these females often had an infant of their own, the visitor was always accepted. On one occasion, a two-year-old infant walked up to an infant-less, though pregnant, female. She reached over and gathered the youngster into her arms, nuzzled its arm and groomed it. A few days later, the same infant approached a female with a month-old infant and sat by her side. On another occasion, a one-year-old infant left its mother, walked 20 feet, and clambered into the arms of a female who was holding her own two-year-old offspring to her chest. The female held both infants briefly, after which the one-year-old returned to its mother.

Two infants spent considerable time with females other than their mothers. In group II a six-month-old infant repeatedly remained for hours with an infant-less female, playing with her, being carried by her, and even suckling on her. Twice this female carried the infant to her chest while the group traveled. Once this infant hurried to her at the beginning of a rainstorm. She, however, pushed it away with the back of her hand, but it persisted and gained her chest, only to return to its own mother a few minutes later. The mother of the infant never responded in any way to its periodic absences.

One infant, one and a half years old, in group VIII, spent most of its time from November, 1959, through May, 1960, with a female and small infant, returning to its own mother apparently during feeding periods and nesting time only. Although the infant never rode on the back of the second female, it held on to her rump with both hands when the group moved slowly, and it usually sat by her side. Once the infant's mother and the second female sat next to each other. When the mother ambled off, her offspring left the second female to hurry after her.

JUVENILE-JUVENILE INTERACTIONS Juveniles interacted quite commonly in play, in grooming, and less frequently in dominance. Two juveniles slept together in the same night nest several times. Twice, when contacted during a heavy rainstorm, two juveniles sat face to face huddled close to each other. Although juveniles often seemed to make definite attempts to initiate contact with each other, and their interactions were always peaceful, most of their direct social intercourse was of brief duration, usually lasting only for the length of the rest period or for one night.

JUVENILE-INFANT INTERACTIONS Juveniles and infants came into frequent and peaceful contact with each other, predominantly in play and in grooming. One large infant, about two and a half years old, at least once stayed the whole night with a juvenile in group

VII. Sometimes juveniles sat beside infants and draped an arm over their shoulders. Although dominance was occasionally asserted by the juveniles, they were extremely gentle with the infants. Exuberant play never resulted in the injury of an infant or even in a quarrel. When the juvenile was not willing to play, it merely ignored the infant or rebuffed it lightly, as two examples illustrate.

1. *A 1¼-year-old starts to climb onto the back of a sitting juvenile. The juvenile jerks back its head with mouth partially open, but with teeth not showing, and the infant desists.*
2. *A 6-month-old bumbles up to a reclining juvenile and starts to climb up its abdomen. The juvenile reaches out with one hand, and pushes the infant away with its knuckles. Not deterred in the least, the infant persists and climbs all over the juvenile.*

Juveniles occasionally carried infants for short distances. This was observed eight times in three different groups and appeared to be done by both male and female juveniles. Twice juveniles merely held a large infant to the chest and walked less than 5 feet with it. In group II, a juvenile, seemingly a female, transported the same three-month-old infant two different times. Once the infant struggled and squirmed, and, when the mother moved over to the juvenile, the infant reached out and grabbed the hair on her side. The juvenile released it and the mother gathered the infant in. Whenever juveniles carried small infants, the mothers watched carefully and usually retrieved their offspring when they showed discomfort or were carried out of sight.

1. *A juvenile sits, holding a 6-month-old to its chest. The infant squirms and wriggles and tries to leave. Suddenly its mother rises 10 feet away, and the juvenile releases the infant immediately. She gathers her youngster in and continues to rest.*
2. *A female sits on a fallen log, her 3-month-old in her lap. A juvenile reaches over, takes the infant, and walks slowly for 20 feet holding it to the chest. The juvenile then sits, still holding the infant, for about 30 seconds before the female rises and slowly approaches the juvenile. The latter turns its head to look, loses its balance, and falls backwards four feet into the vegetation. The female rushes up silently, climbs down, and retrieves her infant.*

INFANT-INFANT INTERACTIONS Infants interacted primarily in play. Although play between infants was frequently quite wild, it never ended in strife. Dominance behavior was seen only two times and grooming only once. Older infants sometimes held each other chest to chest.

Group Co-ordination and Control

Data concerning the way gorillas maintain their cohesive social units and communicate with each other have been presented in this and other chapters. The means by which the behavior of the group is co-ordinated may be summarized under four headings.

1. Postures and Gestures Gorillas co-ordinate the behavior within the group primarily by employing certain postures and gestures. A dominant male who walks without hesitation in a certain direction indicates that he is leaving and the others follow. An unexplained sudden run by a member of the group communicates danger. In order to be groomed, a gorilla merely presents a certain part of the body to another animal. These examples illustrate that during the daily routine of a gorilla group nearly all behavior is readily communicated directly by postural means, and recourse to other types of communication is usually not essential.

2. Facial Expressions The extent to which facial expressions alone communicate emotions is difficult to assess, for they occur in conjunction with gestures and vocalizations. It is probable that facial expressions emphasize and elaborate information communicated by other means.

3. Vocalizations and Other Sounds Vocalizations serve to draw attention to the animal emitting them. This is frequently an important function in dense vegetation, for not until the group has become attracted to the performing animal is further communication by means of gestures possible. Vocalizations notify the others of a specific emotional state of the performer, alerting them to watch for gestures which communicate further information.

4. Forceful Control This method is most commonly observed and appears to be of importance only in females handling their small infants.

Response of Gorillas to Dead or Dying Members of the Group

Most observations of the response of gorillas to dead or dying members of their groups were made by hunters at the site of a shooting. In the accounts of Barns (1922), Prince Wilhelm of Sweden (1923), Akeley (1923), Bingham (1932), and others, the group always fled during the hunt after one or more animals had been shot. Geddes (1955) reported that a female was ignored by the rest of the group after she apparently died from a snake bite in a large enclosure. However, Merfield and Miller (1956) stated that ". . . gorillas never abandon their wounded until they are forced to do so, and I have often seen the Old Man trying to get a disabled member of his family away to safety." The same authors also related an instance based on second-hand information in which a male attempted to drag a dead female away. Similar accounts were given by Jenks (1911) and Aschemeier (1921). In a more fanciful vein, Owen (1859) claimed that "... when a gorilla dies, his fellows cover his corpse with a heap of leaves and loose earth collected and scraped up for the purpose."

Observations on the response of gorillas to members of the group dying a natural death show some variation. A female carried her dead infant at Kabara for four days before abandoning it. The other group members did not respond to the corpse. In August, 1958, when a male was found dead of unknown causes at Kisoro, the group of which this male had been the leader was not in the vicinity. Similarly two females and a male, apparently all killed by the same leopard at different times, were abandoned by the group (Baumgartel, pers. comm.).

In November, 1959, tourists noted that one animal in a group at Kisoro suffered from diarrhea. On February 6, 1960, Mills (1960) encountered the male of that group accompanied only by a large infant, but the other members were not found. The male and the infant were again seen alive and together on February 22. The following morning a party of tourists found the male dead with the infant sitting near the corpse. According to Baumgartel (pers. comm.) and Mills (1960) the infant, a male later estimated by me to be about two and a quarter years old, was unwilling to leave the body of the big male even when closely approached. Eventually it was driven into a patch of undergrowth and captured. The cause of death of the male was diagnosed as gastroenteritis (Peden, 1960).

Apparently the sick male, accompanied by the infant, parted voluntarily from his group several weeks before his death. The fact that the infant did not rejoin the females suggests that there was probably no contact with the main group during this period.

The Development of Infants from Birth to the Age of Six Months

In various sections of this report, especially in the one on mother-infant relations, I have given examples of the behavioral development of infants from the time of birth to complete independence from the female. The most striking changes occur during the first six months of the infant's life, when it grows from almost complete helplessness to an active, alert youngster.

I observed the behavioral development of several infants from birth to the age of one year, and a composite picture of my data is presented in Table 56. The observations in Table 57 on the infant gorilla Jambo, which is being raised by its mother at the Basel zoo, agree quite closely with those on free-living animals. For comparison, I have included in Table 58 some developmental information on the infants Colo and Goma, which were born at the Columbus and Basel zoos, respectively, and which were raised by human foster mothers.

In the wild I was not able to discern the first appearance of various behavior patterns in infants, and this makes direct comparisons with captive animals difficult. For example, I never observed free-living females to place their young on the ground until they were at least two and a half months old. Thus, I could not observe the earliest age at which an infant first raised its head, began to crawl, or turned from back to belly. At any rate, caution should be exercised in comparing infants raised by human keepers with those raised by their own mothers. In the wild, for instance, infants begin to assume the typical quadrupedal gait at the age of about four months. Goma at the Basel zoo did not progress in this manner until the age of six months (see Tables 56 and 58).

In general, observations both in the wild and in captivity show that gorillas are nearly helpless during the first two months of their life. Their grasping reflex is so weak that they are unable to hang onto their mothers unsupported for more than a few seconds. They respond in no visible way to moving objects in their surroundings during the first month after birth; and they make no attempt to reach for things during the first two months of their life.

By the age of two and a half months there is a marked increase in movement. The upper and lower middle incisors have erupted (Thomas, 1958; Lang, 1959), and the infant reaches for and chews on vegetation. The infant Goma at the Basel zoo first responded to known persons at the age of two months (Lang and Schenkel, 1961*b*). Free-living infants crawl for the first time at the age of two and a half to three months. Their crawl changes to an unsure quadrupedal gait by the age of four months. Development is now rapid as they begin to climb independently and to visit other youngsters in play. By five and six months of age the infants venture farther from the female, play with greater exuberance, and exhibit good control of their limbs.

Table 58 lists some behavioral norms of chimpanzees and humans for comparison with those of gorillas. On the basis of their comparison with the gorilla "Colo," Knobloch and Pasamanick (1958) conclude that the rate of development of gorillas in the early months of life is approximately twice as fast as that found in humans and slightly faster than that found in chimpanzees.

Mating Behavior

Only in the past few years has some information on the mating behavior of captive gorillas become available. Records now exist of six infants born in captivity: one in Columbus, Ohio, one in Washington, D.C., two in Basel, Switzerland, and two at I.R.S.A.C., Lwiro. In addition, various forms of sexual behavior have been described for gorillas in several zoological gardens. Unfortunately many zoos make no sustained effort to study their animals, or to keep notes on such topics as frequency and duration of sexual receptivity and the development of various behavior patterns prominent in sexual behavior. The chimpanzee has been exhaustively studied (Yerkes and Yerkes, 1929; Bingham, 1928; Yerkes, 1943), but very little precise information on the sexual behavior of the gorilla is as yet available.

My work on free-living gorillas yielded remarkably few observations on sexual behavior. In 466 hours of observation I witnessed only two copulations and one invitation to copulate. The only information in the literature on wild gorillas consists of two cursory reports of supposed copulation by Osborn (1957) and Baumgartel (1960).

On the following pages I have briefly summarized the scanty literature and presented my own observations on captive and free-living animals. It should be remembered that the mating behavior of gorillas in captivity may be highly atypical. Although zoo observations are interesting, they are not necessarily indicative of the behavior of free-living animals. In many respects the behavior of chimpanzees is similar to that of gorillas, and I have drawn comparisons whenever possible.

Estrous Cycle

Many of the physiological changes in the monthly cycle of the female gorilla, in sharp contrast to those of the chimpanzee, are not readily discernible. A prominent genital swelling at estrus is absent, although I noted several times that in wild animals the genital area appeared very slightly turgid. Raven (1936) maintained that he saw a free-living mountain gorilla in the mountains west of Lake Kivu with a genital swelling as large as that found in chimpanzees. This observation has never been repeated, and may have been based on a misidentified chimpanzee or a diseased animal. The menstrual flow of blood is so minute that even in captivity there is frequently no evidence of it. The physiological state of the female is, therefore, not determinable in the wild, and rarely in captivity, except when expressed behaviorally.

Captive females exhibit intermittent periods of sexual receptivity during which they initiate contact with the male or with their human keeper. This period of receptivity lasts for several days, probably at the time of ovulation, and then ceases for the duration of that cycle. Reed and Gallagher (in press) found that the female at the National Zoological Park, Washington, D.C., remained in estrus for three days. Quinn (pers. comm.) told me that Oka, a female at the New York (Bronx) zoo, presented to him usually for three to four consecutive days. Chimpanzees show an estrous period of similar length (Harrison, 1958).

The average duration of the menstrual cycle in man is 28.3 days (Harrison, 1958) and in the chimpanzee 36.2 days (Yerkes, 1943). Lang (1959) noted that in the gorilla it is about 30 days, and the data presented by Thomas (1958), although quite variable, also point to a monthly cycle. Reed and Gallagher (in press) observed that the cycle of a female at the National Zoological Park was 31 or 32 days. Noback (1939), on the other hand, observed that the average length of the first eight periods in one female was 49 days.

Further indicative data were kindly given to me by Mr. Quinn of the New York (Bronx) zoo for the female lowland gorilla "Oka." The dates on which she first showed receptivity are as follows, indicating a cycle of about 31 (25–35) days.

1954	May 8	1955	Jan. 12
	June 10		Feb. 11
	July 12		Mar. 6
	Aug. 10		Apr. 10
	Sept. 10		May 8
	Oct. 9		June 12
	Nov. 11		Aug. 12
	Dec. 12		Sept. 13
			Oct. 12
			Nov. 6

After the birth of her infant, the female lowland gorilla "Christina" at the Columbus zoo remained separated from the male, but in late 1958 was allowed to join him when receptive and occasionally when not. The dates of her receptivity and the responses of the animals to each other again suggest a monthly cycle.

Sept. 25	intercourse
26	none
Oct. 23	intercourse
24	intercourse
25	sex play only
27	none
31	none
Nov. 18	sex play
19	none
20	sex play
21	none
24	none
Dec. 16	none
17	intercourse
18	intercourse

Masturbation

I have twice seen male lowland gorillas masturbate in captivity. Mambo, who at the time of observation was about nine years old and housed with an adult female in the New York (Bronx) zoo, stimulated his penis by handling and then by working it back and forth between index finger and thumb. As the act progressed, his eyes grew rigid and seemed glassy. Sampson at the Milwaukee zoo was about seven to eight years old and housed with another male when I watched him. He stimulated himself with one hand for three to four minutes and thrust several times before being chased by the keeper. No ejaculation was observed in either instance.

A receptive female observed by Yerkes (1928) pressed her rump against the bars of the cage, and once placed the hand of her keeper on her genital area. A female at the National Zoological Park played with her clitoris according to Emlen (pers. comm.). I watched a female at the Basel zoo place her three-month-old infant between her legs and in a squatting position thrust at least ten times. Masturbation was not observed in free-living gorillas.

Erotic Behavior

Nearly all information in this section was obtained from captive animals, which indulged in various erotic acts not observed in the wild.

Captive Gorillas My wife observed a two-year-old male infant, belonging to Cordier, sit and pull in a seven-month-old female and make several thrusting movements. This was the earliest indication of sexual behavior noted. A female mountain gorilla about five to six years old and apparently subadult mounted and thrusted against a dog (Yerkes, 1927). The following year she once threw herself on her back, pressed her genitalia against Yerkes' feet, and determinedly tried to pull him upon her. She also presented on all fours, her body shivering (Yerkes, 1928).

Receptive adult females characteristically initiate sexual behavior by presenting their genital area to the male. At the Columbus zoo, the female backed to the bars, which prevented direct access to the male, and the latter carried on digital intercourse (Thomas, 1958). Sumaili, a mountain gorilla female in the New York (Bronx) zoo, made advances to the male in her cage by rubbing her rump against him and by holding his hand pressed against her genital area. Occasionally she turned on her back with legs spread widely. Once she fondled and licked the testes of the male. The male in turn rubbed the breast and genital area of the female with his hand. Such behavior preceded the first attempts at copulation by the male for more than a year (Quinn, pers. comm.). Makoko, a male lowland gorilla at the New York (Bronx) zoo, initiated genital manipulation with a female three times, and he thrust against her once at the age of about eleven years (Riess *et al.*, 1949). Benchley (n.d.), on the other hand, never observed any form of sexual behavior in two adult male mountain gorillas at the San Diego zoo.

FREE-LIVING GORILLAS The instance below is my only observation on sexual behavior which did not culminate in copulation. On September 7, 1959, at 1700, a female stepped in front of the No. 2 male in the group IV hierarchy. She walked slowly ahead of him with her rump about 3 feet from his face. Her head was turned and she looked back at him as she adjusted her speed of movement to his. He, however, was obviously preoccupied by my presence, and after about thirty seconds the female ceased to present and ambled away.

In the wild, not a single observation on play mounting, homosexual behavior, or any other type of sexual behavior was noted, except two copulations and one invitation to do so. Bingham (1928), Mason (1960), and others have pointed out that the efficiency of the copulatory pattern, especially that of the male, is dependent to a large extent on social experience and practice. In rhesus monkeys, for example, Mason (1960) found that males which were raised in a socially restricted environment often mounted in a disoriented fashion, and that such responses as foot-clasping during mounting were less stereotyped than those of males with previous experience. Both free-living baboons (Bolwig, 1959*b*) and rhesus monkeys (Carpenter, 1942) exhibit a considerable amount of mounting and other sexual behavior which probably provides the social experience for fully oriented copulatory acts in adults. Bingham (1928) noted that all basic sexual acts were present and practiced in young captive chimpanzees but that these tended to be unoriented. A quick reorganization occurred at puberty when sexual behavior appeared in its adult form. Free-living gorillas showed no sexual activity while young, and even blackbacked males were never seen to indulge in any behavior during my presence which could be construed as erotic.

Copulatory Behavior

Brief descriptions of copulation in gorillas have been published by Thomas (1958), Lang (1959), van den Berghe (1959), and Reed and Gallagher (in press). I watched the behavior twice in captivity and twice in the wild.

During my observations I found it difficult to judge if the penis of the male actually penetrated the vagina, or if the animal merely went through the motions of copulating. The penis of the male is quite short for the size of the animal. A masturbating male at the Milwaukee zoo showed about 2.5 cm. of penis. The keeper at the Washington zoo told Emlen that the erect penis is 5 cm. long, a figure also given by Lang (1959); Galloway *et al.* recorded a 7.6 cm. penis in a dead male. The keeper at the London zoo told us that the penis of "Guy" is 9 cm. long when erect; and Thomas (1958) recorded the length of the erect, black penis of the male in the Columbus zoo as 14 cm. long.

In Captive Gorillas Quinn (pers. comm.) observed copulation between an adult female mountain gorilla and a nine-year-old lowland gorilla at the New York (Bronx) zoo on January 29, 1959. This was apparently the male's first copulation attempt. Emlen (pers. comm.) and Reed and Gallagher (in press) observed copulation in the pair of lowland gorillas at the National Zoological Park. These observations, as well as my own notes on the Columbus gorillas, are reproduced below.

New York (Quinn) The female stood quadrupedally and the male covered her from behind. He thrust several times and they parted after ten to fifteen seconds. No erection of the penis was discerned. Immediately afterwards the male lay on his back with legs spread, and the female fondled his testes and licked them.

Washington (Reed and Gallagher) "The first observed coitus occurred when the male was lying supine; the female straddled him, crouched down, and copulation took place, lasting two to three minutes. Most of the pelvic thrust was supplied by the female. Subsequently, during the first period, they were observed to copulate in a standing position, and also in abdominal juxtaposition, the male dorsal.

"The position most frequently observed was the female crouching, with her knees drawn up under her abdomen, her arms folded under her chest, and her chest lowered, almost touching the floor. The male would assume a squatting position behind her, sometimes placing his hands on her hips, sometimes on the floor, sometimes with his hands just hanging at his side."

Washington (*Emlen*) The following notes were taken by Emlen in October, 1960: "0910. She backs up to him crouching low. He pushes her head down with his wrist; then throws his left leg over her rump and starts thrusting. She crouches very low. He thrusts harder and more rhythmically and gets directly behind her. After about 30 to 40 seconds, he everts his lips and gives a soft rhythmical hooting. She gives a faster higher hooting. He dismounts after about 1 to 1½ minutes total. (The keeper was right by the cage.)"

Columbus (*this study*) The adult gorillas were housed in separate but adjoining cages, connected by a barred door through which they could watch each other. Between April, 1956, and July, 1958, the adults had rarely been allowed to copulate. In 1957 they were put together once, and between July and December, 1958, fifteen times. During my visit in 1958 they were together on December 16, 17, and 18 for short periods ranging from ten to twenty-five minutes each day.

The animals were obviously greatly excited when put together; in addition several keepers stood around the cage, and this made the gorillas nervous. The behavior of the pair was, therefore, probably not typical of undisturbed creatures which remain together all the time.

The behavior preceding copulation included three activities, two of which were also seen in different contexts in the wild.

1. The strutting walk: The male approached the female while walking quadrupedally with short, abrupt steps. His body was held very stiff and erect, and his head was tipped slightly upward and sideways (see section on displays). The male thus advanced toward the female and sometimes circled her. He always averted his head but watched her from the corner of his eye. When approached closely, the female likewise turned her head to the side.

2. Wrestling: The male and female lumbered toward each other

bipedally, slowly swinging their arms overhead until close enough to grasp each other behind head and back, behavior also noted in chimpanzees by Bingham (1928). Bent at the waist, they pushed and pulled, and grasped at each other's legs, until finally they toppled over. Then they sat and faced one another, mouths wide open, and emitted almost continuous deep growls and grunts. They mock bit each other's shoulders and they hugged. Several times the male cradled the female in his arms. The whole sequence of events was slow and gentle and appeared ritualistic.

3. Running: The wrestling was several times interrupted when either the male or the female broke loose and led off on a march or run around the cage with the other following closely. This circling of the cage has also been observed by Yerkes (1943) in sexually aroused chimpanzees, which in addition exhibit romping, teasing, petting, fleeing, eating, fighting, and tantrums (Bingham, 1928).

The male did not mount the female on December 16, but he did so three times on the 17th and 5 times on the 18th. The eight attempts at copulation involved a maximum of 109 thrusts, varying from one to forty in number. The total time spent in the mounting position was about ninety seconds, or about eleven seconds per mount. The thrusts were delivered rapidly at the rate of two per second with brief periods of rest during attempts of longer duration.

Copulation was initiated either by the male or by the female. The male typically reached out when the two were facing each other, placed his hand on her hip, swung her around, and with both hands pulled her close as he approached from the rear. Several times the female interrupted the wrestling by rolling onto her back, spreading her legs, and thus accepted the male. The male growled loudly during copulation. Thomas (1958) noted a staccato series of grunts which rose to a climax of guttural sounds.

The following copulatory positions have been recorded at Columbus:

Dorso-ventral

Female on all fours, with male standing upright behind her, his hands clasping her hips.

Female squatting, with male squatting behind her.

Both female and male standing bipedally.

Latero-ventral

Female lying on her side with arms and legs tucked in front of her. Male kneeling behind her, his rigid arms, one placed on each side of the female's body, supporting his torso (Fig. 66D).

Ventro-ventral

Female lying on her back with legs spread, and male squatting between her legs (Fig. 66E).

Female lying on her back with her legs clasping the male's back, and with the male standing on all fours over her (Thomas, 1958).

Male standing upright with female lifted entirely off the ground and supported by the arms of the male, her legs clasping his hips (keeper Phil Amorose, pers. comm.).

Lang (1959) observed both dorso-ventral and ventro-ventral copulations in the lowland gorillas at the Basel zoo; van den Berghe (1959) recorded only dorso-ventral ones in a pair of mountain gorillas at the I.R.S.A.C. station.

In my own observations I never saw an erection or a visible orgasm, and, therefore, conclude that mating was incomplete in these cases. According to Thomas (pers. comm.) the erect penis is easily discernible, and the whole body of the male becomes rigid during orgasm. If the male is aroused, penetration of the vagina occurs promptly, the thrusts are few in number, and orgasm is reached within fifteen seconds (Thomas, pers. comm.), a sequence of events similar to that described by Yerkes (1943) for chimpanzees.

The male and the female continue to fondle each other gently if he reaches orgasm (Thomas, pers. comm.), contrasting with chimpanzees which part immediately after ejaculation (Yerkes, 1943). During my observations the male began to treat the female roughly as he wrestled with her aggressively. All gentleness had vanished and growls of both animals were sharp. The keepers separated them at this point.

The copulatory positions in gorillas are highly variable, contrasting sharply with the stereotyped postures in monkeys. All apes in captivity appear to show this variation in copulatory positions. Yerkes (1943) described dorso-ventral copulation in the chimpanzee. The female crouches low in front of the male, "who approaches from the rear, mounts, sometimes lies along her back, with his head just back of the female's shoulder or again stands or sits half upright." Bingham (1928) reported attempted ventro-ventral copulation in subadult chimpanzees. Both Fox (1929) and Brandes (1930) describe copulation in the orang-utan with the female lying on her back and the male mounting her in a sitting or crouching attitude. Thomas (pers. comm.) has seen dorso-ventral copulation in orang-utans, and Zedtwitz (1930) reported orangs copulating while suspended by their hands from the bars at the top of the cage. Similarly, Coolidge (1933) observed dorso-ventral copulation in a pair of

Hylobates leucogenys while suspended by their hands from a bar. I observed copulation in captive *Hylobates lar* in a dorso-ventral position while crouching on a branch, and Mr. Harold Hayes, the late director of the Madison, Wisconsin, zoo, noted ventro-ventral copulation in the same species. Comparable variability in copulatory positions has not been reported in monkeys.

IN FREE-LIVING GORILLAS My two observations of copulation in wild gorillas are described in detail below. These occurred in group IV on September 4 and on September 23, 1959. The female involved appeared to be the same in both cases, but I am not certain of this. However, the female who copulated on September 23, 1959, gave birth to an infant in late December, 1959. (Lang, 1959, noted that the female lowland gorilla in the Basel zoo continued to be receptive for two months after the onset of pregnancy.) Two different silver-backed males were involved, both subordinate individuals. One occupied the number two position in the hierarchy and the other was a peripheral animal. On the whole, the behavior in the wild was much more direct and lacked the extensive preliminary wrestling, chasing, etc., which characterized the zoo animals.

1. *1415–1430. The dominant male rests with several members of his group on a knoll. Some ten feet away and hidden in part by brush are a second male, a female, and several other animals. Suddenly a sound, best described as a series of rapid and abrupt ö-ö-ö-ö-ö, followed by a short audible inhalation and then another series (see section on vocalizations), gives the first indication that a copulation is in progress. Two or three minutes after the sounds began, the vocalizing male and a female appear from behind the shrubs. The female rests on her knees, belly, and elbows, with the male mounted behind, holding onto her hips (Fig. 66C). The male apparently pushes, and what with the slope being steep, the animals advance downhill 40 feet in 15 minutes. The female parts the vegetation with her hands as they progress. The copulation sound is given almost without interruption by the male through pursed lips. While moving the male does not thrust, but three times they stop and he thrusts a total of 14 times. After about 10 minutes the copulatory sound becomes harsher and harder, and the female emits a short piercing scream. The male has her clasped now by the armpits and he is nearly covering her back. They have come to rest against a tree trunk. The male thrusts rapidly and a hoarse, trembling sound escapes his parted and protruded lips, interrupted only by sharp intakes of breath. The sound becomes nearly a roar—and suddenly he sits back. The female lies motionless for 10 seconds, then walks slowly uphill. The male sits a minute and then he too goes.*

During this whole procedure the dominant male lay on his knoll and the copulating pair was in plain sight, but he paid no obvious attention. A female fed within 20 feet of them without interruption.

2. 1135–1240. The group rests scattered over a steep hillside. A peripheral male stands looking down the slope. A female appears behind him, clasps him around the waist and mounts him, thrusting about 20 times. The male at first pays no attention, but after about 10 seconds turns his head and looks at her. Suddenly he swivels around and sits. With his right hand he reaches over, grabs the female by the hip and pulls her to him. She sits in his lap, facing away from him, her body supported by stiff arms propped on the ground (Fig. 66B), as the male thrusts about 10 times.

The dominant male, who has been resting 15 feet away, slowly rises and walks toward the pair. The copulating male immediately desists and ambles 10 feet uphill. The dominant male sits beside the female one minute, then moves and rests 15 feet away. The other male reoccupies his former spot, and the female approaches him. He reaches out with both hands, swivels her around and again pulls her into his lap (Fig. 66A). He thrusts rapidly, about two times per second. After about 70 thrusts he begins the copulatory sound—the rapid ö-ö-ö noted in the previous copulation. The female waves her head slowly back and forth, and, at about the hundredth thrust, suddenly twists sideways and sits beside the male, who then rolls over and rests on his abdomen for 10 minutes.

Soon, however, he sits and again pulls the female to him, and after about 70 thrusts begins the copulatory sound. She turns slightly sideways as if wanting to leave, but he holds her tightly. Yet she breaks away and both rest side by side.

1239. After a prolonged rest, the female rises and stands by the rump of the male. He glances up and they stare at each other. The process of pulling her into a sitting position and thrusting is repeated. At about 75 thrusts he begins his copulatory sound. His eyes are closed, and the thrusts rock her back and forth, a motion not only aided with his hands on her hips but also by her body. His lips are pursed and the sounds grow more rapid, becoming slurred; her lips are also pursed and her mouth slightly parted. At about 120 thrusts the male suddenly opens his mouth with a loud sighing "ahh," and the female opens her mouth at the same time. He relaxes but she rises and leaves.

The dominant male, who had been lying 15 feet away without apparent response, rises and approaches the other male who moves off uphill.

The copulation, between the initial contact and orgasm, was in this instance broken into 3 separate periods and spread over about one hour. However, the actual time of mounting was only about 4½ to 5 minutes and comprised some 300 thrusts.

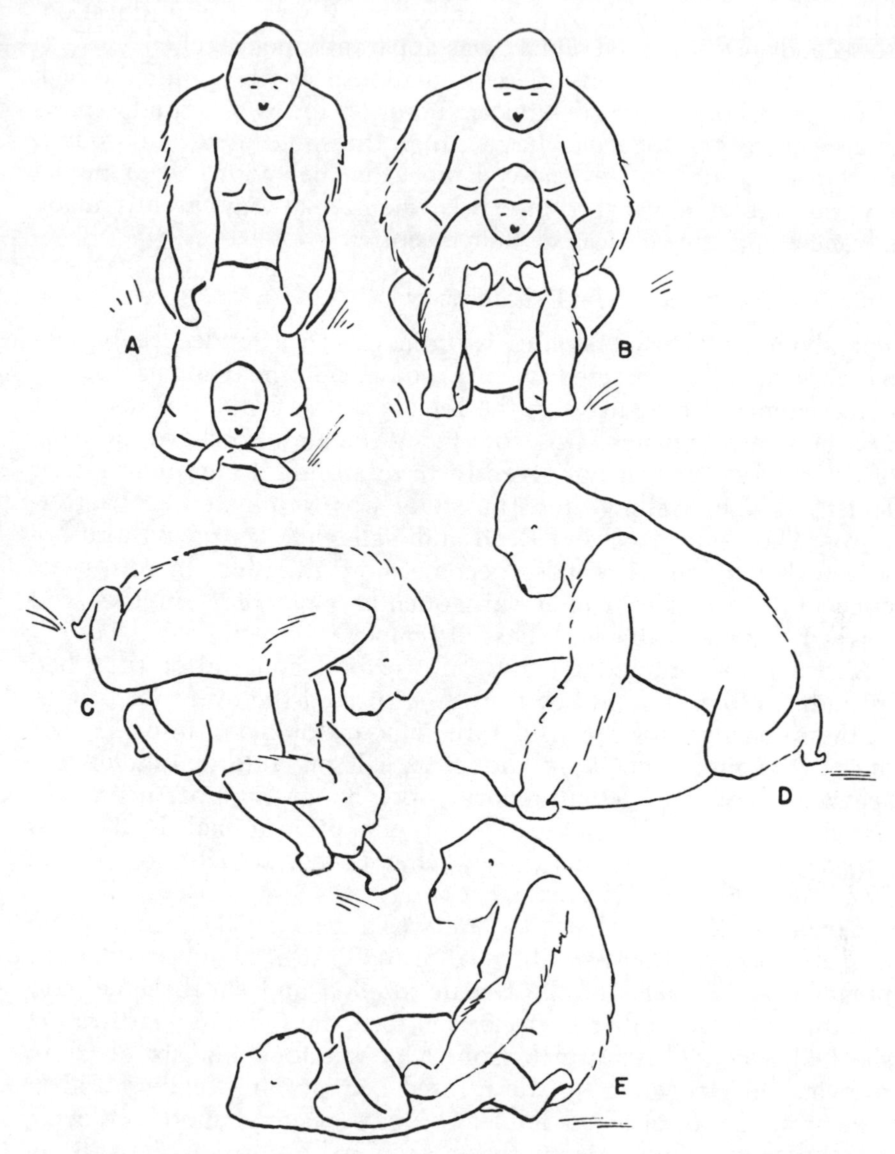

Fig. 66.—Copulatory positions of gorillas at Kabara and in the Columbus zoo. *A–B,* Kabara: the silverbacked male squats, holding the sitting or prone female in his lap (group IV, September 23, 1959); *C,* Kabara: the silverbacked male covers the back of the female (group IV, September 4, 1959); *D,* Columbus zoo: the male covers the female who is lying on her side; *E,* Columbus zoo: the male squats between the legs of the supine female.

In addition to the two copulations, I heard the distinctive vocalization of this activity three more times, but it was of brief duration, 10 seconds or less, and a climax was apparently not reached.

The infrequency of observed copulations, together with the lack of sex play, homosexual behavior, and other forms of sexual expressions, suggests that gorillas have a high threshold of sexual arousal. In the wild, adult males showed no sexual behavior except in the few cases at widely spaced intervals when females apparently made advances and thus indicated their receptivity.

Pregnancy and Parturition

The abdomen of wild females is ordinarily so extended that pregnancy cannot be detected by the body contour. Thomas (1958) noted edema of the ankles in the female at the Columbus zoo after 210 days of pregnancy. One free-living female also showed edema of the ankles, but had not given birth to an infant one month after I first saw the swelling, and the study was terminated soon afterwards. Thomas (1958) and Reed and Gallagher (in press) also observed that pregnant females become more irascible. In retrospect none of the females which subsequently gave birth in the wild showed any noticeable shortness of temper.

Twice I encountered females only several hours after they had given birth to an infant, but I never witnessed the event itself. One of these females lost a considerable amount of blood into her nest during the night following the birth. All my information on the special behavior of females during pregnancy and parturition is, therefore, based solely on the accounts of captive animals by Thomas (1958), van den Berghe (1959), Lang (1959, 1961*b* and *c*), and Reed and Gallagher (in press).

Columbus Zoo Lowland Gorilla (Thomas, 1958) Conception probably occurred on April 6, 7, or 8, 1956. After about 210 days of pregnancy the ankles of the female swelled and she became very irritable. On December 8, 1956, fourteen days before parturition, she had a convulsive attack from which it took her six hours to recover. On December 20 she refused her morning feeding and in the afternoon discharged a clear, serous, vaginal fluid. At 0800, December 22, she stood quietly on all fours, a position she still occupied at 0830. When she was checked again at 0850 the infant lay on the floor still encased in the amniotic sac. She had made no sound, and her face had a dazed expression. The infant was removed and raised by hand.

I.R.S.A.C. Mountain Gorilla *Number 1* (van den Berghe,

1959) The pregnant female became noticeably rounder, and her appetite lessened as parturition drew near. At 1215, on October 26, 1959, she interrupted her meal, lay down on her left side, raised the right thigh, and within five minutes the infant appeared. The amniotic sac ruptured when the head became visible. Then, half sitting and half lying, she took the head of the infant into her hands and pulled it out. The newborn began to move as it lay on her thighs and she broke the umbilical cord. She then raised the infant to her mouth, bit off a foot, then a hand, and finally punctured its head with a canine. The corpse was removed from the female immediately afterwards, and the course of events related to van den Berghe.

Number 2 (Cordier, pers. comm.) The same female gave birth to another infant on April 1, 1961. She cared for the infant this time, but it died about three weeks later.

Basel Zoo Lowland Gorilla *Number 1* (Lang, 1959) The last period of receptivity noted in the female was on August 21, 1957. On March 29, 1958, the female aborted a dead fetus 10.5 cm. long. She became receptive again on May 16. After a copulation on December 8, 1958, she was suspected of being pregnant again, and tests confirmed this on March 2, 1959. During the following five months she gained 7.5 kg., her stomach protruded more, and her breasts grew prominent. On June 7, 1959, she squeezed milk from her breast. The birth occurred during the night of September 22 and 23. By 1100 the umbilical cord had been ripped off near the abdomen. (The umbilical cord of an infant which I found dead at Kabara was torn off 13 cm. from the abdomen.) Although the female carried the infant carefully, she made no attempt to aid it in reaching the nipple. When on the following day the infant was obviously hungry, it was removed from the female.

Number 2 (Lang, 1961*b* and *c*) The female and male were placed together one month after the birth of infant No. 1. The female became sexually receptive in November, 1959, and she continued to show monthly receptivity until August 8 and 9, 1960. A urine test on October 9, 1960, showed that she was pregnant. She was seen to squeeze milk from her breast in January, 1961.

When the keeper entered the cage at 0705 on April 17, 1961, the infant had not been born, but when he returned at 0735, the female held the newborn in her arm, its umbilical cord still reaching her vulva. She carried the placenta around with her until midday, when she detached it from the cord and abandoned it.

The female carried the infant, a male, carefully to her chest, stopping now and then to inspect or lick its hands and feet. She showed less nervousness in handling her second infant than her first one, and it was observed to suckle within 24 hours after birth. The infant continued to receive good care from its mother in the ensuing months, and at the date of this writing (January 20, 1962) remains in good health.

WASHINGTON ZOO LOWLAND GORILLA (Reed and Gallagher, in press) The female showed a normal estrous period beginning on November 28, 1960, but only a brief one, lasting one day, on December 28. On March 21, 1961, a urine test confirmed pregnancy. Her breasts began to enlarge in April; in May her abdomen swelled noticeably and she grew more irritable. The keeper felt the movement of the fetus in late May when placing his hand on the abdomen of the female. She had swollen ankles two months before the birth, but the edema subsided two to three days later. Ten days before parturition, she squeezed a clear fluid from the breast.

On September 9, 1961, at 0658, the keeper discovered the infant, which had been born during the previous hour. The female carried it in her hands rather than holding it to her chest. The umbilical cord and placenta were still attached.

Shortly after 0700 the female placed the infant on the floor, and, in general, showed little maternal interest. Apparently she consumed the fetal membranes somewhat later. Neither the male nor the female showed any "sign of emotional disturbance" when the infant was removed from the cage.

All four of the mothers did not or could not take good care of their first-born, which perhaps indicates, as has been shown in various other mammals, that some of the appropriate responses are learned, and that part of this learning is facilitated by observing other females in the group handle infants. Both females which subsequently gave birth a second time cared for their infants for at least several weeks.

Gestation period

Data on the length of the gestation period of gorillas vary considerably.

Probable Date of Conception	Date of Birth	Probable Gestation Period in Days	Source
Apr. 6–8, 1956	Dec. 22, 1956	251–253	Thomas (1958)
Dec. 8, 1958	Sept. 22–23, 1959	289	Lang (1959)
Aug. 8–9, 1960	Apr. 17, 1961	252	Lang (1961*c*)
Nov. 28–Dec. 28, 1960	Sept. 9, 1961	about 266 (256–295)	Reed and Gallagher (in press)

Similar differences in the length of the gestation period occur in chimpanzees and man. The gestation period of forty-four chimpanzees varied from 202 to 261 days with an average of 231 days (Yerkes, 1943). The average for the orang-utan appears to be about 233 days (Morris and Jarvis, 1959). The average human period is 267 days with a variation of 250 to 285 days (Harrison, 1958).

Breeding Season

Since the recent discovery of a season of preferential breeding in *Macaca fuscata* (Imanishi, 1960) and *Macaca mulatta* (Southwick *et al.*, 1961; Koford, per comm.), and similar suggestive data for *Cercopithecus ascanius* (Haddow, 1952), considerable interest in this phenomenon among primates has been aroused. Burton (1876), who obtained his information from natives, claimed that gorillas in West Africa had a breeding season in December. Sharp (1927) noted that young gorillas were most commonly seen in February and March. Kawai and Mizuhara (1959*b*) inferred that gorillas at Kisoro have a breeding season because six of the young seen by them were of somewhat similar size.

To determine if the young of gorillas are born at certain times of the year at Kabara, I estimated the probable months of birth of all infants of the ten groups in the area. The results are presented in Table 59. The age estimates between about January, 1959, and April, 1960, are, I believe, fairly accurate and vary probably by no more than one month; those for late 1958 show a slightly larger error. But for infants born in late 1957 and early 1958 an error of up to three or four months must be allowed. Notwithstanding the inaccuracy of some of the data, Table 59 indicates that births occur throughout the year. Possible significant differences are not detectable in a sample of this size and variability, but the evidence points to a lack of preferential breeding season in gorillas.

Aggressive and Submissive Behavior

Although I have already described various instances of aggressive and submissive behavior, especially in the sections on intra-group behavior and interactions between groups, the behavior itself is treated here in greater detail.

Aggressive Behavior

The intensity of aggressive behavior which a gorilla exhibits in response to a particular situation varies considerably. In the order of increasing intensity, the responses include: (1) fixed staring or a jerk or snap of the head in the direction of the opposing animal; (2) a short forward lunge of the body; (3) a direct charge; and (4) physical contact in the form of biting or wrestling.

1. STARING An unwavering direct stare is a form of threat. Sometimes a furrowed brow is noticeable, and the lips are occasionally parted slightly or even pursed. Short, abrupt grunts, indicating annoyance, are frequently emitted in conjunction with the stare, especially if directed at a man. The stare occurs commonly during intra-group dominance interactions, where it usually is of brief duration. More prolonged staring was twice observed between males of different groups, and once between the leader of a group and a lone male.

The animal may give this threatening stare while sitting, standing quadrupedally, or while walking in the direction of the other animal. It is most commonly shown by males, although juveniles and females occasionally stared at me and at other gorillas in what appeared to be threat.

Threatening by means of an unwavering stare is common in a wide variety of mammals including other primates and man.

Another threatening gesture consists of jerking the head in the direction of the opposing animal. Sometimes the mouth is half open but with neither teeth nor gums visible. The head-jerk occasionally ends with an audible snap as the jaws are shut.

2. FORWARD LUNGE OF BODY Gorillas show incipient charges by a slight lunge of the body. Sometimes the animals do this without moving the feet, but usually they take one or two abrupt steps. I saw such lunges about ten times in free-living gorillas, and none were accompanied by a sound. However, if the lunge is directed at man, the male may roar and sweep one arm in front of him as if slapping at something. Merfield and Miller (1956) describe such behavior graphically: ". . . the gorilla was standing on all fours behind me, giving vent to short, sharp coughs that turned into prolonged screams of rage as waves of anger overwhelmed him, and making sweeping lunges at me with his right arm."

3. BLUFF CHARGES Definite aggressive bluff charges over distances of 10 to 80 feet toward another animal or a human observer were rarely observed. On one occasion the dominant male of group VII charged the dominant male of group XI. His charges were quadrupedal, silent, and at a fast run, but he stopped before making contact (see section on interactions between groups).

When excited by my presence, group members would occasionally run at and slap each other in what appeared to be redirected aggressiveness, which, although elicited by me, was vented at another animal. I once saw two silverbacked males, and several times two females, run at each other, slap with one hand, and continue without further contact. However, such slapping sometimes provoked quarrels.

Bluff charges by male gorillas at man are frequent and intense if the extensive literature on the subject is indicative (Du Chaillu, 1861; Barns, 1922; Akeley, 1923; and many others). Merfield and Miller (1956) describe a typical attack: "When I showed signs of advancing, he dropped on all fours and charged, screaming and showing his teeth as he came." On the other hand, a charge experienced by Bingham (1932) was silent. A silverbacked male once ran roaring 20 feet in my direction, but stopped 80 feet away. This was the most intense charge to which I have been subjected. When the tendency to charge is combined with the conflicting tendency to retreat, it may take the form of a lateral run. The special behavior commonly observed when gorillas are hunted or otherwise seriously disturbed is discussed in the section on response to man.

A form of charge, often seemingly playful, is sometimes seen in zoos, when the animal runs and actually bumps the wall or another gorilla with the shoulder in passing. I have seen the males at the Columbus and the New York (Bronx) zoos charge in this fashion. A male infant mountain gorilla, about two years old and belonging to Cordier, also ran at and bumped into my wife several times during their first few meetings. Benchley (1942) noted similar behavior by two males in the San Diego zoo. In the wild it was observed once between two infants in play.

4. Physical Contact I have not witnessed serious aggressive contacts between gorillas. Although some animals appeared to quarrel violently on several occasions, the grappling, screaming, and mock biting never resulted in a discernible injury. Only females appeared to be involved in these quarrels directly. In this context it is perhaps significant that females seem to lack a definite dominance hierarchy. On at least three occasions several females not directly concerned with the quarrel entered into the scramble, a trait also noted in baboons (Zuckerman, 1932; Bolwig, 1959*b*). Twice the dominant male stopped the squabbling by merely walking toward the fighting animals and emitting annoyed grunts. One example from group VII describes a typical quarrel:

A female walks leisurely past another one sitting by the trail. The latter slaps her on the back for a reason unknown to me. She in turn wheels around and runs with open mouth straight at the female who swatted her. This female cowers down with legs and arms tucked under, but with head raised screaming loudly. Her lips are curled up and the teeth and gums show. The two females then grapple briefly and mock bite each other's shoulder.

As the two fight, two other females run up and join the melee. All four then scream, grapple with each other, and run around with teeth bared. The rest of the group watches; that is, all but the silverbacked male who sits 5 feet from the nearest combatant and does not even turn his head toward them. After about 15 to 20 seconds three females cease fighting and walk away. Only one female remains in the battle area and emits short screams. Suddenly she takes two steps after one of the retreating females and slaps at her hind leg, whereupon the latter one turns and advances screaming. The former one backs away and collides with the silverbacked male, who gives an annoyed grunt. The two females meet and wrestle briefly as a third female runs up to join the hassle. Finally all part, the whole sequence having lasted about one minute.

Direct attacks on humans are described in the section on response to man.

Submissive Behavior

In response to a threatening stare or a bluff charge other animals react with three characteristic gestures: (1) they turn their head to the side; (2) they shake the head back and forth; and (3) they cower down.

1. Turning of the Head Gorillas in the wild and in captivity frequently use head-turning to indicate submissiveness in response to a threatening stare by another gorilla or a human being. The gesture is also used to indicate that aggression is not intended. Thus, for example, the male at the Columbus zoo did not face the female directly as he approached her before copulation, but looked at her out of the corner of his eye while his head was averted.

The male at the Columbus zoo once responded to my threatening stare by mock biting his wrist, and another time by placing his hands over his eyes. Similar behavior was not observed in the wild.

2. Head Shaking Head shaking is a relatively rare gesture, apparently signifying submissiveness and appeasement in situations in which the animal feels unsure of itself. I saw it about twenty times and then only in response to me. The gorilla characteristically sat or stood quadrupedally and shook its head fairly rapidly back and

forth. The behavior was noted three times in silverbacked males, twice in females, and the remaining times in blackbacked males. They usually shook their heads only when quite close to me, within 60 feet or less.

I conducted several crude tests to ascertain which situations would elicit head shaking. One silverbacked male, for example, sat within 30 feet and watched me. I fixed my gaze upon his face, and he became obviously uneasy, finally jerking his head to one side. When I continued to stare, he began to shake his head from side to side. Finally he rose, beat his chest and walked away. Similarly, a blackbacked male stared at me from a distance of 30 feet. When I shook my head, he averted his gaze. Then, when I in turn stared at him, he began to shake his head. We continued this for ten minutes. Blackbacked males sometimes shook their heads as they approached me very closely. I too employed head shaking when inadvertently meeting the animals at close range; it appeared to reassure them. In general, the gesture appeared to signify "I mean no harm."

A curious modification of head shaking was noted twice in two different blackbacked males. While shaking the head, they relaxed the muscles of the lower jaw so that the mouth hung open. The lateral movement shook the jaw back and forth, producing a rattling sound.

3. COWERING Gorillas cower when attempting to escape the slap of displaying males and when indicating submissiveness in the presence of a dominant animal. The animals crouch on the abdomen, head lowered and arms and legs tucked under, presenting only the broad back. One juvenile held its hands over the back of its head while cowering.

Cowering protects the vulnerable parts of the body, and it also appears to inhibit further aggressiveness. Two infants wrestled vigorously, but when the smaller crouched down the larger desisted. Similar behavior was observed between two infants in the Columbus zoo.

This submissive posture in gorillas has a striking parallel in man. As noted by Lorenz (1957), captives begging for mercy crouch down. Among the Nyakyusa of southwestern Tanganyika a woman greeting a man must crouch and look away from him (Wilson, 1951). This submissive gesture has apparently become ritualized as a greeting. The deep bowing of Japanese or the tipping of the hat and nodding of the head as practiced in Western society was possibly derived from a submissive act which formerly was intended to convey upon meeting that no aggressiveness was intended.

CHAPTER 7

RESPONSES TO ENVIRONMENT

Response to Environment

Various environmental factors influence the behavior of gorillas. The daily fluctuations in the weather have an effect on their activity as noted in various chapters throughout this report. Gorillas also share the forest with numerous other creatures, most of them innocuous, but some, like leopard and man, large and potentially dangerous. The purpose of this chapter is first to describe in detail my observations on the gorilla's response to the vagaries of weather, and second to relate the interactions between gorillas and other species of animals.

Response to Weather

Over most of its range, the gorilla is not subjected to extremes of weather. However, around Kabara, where I made most of the observations, ground temperatures may drop below the freezing point, and rain may turn to hail.

Response to Sun   The gorilla is often erroneously pictured as a creature of the deep forest, lurking in the shadows, avoiding the sun (Geddes, 1955; Schäfer, 1960). Actually the animals readily utilize several vegetation types, like young secondary forest and the giant senecio zone, in which they are subjected to direct sun, sometimes for prolonged periods. At Kabara the gorillas seemingly enjoy the sun when it appears. Their reaction is often immediate: they roll onto their backs, spread their arms above the head or to the side, and expose their chests to the rays. I have observed animals lie in the direct sun for more than two hours, with beads of sweat forming on their upper lip and rivulets of it running down the chest. I have not made a single observation in which a gorilla avoided the sun, but I have watched one or several animals move to and sit in a spot of sun as far as 20 feet from where they had been resting.

RESPONSE TO RAIN In the Kayonza Forest my one encounter with gorillas during a rain revealed that the animals were merely sitting in it. One group, which Emlen and I met during a rainstorm at Kisoro, had taken refuge under a rock overhang, thus keeping dry. Schäfer (1960) claimed that gorillas in the Utu region became especially active during rains, but no evidence in support of this statement was obtained.

The response of gorillas to heavy rain was inconsistent at Kabara. I made twenty-five observations on various groups at the onset of a heavy downpour. On ten occasions all or most of the animals moved as much as 80 feet to sit against the trunks of trees, where they remained dry. On fifteen occasions, the gorillas merely sat in the rain.

Light rain affects the behavior of the animals very little. They rarely interrupt their feeding, but if they have been resting on their backs, they often turn over on their bellies or sit up. At the beginning of a heavy rain all feeding activity ceases, animals which have been sitting in trees descend, infants return to their mothers. The gorillas either move under the canopy of a tree or remain in the open. If they choose the latter, they sit hunched over, head held low, with the chin sometimes resting against the chest or on the folded arms. Arms are either crossed on the chest, or the left hand is placed on the right shoulder and vice versa. Large infants sit close to their mothers, facing their chests. Often the female enfolds her large infant with her arms and leans so far forward that the offspring remains dry. Twice I encountered two juveniles as they held each other chest to chest. The animals sit motionless, soundless, letting the water stream off their shoulders and drip from the brows. They give the impression of being thoroughly miserable. At such times gorillas are not very excitable. I once walked inadvertently through a group of huddled animals and only one raised its head. On another occasion I arrived in full view of the gorillas and sat on the ground under the partial overhang of a tree trunk within 20 to 30 feet of them. Although they glanced at me, they did not move.

A group which remains in the rain one day, may on a subsequent day move under the canopy of a tree, either as soon as the rain starts or within five minutes after its beginning. On at least five occasions the silverbacked male initiated the movement to a shelter; once a female headed for a tree first; and the remaining times several animals appeared to move together. If the tree is large, the whole group may crowd closely together; if several trees or fallen logs provide shelter in the vicinity, the group may split up. Sometimes one or more of the gorillas remain in the rain while others seek shelter.

Activity usually does not resume until the rain subsides. Occasionally the animals wait five to fifteen minutes after the rain has ceased before venturing forth. If, however, the downpour is prolonged,

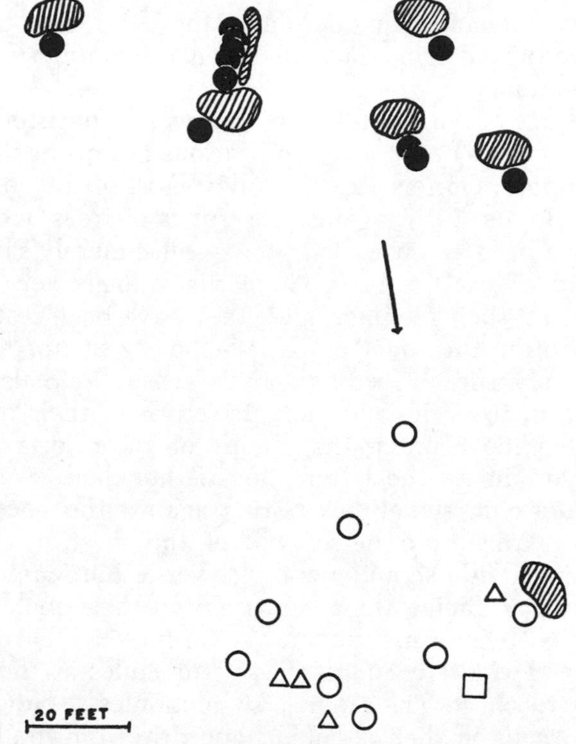

FIG. 67.—Nesting response of group VII to rain. The animals rested in the shelter of tree trunks (hatched areas) during a rainstorm in the late afternoon of March 19, 1960. Later, toward evening, they moved away from the trees to nest in the open in spite of the continuing downpour. Slope of terrain, flat.

black symbols resting spots
white symbols night nests
→ direction of movement
□ silverbacked male
○ blackbacked male or female
△ juvenile

they may feed in the rain. One group, for example, was resting at 1445 when heavy rain began to fall. At 1615 the rain lessened somewhat and the male, followed by his group, began to feed immediately. On another occasion the group sat in a downpour from 1030 to 1430, then slowly began to feed in spite of the rain.

The irregular response of the animals to rain appeared again at

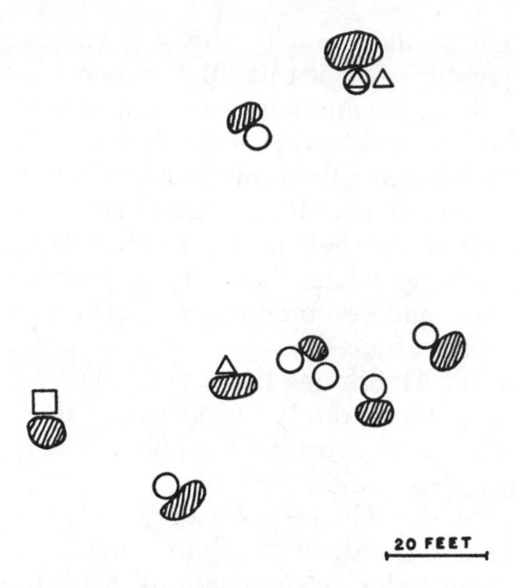

FIG. 68.—Nesting response of group VII to rain. Heavy rains fell during the late afternoon of October 28, 1959, and the gorillas built their night nests in the shelter of tree trunks (hatched areas). Two nests were not found. Slope of terrain, flat to 10°.

□ silverbacked male

○ blackbacked male or female

◬ female with juvenile

△ juvenile

the time of building the night nest. On one occasion, during a long rain, the animals rested under the canopy of trees in the late afternoon but moved out into the open to nest (Figure 67). On another occasion, when it rained at bedtime, the gorillas nested under the shelter of trees (Figure 68). To determine if gorillas tended to nest more frequently under the canopy of a tree trunk when it rained than when it did not, I recorded every nest which was placed in such a way as to keep the animal dry in case of rain. In addition, I noted the time and duration of every rain between the general nest building hours of 1630 and 1830. Only the nests of groups IV, VI, VII, and VIII are included in the calculations.

		Sheltered Nests	
		No.	Per Cent
Total number of nests built when it was not raining during the nest building period	1,502	74	4.7
Total number of nests built when it was raining during the nest building period	335	62	18.5

The data show that gorillas tend to choose a sheltered location to nest in more frequently when it rains than when it does not. However, over 80 per cent of the animals at Kabara slept in the rain even though shelter was almost always available within a few hundred feet. I obtained no evidence to show that gorillas ever abandoned their night nest if rain fell after it had been built.

Wet gorillas never shook their bodies in the manner of dogs and gibbons. Once a female shook her head, apparently to dislodge some water from her brow, and several animals wiped their hands down the sodden hairs of the arm.

Response to Hail Hailstones the size of marbles fell occasionally at Kabara. Such storms rarely lasted longer than half an hour. During two hailstorms the gorillas remained in the open, sitting hunched over; once they sought shelter.

Response to Thunder The response of gorillas to thunder varied with the intensity and suddenness of the sound. Thunder which merely rumbled around the rocky summit of Mt. Mikeno, as it was wont to do, elicited no response in at least three observations. But twice a violent thunderclap startled the gorillas as well as myself. Once the male jerked up his head and looked at me; another time several animals looked around.

Response to Fog Fog is almost a daily occurrence, especially above 10,000 feet. Gorillas respond in no visible way to it. The frequency of vocalizations does not increase.

Response to Wind Winds rarely blow hard under the canopy of trees, but on the more open slopes strong gales commonly bend the trees, and intermittent blasts funnel up the canyons. Gorillas respond in no definite way to wind, although it was my general impression that they were somewhat nervous on windy days.

Response to Water

Gorillas appear to possess a strong antipathy to entering streams. Cordier told me that one group refused to escape across a stream 2 to 3 feet deep and 20 feet wide when natives chased it with spears. The natives themselves repeatedly showed me natural log bridges which, according to them, gorillas used regularly to cross narrow streams, rather than entering the water directly. Sabater Pi (1960) noted a similar avoidance of water in West Africa.

In the Kayonza Forest, a group crossed the same creek three times in one day. It was a small stream, very clear, 7 to 8 feet wide, and at most 1 to 2 feet deep. Once the group crossed on a fallen log which spanned the creek; once they walked across on three tree ferns which they had pushed over; and a third time I watched them as they jumped onto a rock in midstream which was submerged 2 inches before gaining the opposite bank with another leap.

Interactions with Other Animals, Excluding Primates

Throughout their range, gorillas share the forests with numerous other animals, large and small. At the lower end of the size scale are various insects, some, like blackflies and mosquitoes, bothersome to man; at the other end such large creatures as the African buffalo (*Syncerus caffer*) and the elephant (*Loxodonta africana*).

Interactions with Insects There were no bothersome biting insects at Kabara, and I observed no responses of gorillas to insects of any kind.

Interactions with Reptiles Snakes and lizards are not conspicuous in the forests. Only *Chamaeleo bitaeniatus* and *C. johnstoni* occur in the Virunga Volcanoes above an altitude of 8,000 feet. I observed no encounters between reptiles and gorillas. Geddes (1955) relates how a captured group refused to enter a cage in which a snake lay hidden, and he describes the death of a female gorilla presumably bitten by that snake.

Interactions with Birds The variety of birds in the *Hagenia* woodland around Kabara is relatively small. Most birds, like the southern double-collared sunbird (*Cinnyris chalybeus*), are inconspicuous, and gorillas did not respond to their presence. All my observations of the reactions of gorillas to birds are listed below.

Buteo oreophilus A hawk of this species sailed low over a group. As its shadow skimmed over two silverbacked males, they both ducked and then looked up. On another occasion a silverbacked male and a blackbacked male followed the soaring flight of a hawk with their eyes. On a third occasion a buteo sat quietly in a tree above the resting group. Suddenly it launched itself and flew to within 30 feet of the animals. One female gorilla screamed, the male roared, and the bird landed in a tree 50 feet away. One female then shifted her position and watched the hawk.

Columba arquatrix An olive pigeon flew with loud flapping wings out of the foliage of a tree. Two gorillas ducked.

Ruwenzoronis johnstoni Once a blackbacked male, who watched me intently, did not respond to a turaco which fluttered by within 15 feet of him. On another occasion a turaco suddenly emitted its piercing squirrel-like chatter directly above a sitting group; several gorillas jerked visibly, but none looked up.

Francolinus nobilis A francolin sat on a log within 80 feet of a group. No gorillas responded to its presence.

Woodpecker (probably Mesopicus griseocephalus) A bird hammered loudly on a tree, whereupon a male and a female gorilla searched for the sound with their eyes.

Corvultur albicollis A pair of white-necked ravens nested on the bluffs of Mt. Mikeno. Through constant feeding they had become tame enough to land on my wife and to accept food from my hand. Once, as I was eating my lunch in the presence of the gorillas, the ravens flew over. I whistled, which to the birds signified food, and both descended. As they approached, their shadows fell on the gorillas which ducked. Then the male roared and females screamed, some facing me, others the birds. As if in play, the ravens swooped directly at the gorillas several times, causing the male to roar repeatedly and the females to mill about. After accepting food from me, the birds departed, but both returned an hour later to again fly in unison at the gorillas. The silverbacked male stood on four legs, faced the ravens, and roared.

Most of the responses of gorillas to birds were elicited by their sudden appearance. In two of the examples given, gorillas were startled by loud noises which affected me similarly. Dark moving shadows caused gorillas to duck, and I found myself occasionally behaving likewise in comparable circumstances. The most violent reactions occurred in response to direct rapid approach by birds.

INTERACTIONS WITH MAMMALS Although the Virunga Volcanoes support many kinds of mammals, gorillas probably have only fleeting contacts with most of them. At least a dozen species of mice, voles, and shrews scurry through the undergrowth. The mongoose (probably *Mungos*), genet (*Genetta*), and hyena (*Crocuta crocuta*) are primarily nocturnal. I rarely saw the giant forest hog (*Hylochoerus meinertzhageni*) and red forest buffalo (*Syncerus nana*). Lions (*Felis leo*) have not been recorded in recent years, but they formerly ranged widely through the mountains (Prince Wilhelm of Sweden, 1923; Philipps, 1923; Hoier, 1955). At Kabara no bushbuck (*Tragelaphus scriptus*) were seen above an altitude of

9,300 feet. My observations of interactions between gorillas and the more conspicuous mammals are summarized below.

Tamiscus sp. The small gray squirrels were shy and quiet. One juvenile gorilla was seen to follow a running squirrel with its eyes.

Dendrohyrax arboreus Most hyraxes became active at dusk, but some individuals, especially young ones, ventured forth during the day or peered out from their dens in hollow trees and under rocks. They were shy and ducked out of sight at the first sign of possible danger. Once a hyrax scrambled out of a fallen log on which a gorilla sat. Although several gorillas probably saw it, none responded.

Loxodonta africana Lone elephants and small herds occurred throughout the lower altitudes of the Virunga Volcanoes. On Mt. Gahinga they penetrated upward to the 11,400-foot summit, but I noted them only once above an altitude of 9,500 feet in the Kabara area. By localizing their activity, elephant herds may cause considerable damage to the vegetation. No fresh gorilla sign was seen in areas heavily trampled by elephants in the Virunga Volcanoes and in the Kayonza Forest, and I received the impression that gorillas avoid such greatly disturbed areas. At Mt. Tshiaberimu, where elephants and gorillas both fed on bamboo shoots, we commonly saw fresh sign of the two species on the same slope.

I noted no direct encounters between gorillas and elephants. Once a silverbacked male lifted his head at the distant trumpeting of an elephant. One elephant apparently walked past group V at night, with the nearest gorilla only 50 feet from the pachyderm. There was no evidence that the gorillas abandoned their nests, and the elephant did not deviate from its course. Merfield and Miller (1956) observed gorillas and elephants feeding peacefully in close proximity to each other; once the roaring of a male stampeded the elephants.

Cephalophus nigrifrons Red forest duikers were numerous from the park boundary to an altitude of over 13,000 feet. They were timid creatures: I repeatedly observed them quietly moving away from an approaching gorilla group. When startled they emitted a peculiar pshi-pshi sound at which the gorillas occasionally raised their heads but showed no other reaction.

Syncerus caffer Lone animals and herds of two to five buffalo were common around Kabara. Large herds of up to twenty animals frequented the bamboo zone and the Rukumi meadow, with some animals penetrating upward to the summit of Mt. Karisimbi. Although the buffalo were primarily nocturnal, a few fed and moved about at all times of the day.

Buffalos trampled the vegetation considerably at times, and their well-worn trails criss-crossed the whole area, but damage to the ground cover was very local, apparently having no effect on the behavior of gorilla groups.

Bolwig (pers. comm.) observed a gorilla group feeding peacefully near a buffalo herd at Kisoro. At Kabara, a lone buffalo and a gorilla group rested 150 feet apart but out of each other's sight. A buffalo bull fed leisurely within 250 feet of a group while the silverbacked male roared and beat his chest at me. Once group VII fed slowly as a buffalo bull stood at the periphery of the group, only 70 feet from a female with infant. The buffalo and the female could easily see each other. Suddenly the female ambled directly toward the buffalo, which crashed away when she approached to within 30 feet and continued to advance. The female then sat in the place vacated by the animal.

Bos taurus (domestic cattle) During a time of heavy cattle influx into the Kabara area following Congo independence, I found gorillas only in those areas where cattle had not destroyed or trampled the vegetation. Emlen was told by a cattle rancher in the Mt. Tshiaberimu region that gorillas are afraid of cattle. But Mills (pers. comm.) observed a lone male gorilla watching a herd of cattle near Kisoro.

Felis pardus Leopards, according to the literature, stalk young gorillas persistently, killing large numbers of them. "But there is a spotted menace, a potent factor too, in the leopard, who destroys numbers of young animals," wrote Burbridge (1928), and Geddes (1955), Zahl (1960), and others repeated similar statements. Johnson (1931) mentioned that his party heard gorilla screams during the night at Kabara, and, when they investigated the next morning, they found that gorillas had left their nests and that the tracks of two leopards were nearby. Burbridge (1928) related a story told to him by a native in which a gorilla and a leopard battled at night, the leopard being found dead in the morning. Merfield and Miller (1956), on the other hand, stated: "Gorillas, however, seem to have no fear of leopard. . . ." They described a first-hand incident in which they watched a leopard jump to the ground in full view of a group of gorillas without eliciting any response from it.

Between 1955 and 1960, Mr. Baumgartel of Kisoro obtained no evidence that leopards preyed on the gorillas in the volcanoes near his hotel, even though several leopards frequented the forest. However, in September, 1960, a female gorilla was found dead of unknown causes. Then, in February, 1961, the first evidence of leopard predation on gorillas was obtained. The circumstances are described by Baumgartel (1961) in a report to the Uganda Game Department.

> On February 14th 1961 guide Reuben Rwanzagire and his two trackers noticed excited and unusual behavior of birds on the wooded slopes of Mt. Muhavura, a few hundred yards above the saddle camp. On reaching the spot they heard noises from behind a bush such as leopards make, but the animal disappeared and could not be seen. It had been disturbed at the very moment of a kill, for a little red duiker, still breathing, was lying there in its blood. Looking around Reuben found a dead male gorilla nearby and there was evidence of a fight having taken place there shortly before.
>
> Following the trail over flattened vegetation 50 yards uphill, Reuben and his men came to the place where a gorilla family had spent the night. According to Reuben the victim had been taken unawares, lying in bed when the leopard jumped at him, and both together had rolled down the slope to the spot where the body was found.

"A gash in the right groin had laid bare the intestines and there were severe injuries on the neck" (Tobias, 1961), but the gorilla had not been eaten. "Three days after finding the body of this male gorilla, Reuben and his assistants disturbed a huge leopard eating the putrid flesh of another gorilla, apparently a female, which must have been killed shortly before or after it had left its other victim" (Dart, 1961).

A black leopard, the same animal which had apparently caused the death of the other gorillas, was seen to stalk a group unsuccessfully by Dr. Dale Zimmerman, professor of biology, New Mexico Western College. He kindly permitted me to quote from his letter (Sept. 23, 1962) relating the incident.

> Late on the cool, cloudy morning of July 3, 1961, guide Reuben and I had reached an elevation of roughly 12,000 feet on Mt. Muhavura. There, we watched for nearly two hours a pair of adult gorillas and their two young as they fed, rested, and moved about on the opposite side of a deep, vegetation-choked ravine. They evidently neither saw nor heard us, for they behaved as if no potential danger were anywhere in the vicinity. The female and young disappeared behind vegetation farther up the ravine to my left, but the huge male remained in full view resting on his back. As I was about to arise I noticed a conspicuous black mass lower in the ravine and some distance to the right of the male gorilla. With the glass I saw immediately that it was no primate but a rather large cat, apparently a melanistic leopard. It was not aware of our presence, but gazed intently in the direction of the resting gorilla. The cat was on its feet when I first viewed it. Still watching through the binocular I saw it advance in the direction of its gaze, moving rather

deliberately but not particularly slowly. It was then perhaps 600 feet or more from the gorilla and lower down on the side of the ravine. It continued to advance, still seemingly staring toward the gorilla and not once glancing in our direction. After it had moved to within perhaps 300 feet of the anthropoid it stopped, momentarily looked in our direction, then back toward the gorilla. It crouched low and began to move forward with, apparently, considerably more stealth than before.

After watching the leopard's extremely slow advance for some minutes I lowered my binocular to rest my arms and to examine the gorilla. However, the latter had vanished. Reuben spotted him again a few minutes later at a position much higher and much farther away climbing upward. We did not again see the leopard although we remained watching for a while.

I made a special effort to obtain some information on the relations between leopard and gorilla in the Kabara area. Although no leopards were seen, it appeared from the number and size of tracks that about three animals ranged the region, from the giant senecio zone downward into the bamboo. I examined all dung, and found that *Cephalophus* and *Dendrohyrax* represented almost the entire diet of the predator. Gorilla hair or other gorilla remains were never noted in the dung.

To my knowledge only one infant gorilla vanished from the groups which I had under frequent observation during the period of study, and the cause of its disappearance was not determined. In other words, I found no evidence that leopards habitually prey on gorillas. In an environment which contains a bountiful supply of defenseless prey, leopards need not venture combat with gorillas. It is possible that some leopards become gorilla killers just as some large cats show a preference for humans.

Interactions with Other Primates

Various monkeys occur throughout the range of the mountain gorilla, and chimpanzees (*Pan troglodytes schweinfurthi*) too are widespread. Man ventures into all areas frequented by gorillas, and he is probably the only mammal to come into frequent open conflicts with them. I observed gorillas and monkeys near each other several times, but I never saw contact between gorillas and chimpanzees. Descriptions of the responses of gorillas to man are often presented in the literature, and, in addition, I recorded the reactions of gorillas to myself about 300 times.

Interactions with Monkeys The golden monkey, *Cercopithecus mitis kandti*, is the only primate, other than the gorilla, which habitually frequents the bamboo and *Hagenia* woodland of the Virunga Volcanoes. It is also the only monkey which I observed in the immediate vicinity of gorillas. Although the primary habitat of the golden monkey is the bamboo, stragglers, lone animals or groups of two, penetrate upward into the *Hagenia* woodland, where they feed on *Galium* and other vines instead of their customary bamboo leaves. I saw or heard golden monkeys twelve times near Kabara above the bamboo zone.

Once I encountered a group of gorillas feeding on the ground in the bamboo near Rweru while a band of about ten golden monkeys climbed around in the bamboo directly above the apes. Another time my wife and I contacted group VIII as it rested in the *Hagenia* woodland. A golden monkey fled through some shrubs about ten feet above the gorillas upon seeing us. At the same time another monkey emitted its typical sharp "tshio" call 150 feet from the group. At least one of these monkeys remained in the vicinity of the gorillas for one and a half hours. I flushed a lone monkey near a gorilla group on another occasion. In addition, monkeys were heard twice within about 500 feet of gorillas, and twice monkeys were seen on or immediately above gorilla trails made on the previous day. The fact that golden monkeys in the *Hagenia* woodland were encountered six times near gorillas or their sign suggests that the former are attracted to the latter. The gorillas, on the other hand, were never seen to respond to presence of the monkeys.

INTERACTIONS WITH CHIMPANZEES Gorillas and chimpanzees utilized the same parts of the forest in many areas, but the latter apes were unfortunately so scarce in my gorilla study areas that I never encountered them there. In the Kayonza Forest, gorilla and chimpanzee nests of varying ages frequently occupied the same ridge. Bingham (1932) noted the nests of both apes at the base of Mt. Mikeno.

Whenever possible I attempted to determine ecological differences between these two closely related apes. Both are terrestrial, but the chimpanzee less so than the gorilla. Although chimpanzees tend to nest higher in trees than gorillas, there is some overlap between the nesting locations of the two species (see Table 41). In West Africa chimpanzees are largely frugivorous (Nissen, 1931), and judging from dung and direct observation, they subsist largely on a similar diet in Central Africa. In the Budongo Forest, for example, I saw them eat the fruits of *Maesopsis eminii* and *Kleindora gabonensis,* and in the Maramagambo Forest the fruits of *Cola cordifolia.* However, they also forage readily on other plant parts. In the Semliki

Forest, chimpanzees ate the young leaves of *Megaphynium* sp., and in the Ruwenzori Mountains the pith of *Ensete* sp. Thus, it appears that in the rain forests visited by me chimpanzees are primarily but not exclusively frugivorous. As seen in the chapter on food habits, gorillas consume mainly the leaves, barks, and piths of various plants, although they too ingest some fruits. This suggests that the two apes are ecologically separated at least in part by their food habits.

INTERACTION WITH MAN This section has been divided into two parts: (1) a brief discussion of the gorilla's responses to being pursued, as presented in the literature and as gathered from interviews, and (2) a detailed account of the responses of gorillas to one or more observers watching them quietly. My discussion of the first part is rather brief, its purpose being primarily to put the literature in the proper perspective and to relate several interesting aspects of gorilla behavior with which I possess only very limited personal experience.

Response of gorillas to being pursued Hunting, collecting, and photographing are occupations which require persistent tracking and close approach. The displays and bluff charges elicited by these intrusions, elaborated and embellished by such undisciplined word pictures as found in Du Chaillu (1861), Gatti (1936), Cotlow (1957), and others, have created the impression that the gorilla is by nature ferocious, attacking without provocation. Owen (1859), for example, wrote: "Negroes when stealing through the shades of the tropical forest become sometimes aware of the proximity of one of these frightfully formidable apes by the sudden disappearance of one of their companions, who is hoisted up into the tree, uttering, perhaps, a short choking cry. In a few minutes he falls to the ground a strangled corpse."

Although seemingly unprovoked charges undoubtedly occur, nearly every misadventure with gorillas can be traced to careless approach or to active pursuit of the animal. The Japanese investigators at Kisoro, for example, were charged by a male when they approached the resting group too closely (Baumgartel, 1960). Other expeditions to the Virunga Volcanoes have also been attacked under similar circumstances as tabulated below:

Source	Purpose of Visit	Response of Man to Charge
Akeley, 1923	Collecting gorillas	Shot male
Prince Wilhelm of Sweden, 1923	Collecting gorillas	Shot male
Derscheid, cited by Akeley, 1932	Observing gorillas	Wounded male
Burbridge, 1928	Capturing gorilla infants	Shot male
Maxwell, cited by Hoier, 1955	Photographing gorillas	Shot male
Bingham, 1932	Observing gorillas	Shot male

Most of the attacks were undoubtedly made in bluff, but a person with a rifle rarely has the patience to await the course of events.

Natives too occasionally come into close contact with gorillas. Usually the animals are driven away by shouting and throwing of rocks, but, with the double incentive of eliminating crop raiders and providing meat for the larder, gorillas are hunted whenever possible. Such hunts usually entail a community effort in that large groups of men encircle a group of gorillas and shoot, spear, and net anything that moves. After charging repeatedly, the male usually flees or is killed (Cordier, pers. comm.). Merfield and Miller (1956) describe the plight of the group without the male: "I have seen the native hunters, having dispatched the Old Man, surround the females and beat them over the head with sticks. They don't even try to get away, and it is most pitiful to see them putting their arms over their heads to ward off the blows, making no attempt at retaliation."

During communal hunts and chance encounters, natives are sometimes injured by gorillas, but the precise circumstances of the incident are rarely available. A nurse in a mission hospital near Lubero told Emlen that natives are sometimes bitten when attempting to rescue their dogs during a hunt. Gorillas apparently show a special antipathy toward dogs, a fact also noted by Merfield and Miller (1956).

Between 1950 and 1959, according to Emlen, the Kitsombiro mission hospital near Lubero treated nine gorilla injuries. Three were minor, and six required prolonged hospitalization. During a roundup of a lone silverbacked male three Africans were bitten—in the thigh, calf, and hand, respectively. In May, 1959, a lone blackbacked male was surrounded by natives. One fellow fled but slipped, and the gorilla grabbed him by the knee and ankle and bit the outside of the calf, stripping off a large piece of the muscle 7 to 8 inches long. In the Kayonza Forest, the Batwa still laugh about a Bantu who was bitten in the buttocks several years ago and could not sit for a long time. Near Mt. Kahuzi a native was bitten in the leg. Between 1938 and 1940, fifteen Africans reported injuries in one district in the Utu region, and four of them died according to Derochette (1941).

Attacks are usually made by males, but a few records indicate that females may also do so on occasion. According to the 1959 records of Albert National Park, a female with infant attacked and tore off the finger of one member of a group of natives who had pursued the gorillas with spears. Chorley (1928) maintained that he was charged by a female in Uganda; and a native also told Jenks

(1911) of having been attacked by one. Baumgartel told me that a female placed her hands around the throat of his guide as if to choke him.

Attacks usually consist of a lunge forward, a brief contact, and retreat. There is apparently no prolonged mauling, and the wounds received, although often serious, are usually not fatal (Cordier, pers. comm.).

Brehm (1916), Baumgartel (1960), and others have observed that a gorilla will not attack a person who stands his ground and faces the advancing animal. But as soon as man turns and flees, the gorilla pursues on all fours, and, like a dog, bites. Thus, among the Mendjim Mey of the Cameroons it is a disgrace to be injured by a gorilla, according to Merfield and Miller (1956), for "the people know that the gorilla would not have attacked if the man had not taken fright and run away."

In spite of the occasional attacks which gorillas make on man, reputable authors who have encountered these apes a number of times—von Koppenfels (1877), von Oertzen (1913), Barns (1923), Maxwell (1928), Burbridge (1928)—agree that gorillas are basically not aggressive and that they tend to retreat if given the chance. Reichenow (1920, quoted in Yerkes and Yerkes, 1929) described the typical response of lowland gorillas:

> When the gorilla notices the approach of human beings, he gives a short bellow, usually twice, which makes the females and young ones attentive, for they draw back toward the other side. At the same time the bellow apparently means a warning for the one who is approaching, for the gorilla himself remains quietly in his place. . . . I have, however, in no case seen an immediate attack, that is, in the sense that the gorilla would suddenly run toward me. The animal rather breaks into flight when I again approach nearer, with continuous roaring.

Similarly, Akeley (1923) expressed his opinion concerning the mountain gorilla:

> I believe that the gorilla is normally a perfectly amiable and decent creature. I believe that if he attacks man it is because he is being attacked or thinks he is being attacked. I believe that he will fight in self-defense and probably in defense of his family; that he will keep away from a fight until he is frightened or driven into it.

Response of gorillas to observers watching them quietly Most of my observations on gorillas were made at Kabara, an area where the animals have had few or no contacts with man. Since I usually approached groups alone, the majority of the recorded responses refer to my presence sitting quietly and in full view of them.

Response of unhabituated groups to a single observer: Numerous factors influenced the response of gorillas to my presence, but the most consistent part of their behavior was that they rarely fled. The dominant male tended to face me when I first arrived, and perhaps roar a few times as the females and youngsters gathered near him to watch me intently, sometimes from the vantage points of trees and stumps (Pls. 33, 34). Soon, if nothing occurred to incite them to further action, they fed and rested, displaying only intermittently. If the group retreated, it rarely did so precipitously, and it usually did not move farther than about 600 feet before settling down again. Three condensed examples from my field notes illustrate typical behavior of groups, two of which had never seen me before and one of which had seen me previously only once.

Group III. My first encounter with this group.

0815. I contact the animals feeding quietly at 130 feet. When the silver-backed male sees me he emits a harsh grunt, looks intently, backs away 30 feet, then sits. The other three members of the group move silently until they are 300 feet from me, climb a fallen tree, and then watch. The male joins them and beats his chest at least 6 times and thumps the ground with his hand.

After watching me for about 15 minutes they resume feeding.

0855. The group, led by the male, moves rapidly some 250 feet to the base of a steep slope where I recontacted them later the same day.

Group VIII. My first encounter with this group.

0835. I sit on a steep slope while the animals feed 90 feet distant. A female spots me and slaps her chest; a juvenile too looks up and beats its chest. The silverbacked male, who feeds about 150 feet away, suddenly moves toward the female. When he sees me, he roars about 5 times, and the whole group crowds around him. He continues to roar and beat his chest intermittently for about 10 minutes, but the other gorillas show little excitement—none vocalize and only a few beat their chests. Abruptly the male turns and descends into a narrow ravine and out of my sight. All members of the group follow him, only to reappear one by one about two minutes later. This is repeated twice. A juvenile approaches me to within 50 feet and slaps the ground.

0900. Rain begins to pour and most animals merely sit and look at me. The silverbacked male lies quietly behind a ridge and out of my sight. One by one the others join him.

0915. All, except a blackbacked male, rest quietly in the downpour hidden from view. I leave.

Group VII. My second encounter with this group. I saw the group briefly the previous evening, but this was the first prolonged contact.

0810. A juvenile spots me at 120 feet and beats its chest. Another juvenile followed by a female climbs up on a sloping tree trunk, and they also beat their chests. Several others ascend until a total of 8 juveniles and adults, including the silverbacked male, crowd into two trees, all watching me as I sit quietly.

0840. The male, who has not yet displayed, climbs to a height of 35 feet and beats his chest, then slaps a branch with both hands. He descends, beats his chest again, yawns, defecates, nearly falls off the branch, and displays once more. A female has climbed to a height of 30 feet and feeds on ferns; a juvenile sits up 20 feet and is joined by an infant 1½ years old, and both look at me.

The morning sun is warm. The animals are quiet—there is no vocalizing, and only some females and juveniles slap their chests occasionally.

0950. Four juveniles and two females with small infants still sit in the trees, but the others are feeding slowly. A female descends and presents her back to a juvenile and is groomed. Another juvenile snacks on Hypericum *bark.*

1035. Only two juveniles and two females remain in trees; the others feed silently. The silverbacked male suddenly climbs into a Hagenia *to a height of 40 feet, beats his chest, slaps the branches with his hands several times, and roars twice. A female, a juvenile, and a large infant follow the male into the same tree. The other gorillas appear to ignore me.*

1130. The male descends the tree, followed first by the juvenile and then the female, and all feed.

1210. All forage slowly, only the tops of their heads visible. As I leave no one reacts.

In gorillas which have had little contact with man, curiosity is probably one of the main reasons why they remain to observe a quiet intruder. The animals often watched my every action with apparent interest, and sometimes, if I manipulated an object which they could not see clearly, they craned their necks or climbed into a tree until they obtained a better view. Curiosity also seemed to play an important role in inducing single animals, or the whole group, to approach me very closely at times, as two examples illustrate:

Group VI. My third encounter with this group.

0810. The animals forage, spread over an area of about 120 feet. I sit in a low crotch of a Hagenia *tree, about 6 feet above ground. The gorillas remain unaware of my presence although some are within 50 feet of me.*

0825. A juvenile looks intently in my direction for about one minute. The silverbacked male, who sits by the juvenile, peers over at me, rises, beats his chest, and runs into the vegetation. But no further response to my presence is given and no other animals become aware of me.

0905. The silverbacked male suddenly beats his chest again and runs through the vegetation at the end of the display. The momentum carries him into the open within 45 feet of me, whereupon he rapidly backs into cover and peers out. He roars once. The group congregates around him about 60 feet from me. Five animals—a male, a female, a juvenile, and a female carrying a 2-month-old infant—ascend a log 45 feet away in single file. The animals in back push at the ones in front as if to obtain a better view, and the latter, hesitant to approach closer, back up, with the result that there is continuous pushing and shoving. The silverbacked male, who sits surrounded by most of his group, yawns repeatedly and watches me.

0940. The group moves off feeding slowly.

Group VIII. My 12th encounter with this group.

0805. Most gorillas lie in the sun, a few forage slowly. I climb onto a horizontal branch 5 to 6 feet above ground and sit about 100 feet from the animals. A female spots me, beats her chest, and lies down. Two infants play. No one else appears to notice my presence.

0925. A blackbacked male sees me and beats his chest, whereupon the silverbacked male roars once. The other members of the group rise and collect around him.

The whole group slowly advances in my direction and stops at 30 feet, all the while watching me with apparent curiosity. A female, with a 3-month-old infant held to the chest, angles closer, reaches up, and gives the branch on which I sit a sharp pull. She glances at me as if to judge the result of her effort, and then climbs up on the limb. She shows no response to the clicks of my camera, and after one minute descends and rejoins the group. A juvenile approaches the same location, and it too jerks the branch and swings up. It bites off two pieces of bark. When it leaves after about one minute, a female without infant ascends the branch for about 15 seconds.

Meanwhile the other group members have spread out slowly. The silverbacked male beats his chest a total of 4 times, but otherwise he merely sits, without paying obvious attention to me.

1020. The silverbacked male rises and moves until he is about 100 feet from me. There he rests with his group without further response to my presence.

Response of habituated groups to a single observer: Gorillas usually become used to the presence of a single observer sitting quietly and in full view near them day after day. This habituation process—or the "relatively persistent waning of a response as a result of repeated stimulation which is not followed by any kind of reinforcement" (Thorpe, 1956)—cannot readily be measured, for individual variation in response is great and the circumstances under which the animals are contacted change from day to day. Habituation occurs

(1) during a single encounter when at the end of several hours the responses of gorillas are of a lower intensity than at the time of contact; and (2) when the average intensity of response is lowered progressively after repeated encounters. Examples of the former are found throughout this section. A number of subtle changes in behavior pointed to the fact that, after repeated encounters, the animals had more or less accepted my presence:

a) The silverbacked male ceased roaring at the mere sight of me, and his chest-beating displays became less frequent (see Table 60).

b) The animals largely ignored my presence and continued their daily routine without prolonged interruption.

c) Females permitted their small infants to wander away from them.

d) The tolerance distance, as discussed later in this section, was somewhat reduced.

One example illustrates the casual behavior of gorillas after repeated contacts with me:

Group VII. My 78th encounter with this group.

0830. I advance inadvertently to within 30 feet of the group which forages leisurely. After I have settled myself on a low branch 5 feet above ground, a female with infant on her back spots me, looks intently, and then continues to feed. A juvenile notes me too and backs into the vegetation, and when the silverbacked male glances over from a distance of 45 feet he emits an annoyed grunt. However, no one appears excited as all sit in the sun and feed within 25 to 60 feet of my observation post.

0920. The foraging activity slows down and ceases entirely by 0930.

0930. All rest in the sun within 40 to 60 feet of me.

1030. Two animals begin to snack and 10 minutes later the silverbacked male rises, looks at me seemingly startled, and roars 3 times. Apparently he had forgotten my presence (such behavior was noted on several occasions). He begins to forage and the whole group joins him. During the next 25 minutes they move some 500 feet feeding slowly, and then settle into another rest area.

In general, gorilla groups became quite used to me after about ten to fifteen prolonged encounters. They still reacted to my presence, of course, but observations of groups when they were unaware of me showed that their behavior was not affected much by my intrusion.

The difference in behavior between habituated and unhabituated animals was strikingly shown by a female which joined group VII after that group was accustomed to my presence. This female screamed, ducked into the vegetation, and was highly nervous, whereas the other members tended to ignore me.

Unfortunately I have no data on how long gorillas remain habituated. Group IV appeared to remember me after a three-month intermission, but the response was not clear-cut since several animals were new to the group. Group VII was quite placid when I watched it in September, 1960, even though I had seen the group only three times in August, and before that not since May. Yerkes (1928) found that a captive female mountain gorilla recognized him after one year.

I had hoped to determine just how specifically or how generally the animals recognized me by wearing different clothes and masks, and by substituting other white persons or Africans. However, guerrilla activities shortened the study unexpectedly and I obtained no data on the subject.

Response of unhabituated and habituated groups to two or more observers: In general, groups at Kabara were more excited in the presence of two observers than in the presence of one. Whenever my wife or another person accompanied me, the gorillas tended to display more frequently, as is shown by two examples from groups VII and VIII, respectively.

	Number of Persons	Number of Chestbeats by the Silverbacked ♂ During First Hour of Contact	Encounter No.	Approx. Distance of Group from Observer (in feet)
a)	1..........	0	11	110
	2..........	4	12	270
	1..........	1	13	130
	1..........	0	14	160
	1..........	1	15	170
	2..........	5	16	150
	1..........	obs. less than 1 hour	17	...
	1..........	1	18	150
b)				
	1..........	4	25	90
	1..........	4	26	130
	2..........	10	27	150
	1..........	4	28	80
	1..........	4	29	100

Although groups sometimes moved from sight more quickly if confronted by two persons, they usually remained, especially if previously they had become habituated to one observer. The example below is typical.

Group VII. My 16th encounter with this group.

0935. *Kay and I slowly approach the group in full view, walking parallel to it before climbing onto the sloping trunk of a tree 150 feet from the animals. The group, which has been resting, watches us, and the silverbacked male beats his chest twice. A juvenile slaps his chest too. The male walks behind a screen of herbs and apparently lies down. Several animals remain in the open, however, and these seem to pay us little attention except for brief glances. A female and a juvenile climb 10 to 20 feet up into a* Hypericum *tree and feed on bark. Ten minutes later the male suddenly reappears and beats his chest 3 more times. Thereafter the whole group rests except for several animals which snack slowly.*

1025. The silverbacked male rises and moves off leisurely with his group.

1035. The last animals disappear from sight.

Toward the end of the study, group VII remained in the rest area, even though we made no attempt to be quiet and moved about freely quite near the animals.

Group VII. My 93rd encounter with this group.

0935. The group feeds leisurely as Verschuren and I come upon it. The silverbacked male beats his chest, and all move onto an open slope about 100 feet away and somewhat above us. We make no attempt to remain quiet: we walk around to photograph the animals with the movie camera and to record their sounds with a tape recorder. Most females rest quietly, but several, as well as two juveniles, beat their chests repeatedly. The silverbacked male roars at least 10 times and beats his chest 6 times. He is obviously nervous: he lies down to rest only to jump up a few minutes later to roar at us.

1040. The silverbacked male moves 50 feet uphill and lies down out of sight. During the following 15 minutes the other group members join him.

Gorillas were not contacted with more than two persons at Kabara, but one of our encounters in dense bamboo at Kisoro is probably indicative of the animals' responses to a large party of observers. There were seven persons in this excursion, including tourists and trackers.

1355. The first indication that the gorillas are near is a loud but short roar by the silverbacked male, who sits 80 feet away, slightly above us, looking through the bamboo. In the following 5 minutes he roars 5 more times and beats his chest twice.

1400. He moves slightly so that we cannot see him. The African guide advances 10 feet, and the male roars harshly and swats the bamboo. This is followed by a series of soft hoots and the breaking of a branch. The next series of hoots ends with a hollow thump as if he slapped the ground.

1415. The male has disappeared and is moving uphill with his group, none of which we saw.

Moorehead (1959) related a similar experience from Kisoro:

And this void, this nothingness of sound, was suddenly torn apart by a single high-pitched bellowing scream. . . . As I stood there, heart thumping, transfixed with shock, one of the guides grabbed me by the arm and half dragged half pushed me through the undergrowth towards a little rise where the others were standing. I looked at the point where they were staring and I remember calling out loud, "Oh my God, how wonderful!" . . . He was a huge shining male, half crouching, half standing, his mighty arms akimbo. . . . He shifted his posture a little, still glaring fixedly upon us, and he had the dignity and majesty of prophets. He was the most distinguished and splendid animal I ever saw and I had only one desire at the moment: to go forward towards him, to meet him and to know him: to communicate. . . . And now abruptly he rose to his full height. . . . He lifted his head and gave vent to another of those outlandish and terrifying barking-screams. . . . Then he dropped on to his hands and melted away.

Response of lone males to one or more observers: Lone males were characteristically shy. In strong contrast to groups, they usually fled rapidly and silently, except for an occasional single roar at the time of contact. Once I inadvertently met the same silverbacked male three times during the same day, for we both were traveling in the same direction.

1. *1300. Encounter at 150 feet. He moved at a fast walk away from me.*
2 *1500. Encounter at 80 feet. He sat motionless for two minutes with a wad of* Galium *in his mouth apparently sensing my presence. When he saw me he swung around, took several bipedal steps, roared loudly once, and ran off on all fours.*
3. *1530. Encounter at 60 feet. He raced silently away.*

When I surprised a lone blackbacked male at 60 feet, he emitted a sharp grunt and a scream as he fled. Another blackbacked male ran away silently when he saw two of us at 150 feet. However, one blackbacked male visited the slopes behind our cabin several times. He appeared to observe the activity around our hut with interest, even though he roared whenever someone moved about on the meadow.

Factors affecting the responses of unhabituated and habituated groups to a single observer: The responses of gorillas to man are selective, depending on his distance from the animals, on the suddenness of the meeting, on the visibility, and on several other factors. As a result of these variables, the responses usually vary from encounter to encounter.

1. Distance: Donisthorpe (1958) recognized three critical distances in the gorillas' response to the appearance of her party: (*a*) if the gorillas recognize an intruder at 150 or more feet they will quietly slip away; (*b*) between 60 and 150 feet the group retreats, but the dominant male often stays behind and displays until the group has escaped; (*c*) at 60 or less feet the male may charge. These responses do not apply to groups which are being approached alone and in

full view of the animals. At Kabara no group ever slipped quietly away at the mere sight of me, and, although I was approached by or approached gorillas closer than 60 feet on numerous occasions, I was never charged.

I found the concept of flight distance (Hediger, 1950) not readily applicable to the Kabara gorilla population, for the animals rarely fled. Instead the term "tolerance distance"—"that amount of spatial separation or distance between organized groups which, if reduced, will result in disturbed behavior, excitement, and both offensive and defensive activities"—as used by Carpenter (1958) is useful in describing the responses of the gorillas. If the animals were able to see me clearly, the distance of about 120 feet appeared to be a crucial one: gorillas closer than that were often nervous, and even habituated groups sometimes left a rest area and moved until they were 120 to 150 feet from me before settling down again. As a generalization, the following distances appear to hold for unhabituated Kabara groups:

less than 60 feet —Animals usually very uneasy; rarely remain long before retreating.

61 to 120 feet —Animals remain somewhat uneasy; although groups may spend an hour or more within this area, they usually do not relax completely.

121 and more feet—Animals appear to feel relatively secure; they frequently forage and rest for several hours, displaying only intermittently.

Repeated contacts with a group shortens the tolerance distance gradually until habituated animals may remain seemingly at ease for several hours only 60 to 80 feet from the observer. However, such behavior is not consistent, and groups which one day rest close to the observer may on the following day move somewhat farther away.

2. Density of the vegetation: If the vegetation is so dense that the animals are not able to observe the approach of man clearly, the male usually roars and the other group members gather near him before they all retreat. On several occasions the gorillas were able to hear but not see my approach some 100 to 200 feet away. They usually waited silently, facing the direction of the sound, but with bodies half turned in readiness for flight, until they detected me. Habituated groups then merely sat down again, and others responded by moving away and displaying.

3. Suddenness of contact: Startled groups usually flee a short distance, either silently or roaring and screaming. Once, while tracking another group, I approached inadvertently to within 50 feet of group II—my first encounter with it—for the animals were resting quietly in dense vegetation. All members, except the silverbacked male, ran

silently about 100 feet. The male remained behind, hidden behind a tree. When I did not advance, he suddenly jumped up, roared once, and followed his group. On the other hand, several habituated members of group VII merely looked up and then resumed their foraging when by accident I had advanced into the middle of the group within 20 feet of the nearest animal.

On three occasions I suddenly came face to face with single unhabituated gorillas (two females and a blackbacked male) which foraged silently somewhat apart from the main body of the group. Although the distance which separated us was only 5 to 8 feet, the gorillas made no motion to attack. One female screamed once and backed off about 10 feet, but the other two animals merely looked, and when I slowly backed off, paid me no further attention.

4. Behavior of observer: Any action which by the gorillas can be construed as aggressiveness raises their level of excitement: rapid movement, sudden sounds, unwavering stares. An observer directly above the gorillas on a steep slope caused restlessness.

5. Time of day: In general, it appeared that gorillas were most excitable just before building the night nest.

6. Weather: When the animals sat hunched over in pouring rain, they showed the least reaction to my presence. On sunny days they were less excitable than on cloudy, cool ones. High winds made them nervous.

7. Proximity of another group: On three occasions, when another group was in the vicinity, the silverbacked male of a group was more excited than usual.

8. Sex and age of the animal: In general, the silverbacked male was the most alert, nervous, and excitable member of the group; it was he who roared and displayed long after the others had subsided. He also tended to be shy and frequently remained hidden behind a screen of weeds.

Blackbacked males varied more widely than any other age class in response to my presence, being both the most consistently forward and the shyest animals in the various groups.

Females were as a rule much more placid than the adult males. They were the first ones to rest and feed in my presence without obviously reacting to it, and they seemed to ignore the displays of the male much of the time.

Juveniles resembled females in response, and infants rarely reacted directly to the observer.

9. Individual differences: Each gorilla was an individual in temperament, and the responses of the animals to my presence varied accordingly.

a) Silverbacked male: Nearly every male exhibited his own easily recognizable personality. The male of group XI, for example, was so excitable that I had little opportunity to observe the group for extended periods, since he usually moved from my sight after several minutes. The behavior of the male in group VI was completely unpredictable from day to day: either he displayed a curiosity and forwardness matched by no other silverbacked male around Kabara, or he was seemingly nervous as he roared and beat his chest. The male in group VII was a very even-tempered individual, though he had occasional "moods," sometimes lasting several days.

To obtain some measure of the excitability of silverbacked males, I counted the number of chestbeats given by them in response to my presence, alone and fully visible at the edge of the group, during the first hour of contact. Table 60 presents the results for several males. Encounters of less than one full hour are not tabulated. The conditions under which the groups were contacted varied, making standardization of the data impracticable, but this variation occurred with all groups. Of the four groups listed, the males of groups II and VII showed the lowest level of excitability, the male of group VIII the highest.

b) Blackbacked males: I knew eleven blackbacked males of six groups quite well. Of these, four were relatively bold, four were of average temperament, and three were shy. A male of group IV often left the group and rested within 30 feet of me; he paid me little attention, and apparently was at ease. At the other extreme was a blackbacked male in group VII. During my first few encounters with this group I caught only occasional glimpses of him as he scurried through the vegetation, carefully avoiding the open spaces. However, after seventy-three encounters with the group he approached me to within 15 feet.

c) Females: Females showed, on the whole, less extreme variation in response to my presence than the males. One female with small infant in group VIII frequently ventured some 50 or more feet from the others and sat near me. A female in group IV was very shy. Another female, in group VI, tended to be quit bold until she gave birth to an infant. Then, for about three weeks, she became so shy that I rarely glimpsed her sufficiently well to see the baby.

d) Juveniles: None of the juveniles was markedly shy, but two were quite forward and these appeared to be males.

e) Infants: The responses of infants to my presence were too few in number to detect consistent individual differences.

10. "Mood" of the animal: Occasionally a gorilla, which during numerous encounters with me had exhibited a certain fairly consistent

response, suddenly reacted differently for no obvious reason. This was especially conspicuous in the silverbacked males, primarily, I suspect, because their behavior attracted the attention of the observer more consistently than the other group members. Changes in temperament seemed to occur sometimes from one day to the next. These manifested themselves either in a sudden shyness, lasting anywhere from one to three days, or in an unexplained boldness. A gorilla sometimes left a group to sit within 30 feet of me for a few minutes even though it had never done so before nor did so again. Nearly the whole group occasionally crowded around me for no apparent reason, stimulated apparently by the actions of one animal. At such times the gorillas gave the impression of daring each other to go closer and closer. This is best illustrated by the example below, where, since it occurred after seventy-three encounters with the group, curiosity in response to a strange object would not seem to be the reason for the behavior.

Group VII.

0900. I arrive on the edge of a shallow ravine. The group is feeding below me, 45 to 60 feet distant. As I sit on the ground the animals look up and then return to their foraging.

0945. A few still feed, but most sit.

1035. The silverbacked male suddenly rises and walks toward me, followed by the group. He advances to within 30 feet, emits a sharp grunt, wheels around and, as his group scatters, rushes 20 feet farther away. However, the rest of the group remains within 30 to 40 feet of me, and both blackbacked males approach to within 20 to 25 feet to peer at me and then thump the ground with one or both hands.

1105. The silverbacked male, who has rested quietly for the past half-hour, rises and steadily approaches me to within 20 feet, sits, and looks. One of the blackbacked males feeds within 15 feet of me. A female, who has angled up the slope, spots the African park guard, who accompanied me that day and now sits by a tree some 50 feet behind me. The female screams twice and runs. The male roars, jumps back, and is followed by his group, and several females emit barks. But they only move 30 feet, and, within 5 minutes, a blackbacked male and a juvenile again sit 15 feet from me and the others rest quietly. The silverbacked male, however, stealthily circles me and, placing one leg at a time, creeps up quietly. Gorillas are, however, not good at this sort of thing, and branches invariably snap underfoot. When he has advanced to within 20 feet of me, he slaps the weeds with both hands, and beats his chest. No one reacts and the male returns to his group.

1130. The animals suddenly run 30 feet and spread out to forage. Fog rolls in and engulfs them as I leave.

CHAPTER 8

CONSERVATION

Conservation

By international agreement under the London Convention of 1933 the mountain gorilla was given complete protection, for it was feared that the subspecies was on the verge of extinction with only a few hundred animals remaining in small, isolated forest tracts. Our work revealed that both the number of gorillas and the area inhabited by them are larger than previously supposed and that extinction is not imminent. On the whole, however, the total geographic range of the mountain gorilla is small, and history has shown that animals so situated are highly vulnerable. Constant vigilance must be maintained to prevent a disastrous tip of the balance from security to extinction.

What the observed tendencies and trends indicate for the future varies with different parts of the range. Peripherally and in the areas of good arable soils, as in the Virunga Volcanoes, the Kayonza Forest, and the Mt. Tshiaberimu region, range boundaries are being pushed back or threatened by habitat destruction.

Many gorillas are killed directly. Natives hunt them extensively for food, particularly in the Utu and Mt. Tshiaberimu regions. Some Europeans shoot them for sport. Cordier, who lived for several years near Utu, wrote me that "the biggest killers always were the white mining prospectors who had a yen to shoot gorillas." Museum collectors have disturbed the mountain gorilla little in the past thirty years.

The most reprehensible shooting of gorillas is being done by some zoo collectors, who usually slaughter the female, and even the whole group, in order to obtain the infants for export. That such a drastic collecting technique is unnecessary has been shown by Cordier, who has safely netted whole groups, and by Ryhiner (1958), who has captured infants unharmed by using tear gas. One reliable authority told me that in about 1948, officials organized the killing of some sixty mountain gorillas near Angumu to obtain eleven infants, only one of which survived. However, zoo collecting has not yet been as serious a problem in Central Africa as in West Africa.

The newest threat to the gorilla, and for that matter also to the chimpanzee and other primates, is the collecting by medical institutes. One quote from an account by Westley (1951), a collector for a medical laboratory, illustrates the problem facing the gorilla:

> On the other side of a small clearing a female was playing with a small baby. Everything seemed perfect for a good shot. The gorilla wasn't fifty yards away and was unconscious of our presence. But I couldn't help thinking of the other gorillas that we could hear, and couldn't see. . . . But there was nothing for it, I had to shoot. I took plenty of time and when I stopped shaking, I made a clean hit through the skull, killing her instantly. [The male rushed up.] I fired and hit him in the shoulder. He staggered for a second, but kept going. I fired again, and again he staggered. . . .

Westley obtained the infant but it died five days later because he had made no provision for its care.

A brief discussion of the conservation problems in each of my major gorilla study areas is essential for an understanding of the steps that must be taken to preserve the animal.

Virunga Volcanoes—Albert National Park

Albert National Park was established in 1925 and enlarged in 1929 primarily for the preservation of the gorilla in the Virunga Volcanoes. The six dormant volcanoes straddle the border between the Congo and Ruanda-Urundi, both formerly under Belgian rule. The newly independent Congo assumed control of its side on June 30, 1960, with the subsequent co-operation of the Institute of the Parks of the (Belgian) Congo, which had administered the area since 1934. The other portion of the park is now ruled by newly independent Ruanda.

The Virunga Volcanoes rise above a fertile land, which supports one of the densest human populations in Africa, and thus present a continuous temptation to the agriculturalists who need more land to till, the hunters who crave fresh meat, and the pastoral Watutsi who seek grazing for their scrawny cattle.

At present the hunters pose the least threat to the gorilla. Most of their activity is directed at the red forest duiker and the buffalo. I have seen snares set by poachers for duikers to an altitude of 10,000 feet, and, in January and September, 1960, buffalos were killed on the Rukumi meadow. However, gorillas appear to be rarely molested directly, apparently because the tribes in the area do not eat the meat. Barns (1922), for example, shot a male and his hungry porters refused to eat the animal. However, Rousseau told me that he encountered several Batwa in 1956 as they tracked a wounded male with dogs. Lönnberg (1917) noted that the natives hunted gorillas to obtain the skin, which they used for wrapping copper thread. Supposedly eight gorillas were speared to death by a band of natives in Ruanda in 1959 because they had raided beehives, but this story lacks confirmation. These, however, are isolated instances.

Habitat destruction is the greatest danger facing the gorilla and other wildlife in the Virunga Volcanoes. If the mountains were to be turned over to the agriculturalists, the slopes would be denuded within a few years. Only by constant vigilance have the park authorities been able to maintain the boundaries, although in 1958 they were forced to relinquish 17,500 acres of forest in Ruanda-Urundi.

An insidious danger to the habitat, and one sometimes difficult to control, is the invasion of the forest by Watutsi cattle. Van Straelen (1960), who has over the years been the moving force behind nature conservation in the Congo, noted that the social status of the Watutsi depends on the number of cattle they own, with the result that none are killed for consumption. A steady rise in the number of cattle and a shortage of grazing grounds has made the natives push their herds ever deeper into Albert National Park. Von Beringe (1903) found cattle in the bamboo on the slopes of Mt. Sabinio in 1902, and Meyer (1923) encountered them at an altitude of 8,900 feet on the slopes of Mt. Karisimbi in 1911. Today they penetrate upward to 11,500 feet.

By 1959 damage by cattle to some sectors of the volcanoes had grown serious, especially in Ruanda where the government has been completely ineffectual in curbing the incursions. When I visited the Mts. Visoke-Sabinio saddle in July, 1959, at the height of the dry season, I saw no cattle, but numerous deep trails and huts for more than a mile into the Congo made it evident that several hundred cattle had grazed there during wetter periods of the year. In December, 1959, and January, 1960, I counted at least 300 cattle around the base of Mt. Visoke. Herds penetrated as far as four miles into the Congo to the base of Mt. Mikeno near Bishitsi. By that time the park authorities were powerless to administer the Ruanda portion of the park, and, although they attempted to arrest the trespassers on the Congo side, their efforts were thwarted by the Belgian officials who refused to levy substantial fines to discourage the invasions.

On June 2, 1960, cattle grazed at Rweru, and for the following two and a half months, which included independence day, Watutsi roamed with their herds throughout the volcanoes. When I returned to Kabara on August 10, 1960, cattle were foraging on bamboo tops, which the natives had cut so that the animals could reach the leaves. Both sides of the trail from Rweru to Kabara were devastated. The practice of herding the animals in one small area until forage is exhausted leaves the soil churned like a plowed field. I passed three large herds on the way to Kabara and heard several others. The whole Mts. Mikeno-Karisimbi saddle, except the steepest slopes and the nettle fields, was grazed over. Made aware of the problem, Mr. Anicet Mburanumve, the first Congolese head of Albert Park, sent six men with rifles, and twelve cows were shot and fifty-six confiscated. After one more raid in which about fifty head were obtained, the Watutsi withdrew from the Congo side of the park. When I revisited Kabara from September 11 to 17, 1960, I noted no fresh cattle sign, and on January 29, 1961, Dr. Verschuren wrote me that "the population of the border is afraid and it does not seem that there are still cattle in the Congo area of the mountains.* The Congolese government is to be highly commended for the drastic measures it has employed to preserve the habitat and for having maintained Albert Park, in spite of various difficulties, to the present (December, 1962).

Virunga Volcanoes—Uganda Side

In 1930 the Uganda portion of the Virunga Volcanoes, the northern slopes of Mts. Muhavura, Gahinga, and Sabinio, was established as a gorilla sanctuary. The sanctuary was also gazetted in 1939 as a central forest reserve, owned by the crown. Under heavy public pressure, the size of the reserve was cut from 13 to 9 square miles in 1950, raising the boundary from the 8,000 to the 9,000 foot contour line. This was an extremely unfortunate move, for the excised forest was the most suitable habitat for the gorillas in the area. Much of it has already been converted to fields, and cattle graze in the uncut portions.

Human disturbance in the sanctuary is relatively heavy. The

* Verschuren wrote me in a letter dated April 1, 1962, that in December, 1961, cattle had again penetrated the Congo side of the Virunga Volcanoes. "But, of course, there are not so many cattle as in August 1960. On the Ruanda side the situation is not at all good."

guides from the Travellers Rest Hotel frequently roam the slopes, with or without tourists. A smuggling route crosses the Mts. Gahinga-Sabinio saddle between Ruanda and Uganda. Woodgatherers scour the hills; trail crews keep the boundary line open. Green bamboo stems are in heavy demand for basketmaking and the building of huts (Pl. 35). The forest department has divided the area into four annual coupes in which cutting is rotated. Each coupe in turn is open from September 1 to August 31 and then lies fallow for three years. Only eighty basketmakers are allowed to cut and each may take no more than 900 stems annually (Watt, 1956). However, considerable illegal cutting occurs.

In 1960 the game department discussed the possibility of establishing the sanctuary on a more permanent basis. Conservation of gorillas is difficult, however, for the habitat is only mediocre and not extensive, the gorilla population is not permanent, and in recent years the Ruanda-Urundi government has shown little inclination to maintain the adjoining forest in an undisturbed state.

Kayonza Forest

The Kayonza Forest, officially known as the Impenetrable Central Forest Reserve, was first gazetted in 1932, and its final boundaries were established in 1958. The forest was also made an animal sanctuary in 1961.

From the accounts of natives and from personal observation, it appears that the gorillas are rarely molested in the reserve—the animals raid fields infrequently, and they are not sought for food. The Batwa told me that a few gorillas are accidentally killed in deadfalls which are set for forest pigs.

Prospectors, searching for wolfram and other metals, traversed the forest extensively as early as 1930, but the present low prices of metals have discouraged all but one mine.

Timber felling began in 1940, and the number of pitsawing licenses has been stabilized at 100 since 1957 (Leggat and Osmaston, 1961). Mr. A. H. Tothill, district forest officer at Mbarara, wrote the following in a letter dated May 28, 1959:

> No machinery is used and the logging is purely selective and scattered in the 50 square miles of accessible forest. In this area the density of the sawyer troops cannot exceed an average of 3 troops to a square mile. So "disturbance" is small. Far from any expansion as visualized by Dr. Emlen, I foresee a gradual decline in operations, as over mature timber becomes exhausted from the economic area. Expansion could only be maintained by further access roads and I know of no plans for any.

In their working plan for the Kayonza Forest, Leggat and Osmaston (1961) stated that "harvesting is to be controlled by volume within successive compartments. . . . The main yield is limited to 100,000 cu. ft. a year . . . assuming a rotation of 100 years, and a productive area of 100 sq. miles." Two nature reserves, comprising one square mile each, will be preserved in an uncut state within the forest.

Light logging disturbs gorillas temporarily, but, in the long run, it may favor them by creating new supplies of forage.

The Kayonza Forest must remain a reserve if the gorillas there are expected to survive, for if agriculturalists were allowed access, the hills would soon be as denuded as the surrounding country is now. However, it appeared to me that the status of the gorilla during my visit was satisfactory, and that efforts were being made to insure the preservation of the habitat.

Mt. Tshiaberimu Region

The Mt. Tshiaberimu massif, which is included in Albert National Park, is surrounded by cultivation on three sides. Most of the forest outside the park is unprotected and being converted into permanent fields so rapidly that within a few years several of the isolated gorilla populations will undoubtedly have disappeared. In addition, gorillas are heavily hunted for food in the area. Broughton (1932) took part in a hunt, and Emlen saw the remains of a freshly eaten specimen. Unless the forests, including those on Mt. Tshiaberimu, are given greater protection, the combination of hunting and habitat destruction makes the future for the gorillas dark.

Mwenga-Fizi and Utu Regions

Conditions in the vast lowland forest regions west of the rift escarpment and in the extensive mountain forests of the rift west of Lake Tanganyika are totally different from the isolated habitats discussed so far. Here the boundaries of the gorilla range are little affected by pressures of land utilization except around the larger population centers, which are few in number. Conditions of soil and climate will probably continue to enforce shifting cultivation, a type of agriculture which is beneficial rather than destructive to gorilla habitat. Attracted by the increased amount of forage in secondary forest, gorillas frequently concentrate their activity along roads and near human habitations where they sometimes raid fields (the amount of damage is, on the whole, negligible) and injure natives.

The Africans in turn make sporadic communal attacks on the gorillas, killing and eating many animals yearly. However, the survivors are not deterred and soon revisit the tracts of forest from which they have been driven. The lack of modern firearms fortunately limits the extent of the slaughter, and killing rates do not seem to be serious. The gorilla's position appears to be secure in these regions for the present, especially in the great expanses of forest which are almost uninhabited by man. According to Cordier (letter, Nov. 8, 1961), killing rates have not increased since Congo independence.

As a result of our work, I believe that the following recommendations are essential to the preservation of the mountain gorilla:

1. The gorilla should remain on the list of completely protected animals. This means that every effort should be made to prevent the mass killing of gorillas for food by natives; that modern firearms should not be made available to them to facilitate such killing; and that local governments should place and enforce a strict limit on the number of gorillas captured by zoo collectors, medical institutions, museum collectors, and others. The killing of a female to obtain the infant should be outlawed.

2. The habitats in which gorillas live should be preserved, especially those in areas of rich volcanic soils where permanent cultivation is possible. A special effort should be made to prevent incursions by agriculturalists and pastoralists into forests which already are reserves and contain gorillas—the Virunga Volcanoes, the Kayonza Forest, the Mt. Tshiaberimu massif, the Mt. Kahuzi forest reserve.

The creation of large nature reserves in lowland rain forest at present uninhabited by man would aid in the perpetuation of the gorilla and other wildlife in the event that improvement in agricultural methods and a greatly expanding population disrupt the pattern of shifting cultivation. Such reserves would not need to be protected from all human disturbances, some forest cutting being actually desirable, for gorillas find optimum foraging conditions in secondary rather than primary forest.

CHAPTER 9

SUMMARY

Summary

Chapter I. Introduction and Methods This report presents the result of a twenty-month study of the ecology and behavior of the mountain gorilla (*Gorilla gorilla beringei*) in the eastern Congo, western Uganda, and western Ruanda.

The bulk of the data is based on 466 hours of direct observation during 314 encounters with gorillas, primarily in the Virunga Volcanoes. I usually tracked gorilla groups alone, and, after contacting the animals, sat quietly and in full view of them. Through repeated peaceful contacts, six groups became habituated to my presence. All members of these six groups, as well as several other gorillas, were recognized individually.

Chapter II. Distribution and Ecology Indirect evidence suggests that at some time during the late Pleistocene the gorilla existed as one continuous population from West to Central Africa, north of the Ubangi-Uele river system. During a dry climatic period the forest regressed, wiping out the intervening population. Today 650 miles separate the two populations, which have diverged morphologically into two geographic races.

The range of the mountain, or eastern, gorilla comprises an area of about 35,000 square miles (0° to 4° 20′ S, 26° 30′ to 29° 45′ E), in which the animals are concentrated in numerous isolated and semi-isolated population units. Present barriers delineating this range and separating some of these units include grasslands and wide rivers. Gorillas frequently concentrate their activity near roads, villages, and other areas where the young secondary forest growth provides an abundance of forage. Stragglers, lone males and small groups, may occur twenty or more miles from the nearest gorilla population.

Gorillas inhabit three major vegetation types: lowland rain forest below an altitude of 5,000 feet, mountain rain forest between 5,000 and 11,500 feet, and bamboo forest between 8,000 and 10,000 feet. About three-quarters of all mountain gorillas occur in the lowland rain forest west of the rift escarpment, and most of the remaining ones inhabit the mountain rain forests of the rift. Bamboo is a vegetation type of minor importance to gorillas, not only because of its limited distribution but also because it contains relatively little forage during the drier seasons of the year.

The most intense studies were conducted in the Virunga Volcanoes, where gorillas frequent the mountain forest at an altitude of 8,000 feet or less as well as the zone of giant senecios at altitudes of over 13,000 feet. On the upper slopes of the mountains the temperatures drop near or to the freezing point nightly, a sharp contrast to the tropical climate of the lowlands.

Over ten months were spent in studying the behavior of one population of about 200 gorillas, comprising ten groups, in the Kabara area of the Virunga Volcanoes. The *Hagenia* woodland, a subtype of mountain forest growing at an altitude of about 10,000 feet, provided excellent conditions for observing gorillas over extended periods.

CHAPTER III. THE ANIMALS The following age and sex classes were recognized: infant, 0–3 years; juvenile, 3–6 years; female, 6+ years; blackbacked male, 6–10 years; and silverbacked male, 10+ years.

Gorillas vary greatly in appearance, with the nose being the most easily distinguishable feature. Their hands are broad and the thumb opposable. The skin of adults is black on the unhaired portions of the body and gray on the other parts. The length of the pelage varies with the individual and with the altitude, being longer at higher altitudes; the color of the hair is brownish black to blue black, and males develop a gray or silver saddle with maturity. The lips, ears, nose, palms, and soles remain hairless. Ischial callosities were noted in 6.8 per cent of the gorillas checked. The weights of four newborn zoo infants ranged from 1743 to 2403 grams. By the age of one year animals weigh about 15 to 20 pounds; by two years, 35 pounds; by three years, 60 pounds; by four years, 80 pounds; by five years, 120 pounds. Wild adult females weigh about 150 to 250 pounds, and adult males about 300 to 450 pounds.

A total of fourteen recent injuries was recorded in the Kabara population, but most of these were minor, and probably caused by falls or collisions with sharp objects. At least two of the injuries appeared to be bites.

Acuteness of sight, hearing, and smell in gorillas seems to be roughly comparable to that of man.

The personality of gorillas in the wild compares with that observed in captivity and has been characterized as "markedly shut-in or introverted."

Gorillas are primarily quadrupedal and terrestrial. Although they readily climb around in trees, they do so cautiously. Prolonged bipedal walking is rare, with 60 feet being the longest distance noted in the wild. Females and infants ascend trees to a height of 10 or more feet proportionally twice as often as silverbacked and black-backed males, and juveniles four times as often.

Considerable information on the behavior of gorillas can be obtained by studying their spoor. Nests may remain visible for as long as one year, providing conspicuous indicators of local occurrence and data on group size; trails furnish information on food habits and movements; measurements of tracks and dung aid in determining group compositions.

CHAPTER IV. POPULATION DENSITY, STRUCTURE, AND BEHAVIOR

The density of gorillas in the population units varied from about three animals per square mile in the Virunga Volcanoes, and about 1.5 per square mile in the Kayonza Forest, to perhaps one or less per square mile in the Utu region. Roughly 400 to 500 gorillas inhabit the Virunga Volcanoes. The whole mountain gorilla population comprises probably no less than 5,000 and no more than 15,000 animals.

Females appear to reach sexual maturity at the age of six to seven years, and males at about nine to ten years. A female produces one offspring every three and a half to four and a half years unless the youngster dies in infancy. Disease is probably the main cause of mortality. Gorillas may live as long as thirty-five years in zoos.

In the ten groups which were studied intensively at Kabara there was a ratio of two females to one male. However, when lone males are included in the tabulation, the ratio becomes one and a half females to one male. Evidence suggests that this disparity in the sex ratio is due to a higher post-juvenile mortality among males than among females. Infants outnumbered juveniles by 30 per cent. The Kabara population showed a yearly birth rate of 90/1000. About 40 to 50 per cent of the gorillas appear to die during the infant and juvenile stages.

Group size varied from two to thirty animals. The average size of ten groups at Kabara was 16.9 animals, but in the other areas studied groups averaged only about half as large. The average group composition at Kabara was 1.7 silverbacked males, 1.5 blackbacked males, 6.2 females, 2.9 juveniles, and 4.6 infants. Most groups tended to remain stable over periods of many months, with changes caused primarily by the joining and parting of lone males and the births and deaths of infants. Lone males associated with some groups but not with others. In one group a total of six silverbacked males joined or left within the period of one year, the maximum present at any one time being five. Subgrouping, or the splitting of organized groups into two units, was infrequent and temporary. The responses of groups to each other varied considerably, including obliviousness to the other's presence, close approach, brief mingling, and joining for one night. Inter-group antagonism was rare, and was confined to threatening stares and bluff charges.

Each of the gorilla groups at Kabara wandered continuously, covering from 300 to 15,000 feet per day, within a home range of about 10 to 15 square miles. Groups did not defend territory, and as many as six groups frequented the same section of the forest. They exhibited no conspicuous seasonal movements, although certain vegetation types, like bamboo, appeared to be utilized more extensively during certain months. Some groups showed recurrent trends in their pattern of movement. However, daily routines tended to be unpredictable. Morning movements were usually quite leisurely; half of the average daily route was generally covered between mid and late afternoon.

Gorillas are diurnal, and nearly all their activity falls between 0600 and 1800. They usually rise between 0600 and 0800 and feed intensively for at least two hours. Between 1000 and 1400, groups rest, although they sometimes forage briefly during this period. Between 1400 and 1800, gorillas feed, travel, and finally bed down for the night.

CHAPTER V. INDIVIDUAL ACTIVITIES AND BEHAVIOR Gorillas subsist on a wide variety of leaves, barks, piths, and fruits—a total of 100 food plants were collected in the various study areas. The fact that the animals eat some species in one area but seemingly not in another, even though the plant is present, suggests that "cultural" differences in food habits exist. The animals forage leisurely, alternately sitting and walking. The collecting of food is almost entirely manual, and the palatable parts of the plant are extracted by ripping, shredding, and tearing with hands and teeth. Twice gorillas ate a considerable amount of soil, which, upon analysis, proved to have a high salt content. I obtained no evidence that gorillas eat animal matter in the wild.

I have never seen free-living gorillas drink.

Gorillas build crude platforms or nests of herbs and branches on which they rest during the day and sleep at night. The nests of all members of a group tend to be clustered in an area of half an acre or less. The prime requisite for the location of a nest site is the presence of suitable and adequate vegetation for building the structure. At Kabara, 97.1 per cent of all night nests were located on the ground, in the Kayonza Forest 53.5 per cent were on the ground, and in the Utu region only 21.8 per cent. Silverbacked males rarely nest in trees. There is no organization in the placement of individual nests within the site, although juveniles tend to nest near females, and blackbacked and subordinate silverbacked males often nest at the periphery of the group. In constructing the nest, the animal stands in a central position and pulls and breaks in the available vegetation, which it places around and under its body, a process rarely requiring more than five minutes. Infants as young as eight months of age occasionally construct nests, but do not do so regularly until the age of about three years, when they cease sleeping with their mothers. Nests in trees or on steep slopes apparently serve to keep the animals from falling or slipping, but those on level ground have no obvious function.

Gorillas rarely carry vegetation or other objects over distances exceeding 25 feet. In the wild they show little inclination to inspect strange objects, such as a tin can, manually. They were never observed to use tools.

The animals usually defecate while walking or while lying in the night nest; they urinate while squatting or lying. Gorillas sneeze, cough, yawn, hiccup, burp, and scratch in a manner similar to man. All age and sex classes, except infants, groom themselves readily, concentrating their activity on the arms, shoulders, abdomen, and legs.

Although gorillas emit twenty-two more or less distinct vocalizations, only eight of these are heard commonly in their daily routine: a soft grumbling and grunting of contented animals; a series of abrupt grunts, functioning to keep the group together; a series of hoots preceding the chest-beating display; a harsh staccato grunt of annoyance; barks and harsh screams given by quarreling animals; and a high screech by infants when in danger of falling behind the group. In general, gorillas vocalize infrequently, and most of their sounds are of short duration and low frequency. Their most intense sounds, such as the roar of males and high-pitched scream of females, are given primarily in response to man. The reactions of the animals upon hearing a vocalization depend not only on the sound, but also on the condition under which it is given and on the member of the group who gives it.

The chest-beating sequence of gorillas is an elaborate display consisting of nine more or less distinct acts, most of which may also be given alone, independently of the full display. The sequence typically begins with a series of hoots, which may be interrupted as the animal places a leaf between its lips. Afterwards the gorilla, hooting at an accelerating tempo, rises on its hind legs, throws an herb into the air, kicks up one leg, and, at the climax, beats its chest with the hands several times. It then runs sideways while slapping and tearing at the vegetation and finally thumps the ground with its palm. Infants beat their chests and swat at vegetation at the early age of four months. Only large males were observed to hoot. The display apparently functions in intimidation and communication. Its causation seems to be the build-up of tension in any exciting situation. Several acts in the sequence appear to be displacement activities which have become ritualized.

CHAPTER VI. SOCIAL BEHAVIOR The leader of each group is a silverbacked male, who by his actions determines the behavior of the whole group. The order of dominance within a group is to a large extent correlated with body size. All silverbacked males are dominant over all other animals; all females and blackbacked males are dominant over all juveniles and infants not in contact with their mothers; and among and between infants and juveniles size likewise seems to be the main correlate. If more than one silverbacked male is present in a group, there is a linear hierarchy among them. Females, however, seem to lack a definite hierarchy. Dominance behavior was infrequent and was seen only .23 times per hour of observation. It was most commonly shown in asserting the right of way along a trail or to a certain sitting place.

Mutual grooming was observed .28 times per hour of observation. It is rare between adults, and appears to have little or no social significance, merely a utilitarian one. Most grooming involves a female and a younger animal; occasionally a juvenile grooms a female, where the activity apparently functions in establishing social intercourse.

Infants and juveniles readily play by themselves and with each other, but almost all play activity ceases by the time the animals are six years old. Most lone play involves some form of running, climbing, sliding, and swinging; social play consists of wrestling, chasing, and such games as "king of the mountain" and "follow the leader."

Interactions between members of a group vary considerably in kind, frequency, and duration. Associations are close, but interactions involving bodily contact are not prominent between adults. All age and sex classes are readily tolerated by each other, quarrels and other forms of aggressive behavior being uncommon. Females, juveniles, and infants sometimes sit near or rest against the dominant male. Juveniles may remain closely associated with a female until they are four or more years old.

Small infants are completely dependent on their mothers for food, transport, protection, and social comfort. The rate of development in gorillas is rapid, about twice as fast as that found in humans, so that within a period of six months the infant grows from almost complete helplessness to an active, alert youngster. Infants begin to ingest vegetation by the age of two and a half months, and by the age of seven to eight months they probably derive the bulk of their nourishment from forage, although some continue to suckle to the age of one and a half years. Females usually carry newborns held to the chest with one arm, for the infants are unable to hold on securely by themselves. Infants three months and older frequently ride on the back of the female. Crawling in infants was observed by the age of three months; by four and five months they walk in typical quadrupedal fashion; by six and seven months they climb by themselves; and by one and a half years they sometimes travel under their own power if the group moves slowly. When infants are small, females protect them carefully, preventing them from crawling away and rescuing them from possible harm. The female provides a focal point to which the infant can always return and derive comfort and security. One female showed strong concern for her wounded infant; another carried her dead newborn for four days before discarding it. Infants readily leave their own mothers and briefly join other females. On several occasions juveniles were seen to carry infants for short distances.

The sexual drive of gorillas appears to be low. Captive females have an estrous cycle of about thirty days. Masturbation and various forms of erotic behavior, commonly exhibited by captive gorillas, were never observed in the wild. Gorillas copulate in ventro-ventral and dorso-ventral positions. Free-living gorillas were seen to copulate only twice; in both cases the male participant was a subordinate animal, i.e., not the dominant male in the group. The gestation period of four zoo infants varied from about 251 to 289 days. In six births in captivity, parturition was rapid, half an hour or less. There is no evidence of a breeding season.

Gorillas exhibit several forms of aggressive behavior toward other gorillas and human intruders. In the order of increasing intensity, these are: (1) an unwavering stare or a sudden turn of the head; (2) an incipient charge indicated by a forward lunge of the body; (3) a bluff charge over 10 to 80 feet, either silent or roaring; and (4) actual physical contact in the form of biting or wrestling.

In response to the various aggressive acts, gorillas show several forms of submissive or appeasement behavior: (1) they turn their head to the side, facing away from the threatening animal; (2) they shake their head back and forth; and (3) they cower down, head lowered, and arms and legs tucked under the abdomen.

CHAPTER VII. RESPONSES TO ENVIRONMENT Gorillas often sunbathed at Kabara by lying on their backs with chests exposed to the rays. At the onset of heavy downpours the animals either sought shelter beneath the leaning bole of a tree, or they remained in the open, sitting hunched over. They built their night nest in a sheltered position more frequently when it rained at bedtime (18.5%) than when it did not (4.7%).

The apes were sometimes startled by the sudden appearance of such large flying birds as ravens and hawks. No special interactions were observed between gorillas and elephants or buffaloes; buffaloes and gorillas were seen to feed and rest peacefully within 250 feet of each other on three occasions. Leopards do not appear to prey on gorillas to any great extent. Gorillas did not react to the presence of *Cercopithecus mitis,* even though these monkeys were twice seen within 10 feet of the apes. Chimpanzees and gorillas utilize the same parts of the forest in many areas, but no interactions were observed.

Responses to man vary with the circumstances of the encounter. Male gorillas, and occasionally also females, may charge their pursuers when being hunted. Although such charges rarely lead to a direct assault, the animals may bite a persistent attacker severely. When not harassed, however, the gorillas' response to one or two persons observing them quietly consists usually of no more than intermittent roars and displays. At Kabara the groups showed more curiosity than fear, and after ten to fifteen prolonged contacts gradually became habituated to the point where they occasionally approached me to within 15 feet.

Chapter VIII. Conservation Although the extinction of the mountain gorilla is not imminent, constant vigilance should be maintained by conservationists to insure its survival. Many gorillas are slaughtered yearly by natives for food, but killing rates do not appear to be serious so far, and are likely to remain unimportant so long as modern firearms are unavailable to the natives. In areas of good arable soil, as in the Virunga Volcanoes, the Kayonza Forest, and the Mt. Tshiaberimu region, the gorilla's habitat is gravely threatened by agriculturalists and pastoralists.

TABLES

AND

APPENDIXES

TABLE 1

A Partial List of Gorillas Collected in the Virunga Volcanoes

No. of Animals Collected	Year in Which Collected	Method of Collection	Source
2 . . .	1902	shot	von Beringe (1903)
1 . . .	1904	shot	Pretorius (1947)
7 . . .	1913 (received at museum)	shot	Musée Royal de L'Afrique Centrale, Tervuren, Belgium (received from Pauwels)
7 . . .	1913-14	shot	Lönnberg (1917)
1 . . .	1920	shot	Barns (1922)
3 . . .	1920	shot and captured	C. D. Forster quoted in Akeley (1923)
3 . . .	1921	shot and captured	T. Philipps (1923)
5 . . .	1921	shot	Akeley (1923)
14 . . .	1921	shot	Prince William of Sweden (1923)
2 . . .	1922	shot	Musée Royal de L'Afrique Centrale (received from Barns)
9 . . .	1922, 1925	shot and captured	Burbridge (1928)
1 . . .	undated	?	Musée Royal de L'Afrique Centrale (received from Guyaux)
2 . . .	undated	?	Musée Royal de L'Afrique Centrale (received from Schouteden)
1 . . .	1929	shot	Bingham (1932)
1 . . .	1931	shot	Maxwell quoted in Hoier (1955a)
1 . . .	1945	shot	natives, quoted in Hoier (1955a)
1 . . .	1946	shot	Wykes (1960)
1 . . .	1956	shot	Rousseau (pers. comm.)
1 . . .	1958	found dead	Baumgartel (1960)
3 . . .	1960	2 found dead 1 captured	Baumgartel (1960, 1961)
2 . . .	1960	found dead	Schaller (this study)
2 . . .	1961	found dead	Baumgartel (1961)

TABLE 2

Recent Investigations into Gorilla Behavior in Uganda

Name	Sponsoring Institution	Study Location	Period of Study	Publications
R. Osborn....	Coryndon Museum	Kisoro	Oct. 1956 - Jan. 1957	Manuscript report (1957)
J. Donisthorpe.	...	Kisoro	Jan. 1957 - Sept. 1957	Donisthorpe (1958)
N. Bolwig....	Witwatersrand University	Kisoro	Jan. 24 - Feb. 21, 1959	Bolwig (1959)
		Kayonza Forest	Feb. 23-28, 1959	
J. Emlen .... G. Schaller ...	New York Zoological Society	Kisoro	Mar. 26 - Apr. 11, 1959	Emlen & Schaller (1960) and this study
		Kayonza Forest	Apr. 17-27, 1959	
M. Kawai .... H. Mizuhara ..	Japan Monkey Center	Kisoro	May 22 - July 8, 1959	Kawai and Mizuhara (1959a,b)
		Kayonza Forest	July 12-18, 1959	
G. Schaller ...	New York Zoological	Kayonza Forest	June 8-19, 1960	This study
N. Bolwig	Makerere College	Kayonza Forest	June 8-19, 1960	...
J. Itani......	Japan Monkey Center	Kayonza Forest	July - Aug., 1960	...

TABLE 3

Location and Dates of Gorilla Field Work*

Location		Dates
Kisoro, Uganda sector of the Virunga Volcanoes	1959	Feb. 21
Kabara, Mts. Mikeno-Karisimbi saddle area of the Virguna Volcanoes		Mar. 6-15
Kisoro, Virunga Volcanoes		Mar. 26 - Apr. 11
Kayonza Forest, Uganda		Apr. 17-27
Mt. Tshiaberimu region, Congo		May 16-31
Mt. Kahuzi, near Bukavu, Congo		June 9
Mwenga-Fizi Region		June 12-21
Utu region		June 26 - July 9
Mts. Visoke-Sabinio saddle, Virunga Volcanoes		
Rift mountains west of Lake Kivu		July 23
Kayonza Forest, Uganda		July 25-26
Kisoro, Virunga Volcanoes		Aug. 14
Kabara, Virunga Volcanoes		Aug. 19 - Jan. 16
	1960	Feb. 1-29 Mar. 5 - May 6 May 15 - June 2
Kayonza Forest, Uganda		June 8-19
Kisoro, Virunga Volcanoes		July 21-22
Kisoro, Virunga Volcanoes		Aug. 1
Active Virunga Volcanoes		Aug. 5-8
Kabara, Virunga Volcanoes		Aug. 10-17
Utu region		Aug. 27 - Sept. 4
Kabara, Virunga Volcanoes		Sept. 11-17
Kisoro, Virunga Volcanoes		Sept. 22

*Only the actual dates of field work are noted; travel times to, from, and between areas are not included. The same applies to Table 5.

TABLE 4

Number of Visual Encounters* and Hours of Direct Observation of Gorilla Groups

Location	No. Visual Encounters	Total Hours Direct Observation	
Mt. Tshiaberimu region	heard only	. . .	
Utu region	1	1 hour	05 min.
Kayonza Forest	5	7 hours	00 min.
Kisoro (Virunga Volcanoes) . . .	2		30 min.
Kabara (Virunga Volcanoes):			
Group I	4	6 hours	55 min.
II	21	32 hours	35 min.
III	1	1 hour	35 min.
IV	48	96 hours	30 min.
V	16	22 hours	00 min.
VI	56	83 hours	40 min.
VII	96	158 hours	55 min.
VIII	33	40 hours	40 min.
IX	2	4 hours	20 min.
X	1		05 min.
XI	7	5 hours	05 min.
?	3	3 hours	25 min.
Lone males†	18	1 hour	40 min.
Total	314	466 hours	00 min.

*Only one encounter with a group is listed per day, even though the group may have been contacted twice.

†Animals not with group at time of contact.

TABLE 5

Location and Dates of Field Work on Apes Other Than Gorillas

Species	Location	Date	No. of Visual Encounters	Hours of Direct Observation
Chimpanzee (Pan troglodytes schweinfurthi)	Semliki Forest, Albert National Park	May 4, 1959	0	(sign only)
	Ruwenzori Mts., Albert National Park	May 6-7, 1959	0	(heard and sign only)
	Maramagambo Forest, Queen Elizabeth National Park	Aug. 8-11, 1959	2	10 minutes
	Budongo Forest, Uganda	July 2-4, 1960	3	3 hrs. 50 min.
	Gombe Stream Game Reverve, Tanganyika	Oct. 9-11, 1960	0	(sign only)
Orang-utan (Pongo pymaeus)	Sarawak, Borneo	Nov. 17-20, 1960 Nov. 24 - Dec. 4 Dec. 15-24 Dec. 28-30	4	5 hrs. 45 min.
Gray Gibbon (Hylobates moloch)	Sarawak, Borneo	same as above	10	about 1 hour
White-handed Gibbon (Hylobates lar)	near Kuala Lumpur, Malaya	Jan. 3-8, 1961	0	(heard only)
Siamang (Symphalangus syndactylus)	near Kuala Lumpur, Malaya	Jan. 3-8, 1961	1	about 10 min.

TABLE 6

The Virunga Volcanoes

Sector	Name of Volcano	Altitude in Meters (in Feet)	Country in which Located	Volcanic Activity	Presence or Absence of Gorillas
1 Western	Nyamuragira	3,056 (10,023)	Congo	active	none
	Nyiragongo	3,470 (11,381)	Congo	active	none
2 Central	Mikeno	4,437 (14,553)	Congo	dormant	many
	Karisimbi	4,507 (14,782	Congo, Ruanda	dormant	many
	Visoke	3,711 (12,172)	Congo, Ruanda	dormant	many
3 Eastern	Sabinio	3,680 (11,960)	Congo, Ruanda, Uganda	dormant	some
	Gahinga	3,507 (11,400)	Ruanda, Uganda	dormant	some
	Muhavura	4,166 (13,540)	Ruanda, Uganda	dormant	some

TABLE 7

Forage Utilized by Gorillas in Mountain Woodland and Bamboo at Kisoro*

Habitat	Kinds of Forage Eaten										Total No. of Feedings Counted	Total Distance Traveled by Gorilla Group (in Feet)
	Bamboo		Cynoglossum		"Vines"		Peucedanum		Xymalos			
	No. of Feedings	Per Cent	No. of Feedings	Per Cent	No. of Feedings	Per Cent	No. of Feedings	Per Cent	No. of Feedings	Per Cent		
Bamboo	39	54.9	24	33.8	6	8.5	2	2.8	0	0	71	16,200
Mountain woodland with scattered bamboo	49	38.5	31	24.4	23	18.1	20	15.8	4	3.2	127	21,900

*The data are adapted from information collected by Emlen while tracking one group of 7 gorillas for 7 consecutive days in March-April, 1959.

TABLE 8

Weights of Lowland Gorillas in Captivity

Sex	Name of Animal	Location	Approx. Age	Weight in lbs.	Source
Male	Bamboo	Philadelphia	15	435	Conant, 1941
	Bushman	Chicago	21-1/2	542	Riess _et al._, 1949
	Massa	Philadelphia	11	360	Riess _et al._, 1949
	Makoko	New York	11	438	N.Y. Zool. Soc. files
	Mambo	New York	9	351	N.Y. Zool. Soc. files
	Gargantua	Ringling Bros. Circus	17	550	Riess _et al._, 1949
Female	Achilla	Basel	13	154	Lang, 1959
	Miss Congo	Chicago	7	180	Conant, 1941
	M'toto	Ringling Bros. Circus	10	438	Conant, 1941
	Susie	Cincinnati	15	334	Conant, 1941
	Suzette	Chicago	7	180	Conant, 1941
	Oka	New York	19	349	N.Y. Zool. Soc. files

TABLE 9

Weights and Heights of Male Mountain Gorillas Collected in the Wild

Location Collected	Weight in lbs.	Height in cm.	Source
Kisoro, Virunga Volcanoes . .	265	147.3	Galloway *et al.*, 1959
Kabara, Virunga Volcanoes . .	360	171.4	Akeley, 1923
Mt. Kahuzi area near Bukavu .	482	205.7	Gatti, 1936
Kabara, Virunga Volcanoes . .	450(?)	161.9	Barns, 1922
Virunga Volcanoes	292	. . .	Gyldenstolpe, 1928
Virunga Volcanoes	334	. . .	Gyldenstolpe, 1928
Virunga Volcanoes	379	. . .	Gyldenstolpe, 1928
Kisoro, Virunga Volcanoes . .	294	. . .	Mills, 1960
West of Lake Kivu	459	177.8	Gregory and Raven, 1937
Virunga Volcanoes	420	. . .	Hoier, 1955a
West of Lake Kivu	. . .	179.7	Gregory and Raven, 1937
West of Lake Kivu	. . .	173.0	Coolidge, 1930
Kabara, Virunga Volcanoes . .	. . .	182.0	Burbridge, 1928

TABLE 10

The Distribution of Ischial Callosities in Six Gorilla Groups

Group	No. of Females with Callosities	No. of Females in Group	No. of Blackbacked and Silverbacked Males with Callosities	No. of Males in Group	Total No. of Animals in Group
II	0	6	0	4	19
IV	2	10	0	5	24
V	1	3	1	6	15
VI	1	9	0	2	20
VII	2	6	0	3	18
VIII	1	8	0	3	21
Totals	7	42	1	23	117

TABLE 11

Handedness in the Chest-beating Display of Individual Males

	Right Hand Used First	Left Hand Used First
Silverbacked ♂		
Group IV (a)	3	2
IV (b)	5	0
V	6	1
VI	6	0
VII	7	1
Blackbacked ♂		
Group IV	16	0
VIII (a)	8	4
VIII (b)	8	5
Total	59	13

TABLE 12

Recent Injuries Noted in the Kabara Gorilla Population

Sex and Age Class	Group	Date Injury Was First Noted	Type of Injury
1. Silverbacked ♂ . .	IV	Apr. 29, 1960	Swollen eyelid.
2. Silverbacked ♂ . .	IV	Jan. 11, 1960	Eye swollen and closed for two days, and a raw patch of skin in the outer corner.
3. Silverbacked ♂ . .	VII	Oct. 26, 1959	Swollen area around one eye.
4. Silverbacked ♂ . .	lone	Nov. 24, 1959	Small piece of skin gouged from brow.
5. Silverbacked ♂ . .	VIII	Mar. 7, 1960	A deep and clean cut 1-1/2 inches long on the back of the hand.
6. Silverbacked ♂ (old).	IX	Dec. 30, 1959	A section of skin about 4 inches long and 1 inch wide ripped from the neck starting at the ear and running toward the shoulder.
7. Silverbacked ♂ . .	IV	Sept. 1, 1959	A 6 inch long freshly healed scar in the saddle.
8. Blackbacked ♂. . .	V	Nov. 1, 1959	A piece of skin about 1 inch in diameter torn from the rump.
9. Female	VI	Feb. 26, 1960	A small skin abrasion over the eye.
10. Female (old)	IV	Jan. 12, 1960	A raw, grey and scabby area on the neck about 3 inches in diameter (some other parts of her body were nearly hairless—as if she suffered a skin malady, aggravated by scratching).
11. Female	VIII	Nov. 22, 1959	Her whole face was pushed so far sideways that the mouth appeared almost vertical. One cheek was immensely swollen and a yellowish liquid drained from the eyes. Apparently she received a terrific blow to the side of the face and the jaw may have been broken. However, she supported a small infant, and after 4 months the swelling had subsided somewhat.
12. Female (old)	IV	Sept. 1, 1959	She was blind in one eye. The lens was white and a whitish fluid drained continuously. In August, 1959, a swelling was primarily confined to the side of the nose near the injured eye, but by August, 1960, the swelling had advanced along the zygomatic arch.
13. Juvenile	VII	Mar. 27, 1960	A cut in the heel about 1 inch long.
14. Infant (8 mos. old)	VI	May 27, 1960	The muscles over an area more than 4 inches in diameter on the rump were exposed and raw and a section of the ishium seemed visible. The skin was torn away and hung still attached between its legs. The infant was very weak. During the following four days the wound looked better—somewhat drier, but it probably proved ultimately fatal. The study was terminated and I had no further opportunity to determine the fate of this infant.

TABLE 13

The Frequency of Tree Climbing in Kabara Gorillas

Age and Sex Class	A Observed Instances of Climbing	B Total No. of Animals in 6 Groups Sampled during 74 Encounters	Index of Climbing Frequency A/B
Silverbacked ♂	10	129	.77
Blackbacked ♂	10	111	.90
Female (with or without infant)	81	583	1.40
Juvenile	77	235	3.27
Infant (0 - 1-1/4 yrs.).	17 } 69	242 } 530	.70 } 1.30
Infant (1-1/4 - 3 yrs.).	52	288	1.80

TABLE 14

Relative Abundance of Gorilla Nest Sites in Several Areas

Area	Dates Transected	Predominant Vegetation	No. of Hours Transected	Estimated Rate of Walking in mph	No. of Nest Sites Counted	Av. No. Nest Sites per Mile of Walking	Av. No. Nest Sites per Hour of Walking
Kabara	Mar. 6-15, 1959	*Hagenia* woodland	20	1	16	.80	.80
Visoke-Sabinio saddle	July 18-21, 1959	Bamboo, grass meadow	21.5	1-1/2	5	.16	.23
Kayonza Forest . . .	Apr. 17-27, 1959 June 8-19, 1960	Mountain forest	41 28.5	1 1-1/4	27 16	.67 .45	.67 .56
Mt. Tshiaberimu . .	May 16-31, 1959	Bamboo	51	1	34	.66	.66
Utu region	June 29-July 3, 1959	Old secondary or primary lowland forest	35	2	5	.07	.14

TABLE 15

Estimated Population Density of Gorillas in Selected Areas

Location	Approximate Total No. of Square Miles in the Area	Estimated Gorilla Population	No. of Animals per Square Mile
Virunga Volcanoes	155	450 (400-500)	2.9
a. Sabinio-Muhavura .	25	45 (40-50)	1.8
b. Kabara	30	200	6.6
Kayonza Forest	96	150 (120-180)	1.5
Mt. Tshiaberimu	25	35 (30-40)	1.4

TABLE 16

Sexual Development of Gorilla Pairs in Captivity

Zoo	Name of Animal	Sex	Approx. Age at Arrival in Zoo (yrs.)	Wgt. at Arrival (lbs.)	Date of Arrival	Age at First Sexual Presentation (yrs.)	Approx. Date of First Conception	Age at First Conception	Source
Basel. . . .	Christopher	♂	1-1/4	17-1/2	Jan. 1951	. . .	. . .	9	Lang, 1959
	Achilles	♀	1-1/4	18-1/2	Oct. 1948	6 (June, 1953)	Aug. 1957	10	
Columbus .	Baron	♂	4	72	Jan. 1951	. . .	. . .	9-1/2	Thomas, 1958
	Christina	♀	1-1/2	21	Jan. 1951	5 (July, 1954)	April 1956	7	
Washington	Nikoumba	♂	1-1/4	17	Feb. 1955	. . .	. . .	7	Reed (pers. comm.)
	Moka	♀	1-1/2	20	Feb. 1955	First Mating 6-1/4 (July, 1959)	Dec. 1960	7-1/4	

TABLE 17

The Occurrence of Nematode Eggs in Gorilla Dung Samples from Kabara

Sex and Age Class	Group II		Group IV		Group VII		Group VIII		Lone Male		Total	
	No. of Samples	No. with Eggs	No. of Samples	No. with Eggs	No. of Samples	No. with Eggs	No. of Samples	No. with Eggs	No. of Samples	No. with Eggs	No. of Samples	No. with Eggs
Silverbacked ♂	1	1	3	3	1	1	...	...	1	1	6	6
Female	3	1	5	1	5	2	...	...	...	...	13	4
Female or blackbacked ♂	...	...	6	4	2	2	...	...	...	...	8	6
Juvenile	1	0	2	0	4	3	...	...	...	...	7	3
Infant	3	1	5	2	2	1	1	1	...	...	11	5
Totals	8	3	21	10	14	9	1	1	1	1	45	24
Per cent with eggs	37.5		47.6		64.2		100.		100.		53.3	
Date of sample collection	Aug. 23, 1959		Sept. 7, 1959		Oct. 16, 1959		Nov. 27, 1959		Aug. 24, 1959			

TABLE 18

Sex and Age Class Composition of the Kabara Population

Sex and Age Class	A Animals in Groups at the Beginning of the Study*		B Total Kabara Population Including All Known Lone Males, Births, and Animals Who have left Groups or Joined Them by the End of the Study	
	No. of Animals	Per Cent	No. of Animals	Per Cent
Silverbacked ♂. .	17	10.0	25	13.1
Blackbacked ♂. .	15	8.9	18	9.4
Female	62	36.7	65	34.1
Juvenile	29	17.2	31	16.2
Infant	46	27.2	52†	27.2
Total	169	100.0	191	100.0

*Indicates animals present at the time of the first complete count of each group.

†Two infants which died soon after birth are not included.

TABLE 19

The Composition of Gorilla Groups at Kabara

Group	Silverbacked Male	Blackbacked Male	Female	Juvenile	Small Infant (less than 1-1/4 yrs.)	Large Infant (more than 1-1/4 yrs.)	Total
I	1	0	3	2	1	1	8
II	1	3	6	5	3	1	19
III	1	0	2	1	1	0	5
IV	4	1	10	3	3	3	24
V	2	2	3	2	2	0	11
VI	1	1	9	2	3	4	20
VII	1	2	6	4	3	2	18
VIII	1	2	8	3	4	3	21
IX	4	3	9	5	3	3	27
XI	1	1	6	2	4	2	16
Total	17	15	62	29	27	19	169
Per Cent	10.0	8.9	36.7	17.2	16.0	11.2	100

TABLE 20

The Composition of Mountain and Lowland Gorilla Groups

Area	Silverbacked ♂	Blackbacked ♂	Female	Juvenile	Infant	Unidentified	Total	Source
Kisoro	1		1		1		3	Osborn (1957)
	1		2		1		4	Baumgartel (pers. comm.)
	1		2	1	1		5	this study
	1	1	3		2		7	Kawai & Mizuhara (1959b)
	1		2		3	3	9	Kawai & Mizuhara (1959b)
	1	1	4	4	6	2	18	Kawai & Mizuhara (1959b)
Mt. Visoke, Virunga Volcanoes	1*		3	1	1		6	this study
Kayonza Forest . . .	2*		2	4	2	4	14	this study
	1*		5	2	3	4	15	this study
	1			1			2	this study
	1			2		2	5	this study
	1		1		1		3	Bolwig (pers. comm.)
	1	2(?)	3		2		8	white hunter (pers. comm.)
Utu region	1		1	1			3	this study
	1		2		1		4	Cordier (pers. comm.)
	4	1(?)	5		5		15	Cordier (pers. comm.)
Fizi-Mwenga region	1*		3	2	3	4	13	this study
West Africa	1	2	4		3		10	Merfield & Miller (1956)
	1	1	3		2	2	9	Merfield & Miller (1956)
	1		1				2	Merfield & Miller (1956)
	1	1	2				4	Merfield & Miller (1956)
	3		several		4-5		10+	Merfield & Miller (1956)
	2	1	4	2		1	10	Geddes (1955)

*May be incomplete.

TABLE 21

Instances of Inter-group Contacts at Kabara

No.	Date	Groups Involved	Nature of Evidence	Figure No.	Comments
1	Aug. 22-23, 1959	I,II,III	Indirect	. . .	The nest sites of all groups were within an area 100 yards in diameter.
2	Aug. 24-25	II,III	Indirect	. . .	The two groups apparently nested together.
3	Aug. 25-26	I,II	Indirect	. . .	The two groups nested 150 feet apart and probably mingled.
4	Aug. 26-27	I,II	Indirect	. . .	The two groups nested 150 feet apart.
5	Oct. 19-20	VII,III	Indirect	24	The two groups joined and nested together.
6	Oct. 24.	VII, ?	Indirect	. . .	The groups rested side by side and then parted, perhaps after having mingled.
7	Nov. 1	VII,V	Direct	22	The two groups approached each other to within 150 feet before parting.
8	Nov. 22-24	VIII,VI	Direct	19,20	The two groups remained near each other for 3 days and finally joined briefly.
9	Dec. 2	VIII,VI	Indirect	. . .	The groups apparently rested within 80 feet of each other. They may have mixed.
10	Mar. 24-25, 1960	VII, ? (probably XI)	Direct	23	The groups nested 250 feet apart and approached each other to within 100 feet without joining.
11	Apr. 18	XI,VII	Direct	21	The groups rested side by side; there was some antagonism between the males.
12	May 22	IV,VIII	Indirect	. . .	The two groups apparently rested side by side.

TABLE 22

Seasonal Variation in the Presence or Absence of Bamboo Shoots and Gorilla Sign in the Bamboo Zone along the Trail between Rweru and Kabara

Date	Abundance of Shoots	Gorilla Sign
Mar. 6, 1959	xx	fresh dung
Aug. 19, 1959	0	0
Sept. 14, 1959	x	group IV seen
Oct. 14, 1959	xxx	gorillas seen
Oct. 27, 1959	xxx	gorillas seen (by visitor)
Dec. 2, 1959	xxx	fresh dung
Jan. 16, 1960	x	0
Feb. 1, 1960	0	sign about 5 days old
Feb. 29, 1960	0	0
Mar. 5, 1960	0	0
May 6, 1960	0	0
May 15, 1960	x	0
June 2, 1960	x	0
Aug. 10, 1960	0	0
Aug. 17, 1960	0	0
Sept. 11, 1960	0	0
Sept. 17, 1960	x	0

xxx: many
xx: medium
x: few
0: none

TABLE 23

Arrival Dates of Gorilla Groups in the Study Area

Group V	Group VI	Group VIII
Sept. 18, 1959	Sept. 24, 1959	Nov. 18, 1959
Oct. 28, 1959	Nov. 13, 1959	Jan. 9, 1960
Jan. 6, 1960	Jan. 9, 1960	Mar. 3, 1960
Mar. 29, 1960	Feb. 11, 1960	May 20, 1960
May 17, 1960	Mar. 7, 1960	
	Apr. 9, 1960	
	May 24, 1960	

TABLE 24

Extent of Daily Movement by Gorilla Groups at Kabara

Group	No. of Observations	Mean Distance Traveled (in feet)	Standard Deviation (s)	Range of Variation in Distance Traveled
I	3	1750	1376	350 - 3,100
II	14	1625	583	1,000 - 2,700
IV	16	1553	967	550 - 4,240
V	8	1106	218	900 - 1,500
VI	16	1242	660	300 - 2,600
VII	48	2177	353	700 - 6,000
VIII	9	1400	695	500 - 2,500

TABLE 25

Distances Traveled along the Route from One Nest Site to the Following Nest Site by Groups at Hourly Intervals during the Day at Kabara

Time of Encounter	No. of Records	Mean Distance Traveled from Nest Site (feet)	Mean Distance Advanced during Hour (feet)	Time of Leaving	No. of Records	Mean Distance Traveled to Nest Site (feet)	Mean Distance Advanced during Hour (feet)
0801-0900	38	200	. . .				
0901-1000	56	290	90				
1001-1100	37	408	118	1001-1100	24	1,688	. . .
1101-1200	11	481	73	1101-1200	28	1,467	221
1201-1300	9	483	2	1201-1300	14	1,454	13
1301-1400	9	717	234	1301-1400	11	1,268	186
				1401-1500	9	1,128	140
				1501-1600	2	650	478
				1601-1700	6	167	483

TABLE 26

Gorilla Group Activity at Various Times of Day

Time of Encounter	Feeding Intensively and/or Moving Rapidly		Some Feeding Leisurely; Others Resting		Resting		Total No. Obs.*
	No. Obs.	Per Cent	No. Obs.	Per Cent	No. Obs.	Per Cent	
0700-0729	4		0		0		4
0730-0759	5		0		0		5
0800-0829	13		2		0		15
0830-0859	22		2		2		26
Totals	44	88.0	4	8.0	2	4.0	50
0900-0929	24		11		9		44
0930-0959	15		14		15		44
Totals	39	44.3	25	28.4	24	27.3	88
1000-1029	5		8		21		34
1030-1059	4		3		14		21
1100-1129	5		2		11		18
1130-1159	0		0		6		6
1200-1229	5		2		10		17
1230-1259	0		0		7		7
1300-1329	5		0		3		8
1330-1359	0		0		5		5
Totals	24	20.6	15	13.1	77	66.3	116
1400-1429	3		0		2		5
1430-1459	1		0		1		2
1500-1529	2		0		0		2
1530-1559	1		0		0		1
1600-1629	4		1		3		8
1630-1659	4		0		1		5
1700-1729	2		0		0		2
1730-1759	0		0		0		0
Totals	17	68.0	1	4.0	7	28.0	25

*A total of 279 observations was made.

TABLE 27

Food Plants Utilized by Gorillas at Kabara

Name	Relative Degree of Utilization*	Part Eaten
Grass-Sedge		
Arundinaria alpina	xx	Young stem, shoot
Carex petitiana	x	Base of leaves
Fern		
Polypodium sp.	xx	Whole plant
Herb		
Lobelia wollastonii	xx	Root, base of leaf cluster
Lobelia giberroa	x	Stem
Peucedanum linderi	xxx	Inside of stem
Peucedanum kerstenii	xx	Lower part of stem
Chaerefolium silvestre	xx	Root, lower part of stem
Cynoglossum geometricum . .	xx	Root, lower part of stem
Cynoglossum amplifolium . . .	xx	Root
Rumex ruwenzoriensis	xx	Root, lower part of stem
Laportea alatipes	xxx	Bark, stem, leaves
Carduus afromontanus	xxx	Leaf, flower, stem
Helichrysum (probably H. formosissimum)	xx	Leaves, stem
Vine		
Droquetia iners	x	Whole plant
Clematis sp.	x	Bark
Galium simense	xxx	Whole plant
Stephania abyssinica	x	Fruit
Urera hypselendron	x	Bark
Shrub		
Rubus runssorensis	xx	Fruit, leaves, small branches
Pycnostachys goetzenii	x	Bark
Vernonia adolfi-frederici . . .	xxx	Flower, pith
Tree		
Pygeum africanum	xx	Bark, fruit
Rapanea pulchra	x	Bark
Erica arborea	x	Bark
Hagenia abyssinica	x	Pith, bark, root
Hypericum lanceolatum	x	Bark, root, rotten wood
Senecio erici-rosenii	xx	Pith, base of leaf cluster
Senecio alticola	xx	Pith, base of leaf cluster

*xxx: heavily used
xx: moderately used
x: infrequently used

TABLE 28

Food Plants Utilized by Gorillas at Kisoro

Name	Part Eaten	Relative Degree of Utilization[†]	Source[‡]
Grass-Sedge			
Arundinaria alpina	Young stem, shoot	xxx	
Carex petitiana	Base of leaves	x	
Cyperus sp.	Base of leaves	?	**
Herb			
Lobelia wollastonii	Base of leaves, root	xx	
Peucedanum linderi	Inside of stem	xxx	
Peucedanum kerstenii	Lower part of stem	xx	
Chaerefolium silvestre	Lower part of stem	xx	
Cynoglossum geometricum	Root	xxx	
Cynoglossum amplifolium	Root	xxx	
Rumex nepalensis	Stem and leaf base	?	*
Rumex usambarensis	Stem	?	*
Carduus afromontanus	Stem and leaf	xx	
Senecio trichopterygius (?)	Inner part of stem	?	**
Laportea alatipes	Leaves, stem	xx	
Kniphofia grantii	Stem	x	
Vine			
Crassocephalum bojeri	Bark	xx	
Stephania abyssinica	Fruit	x	
Droquetia iners	Leaves, stem	?	*
Basella alba	Leaves	?	*
Piper capense	Bark, stem	?	*
Shrub			
Vernonia adolfi-frederici	Pith	xxx	
Pycnostachys goetzenii	Bark	x	
Rubus runssorensis	Fruit, leaves	x	
Tree			
Hypericum lanceolatum	Bark	x	
Xymalos monospora	Fruit, bark	x	
Senecio erici-rosenii	Pith, base of leaves	xx	
Senecio alticola	Pith, base of leaves	xx	

[†] xxx: heavily used
xx: moderately used
x: infrequently used

[‡] *: Donisthorpe (1958)
**: Bolwig (pers. comm.)

TABLE 29

Food Plants Utilized by Gorillas in the Kayonza Forest

Name	Part Eaten	Relative Degree of Utilization†	Source‡
Sedge			
Cyperus sp.	Base of leaves	?	*
Fern			
Cyathea deckenii	Shoot, pith of frond	xxx	
Conchitis sp.	Inside of stem	xx	
Dryopteris sp.	Inside of stem	xx	
Herb			
Aframomum milbraedii . . .	Fruit	xx	
Rumex usambarense	Stem	xx	
Pilea bambuseli	Leaves	x	
Brillantaisia nyanzarum . .	Bark	x	
Vine			
Senecio syringifolius	Leaves	x	
Culcasia scandens	Bark	x	
Momordica foetida	Leaves	xxx	
Urera hypselendron	Bark	xxx	
Fleurya ovalifolia	Stem	xx	
Crassocephalum bojeri . . .	Bark	xx	
Piper capense	Bark	xxx	
Basella alba	Leaves	xxx	
Kigelia sp.	Fruit	?	*
Mikania cordata	Leaves	xxx	
Shrub			
Rubus pinnatus	Leaves	xx	
Mimulopsis arborescens . .	Bark	x	
Pycnostachys goetzenii . . .	Bark	x	
Acalypha sp.	. . .	?	***
Tree			
Xymalos monospora	Bark	x	
Myrica kandtiana	Bark	x	
Ensete sp.	Pith of stem	?	**
Myrianthus arboreus	Leaves, bark, fruit	xx	
Cultivated			
Musa sp. (banana)	Pith of stem	x	

†xxx: heavily used
xx: moderately used
x: infrequently used

‡ *: Batwa guide
**: Bantu assistant
***: Pitman (1942)

TABLE 30

Food Plants Utilized by Gorillas at Mt. Tshiaberimu and Vicinity

Name	Part Eaten	Relative Degree of Utilization†	Source‡
Grass			
Arundinaria alpina	Young stem, shoots	xxx	
Pennisetum purpureum	Stem	?	*, **
Herb			
Rumex bequaertii	Stem	xx	
Rumex ruwenzoriensis	Stem	xx	
Peucedanum linderi	Stem	?	*
Sonchus schweinfurthii	Leaves	?	***
Gunnera perpensa	Leaves	?	***
Vine			
Smilax kraussiana	Leaves	xx	
Urera hypselendron	Bark	xx	
Shrub			
Rubus runssorensis (?)	Leaves, bark, small branches	xx	
Pycnostachys goetzenii	Bark	x	
Mimulopsis arborescens . . .	Bark	x	
Tree			
Pygeum africanum	Bark	xx	
Macaranga kilimandscharica	Bark	x	
Maesa lanceolata	Bark	xx	
Galiniera coffeoides	Pith	x	
Grumilia megistosticta	Bark	x	
Cultivated			
Zea sp. (maize)	Fruit	?	*, ***
Pisum sp. (peas)	Fruit	?	*

†xxx: heavily used
xx: moderately used
x: infrequently used

‡ *: Bantu assistant
**: white settler
***: J. T. Emlen

TABLE 31

Food Plants Utilized by Gorillas in the Mt. Kahuzi Region West of Lake Kivu

Name	Part Eaten	Source
Grass		
Pennisetum purpureum . .	Stem	Rahm (pers. comm.)
Arundinaria alpina	Young stem, shoot	This study
Herb		
Brillantaisia nyanzarum .	Bark	Christiaensen (pers. comm.)
Aframomum sp.	Pith	Agricultural official (pers. comm.)
Vine		
Urera hypselendron	Bark	Africultural official (pers. comm.)
Gynura scandens	Bark	Christiaensen (pers. comm.)
Ficus thonningii	Bark	Christiaensen (pers. comm.)
Shrub		
Mimulopsis arborescens .	Bark	Christiaensen (pers. comm.)
Vernonia kirungae	Pith	This study
Tree		
Galiniera coffeoides	Pith	This study
Conopharyngia holstii . . .	Bark	This study
Cultivated		
Daucus sp. (carrot)	Root	White planter

TABLE 32

Food Plants Utilized by Gorillas in the Mwenga-Fizi Region

Name	Part Eaten	Collecting Station†	Source‡
Grass			
Arundinaria alpina	Shoot, stem	F,M	*
Pennisetum purpureum	Stem	F	*
Fern			
Cyathea deckenii	Inside of frond	N	*
Dryopteris sp.	Stem	N	
Marattia fraxinea	Inside of stem	N	
Herb			
Aframomum sanguineum	Pith	F,N	
Brillantaisia cicatricosa	Bark	N,M	
Triumfetta annua	Leaves	M	*
Hypoestes verticillaris	?	M	*
Palisota ambigua	Leaf base	N	
Vine			
Urera hypselendron	Bark	F,N,M	
Piper capense	Bark	F	
Adenia cissampeloides	Bark	M	
Lasianthus kilimandscharica	Bark	N	
Oncinotis glabrata	Bark	N	
Cissus petiolata	Leaves	N	
Shrub			
Triumfetta macrophylla	Bark	F	*
Alangium chinense	Bark	M	
Tree			
Xymalos monospora	Bark	M	
Galiniera coffeoides	Pith	M	
Maytenus acuminata	Bark	M	
Trema guineensis	Bark	M	*
Polyscias fulva	Bark	N	*
Myrianthus arboreus	Fruit	N	*
Cultivated			
Musa sp. (banana)	Pith of stem	near N	**

†F: Fizi
M: Mulenge
N: Nzombe

‡ *: Bantu guide
**: J. T. Emlen

TABLE 33

Food Plants Utilized by Gorillas in the Utu Region

Name	Part Eaten	Relative Degree of Utilization†	Source‡
Fern			
Dryopteris sp.	Base of stem	x	
Marattia fraxinea	Inside of stem	xx	
Herb			
Marantochloa leucantha . . .	Young shoots	xx	
Megaphynium macro-stachyum	Fruit	?	*
Palisota sp.	Base of main stem	xx	
Aframomum sp.	Pith, fruit	xxx	
Costus affer	Pith	xx	
Tree			
Musanga cecropioides	Fruit, leaf	?	*
Ficus exasperata	Leaf	xx	
Ficus vallis-choudee	Leaf	?	
Raphia sp.	Base of frond	?	*
Pycnanthus angolensis . . .	Leaves	x	
Myrianthus arboreus	Fruit	xx	
Uapaca sp.	Pith	x	
Cultivated			
Musa sp. (banana).	Pith of stem	xxx	
Manihot sp. (manioc).	Tuber, bark of stem	xx	
Colocasia esculenta (taro) .	Root	?	*

†xxx: heavily used
xx: moderately used
x: infrequently used

‡*: Bantu guide

TABLE 34

Food Plants Utilized by Gorillas in West Africa as Noted in the Literature

Scientific Name	Common Name	Source
Wild		
Pennisetum sp.	Elephant grass	Merfield & Miller (1956)
Musanga smithii	Umbrella tree	Sabater Pi (1960)
Uapaca guiniensis		Sabater Pi (1960)
Prevostea africana.		Sabater Pi (1960)
Aframomum sp.		Sabater Pi (1960), Merfield & Miller (1956), Bates (1905)
Ficus exasperata		Merfield & Miller (1956)
Fleurya aestuans		Sabater Pi (1960)
Parinarium excelsum . . .		Owen (1859)
Mangifera (?) sp.	Wild mango	Malbrant and Maclathy (1949)
Cultivated		
Carica sp.	Paw-paw	Owen (1865)
Ananas sp.	Pineapple	Geddes (1955), Owen (1865)
Manihot utillissima	Manioc	Sabater Pi (1960)
Musa sp.	Banana	Sabater Pi & de Lassaletta (1958), Bates (1905)
Saccharum officinarum . .	Sugar cane	Dekeyser (1955), Merfield and Miller (1956)
Arachis sp.	Peanut	Malbrant and Maclathy (1949)
Cola sp.	Kola nut	Malbrant and Maclathy (1949)
Elais guiniensis	Palm nut	Owen (1859)

TABLE 35

Chemical Analysis of Peucedanum linderi*

	Leaves	Stem
Dry matter as received	92.30†	95.10
Nitrogen free extract	46.05	39.88
Total ash	12.90	9.44
Silica free ash	11.40	9.25
Silica	1.50	.19
Ether extract	2.32	2.05
Crude fiber	19.94	35.38
Crude protein	18.79	13.25

*Analyzed by the Veterinary Department, Entebbe; courtesy of Mr. Baumgartel.

†Numerical values are given as per cent oven dry matter.

TABLE 36

Similarities in Food Habits between Study Areas*

Areas Compared	Total No. of Food Species Found in Both Areas	No. of Shared Food Species Eaten in Both Areas	Per Cent of Shared Food Species Eaten in Both Areas
Kabara and Kisoro	38	18	47.3
Kabara and Kayonza Forest . .	54	2	3.7
Kisoro and Kayonza Forest . .	47	7	14.9
Virunga Volcanoes—Kayonza Forest	57	9	15.7
Virunga Volcanoes—Mt. Tshiaberimu	50	8	16.0
Virunga Volcanoes—Mwenga-Fizi region	60	4	6.6
Virunga Volcanoes—Utu region	55	0	0
Kayonza Forest—Mt. Tshiaberimu	43	3	7.0
Kayonza Forest—Mwenga-Fizi region	46	7	15.2
Kayonza Forest—Utu region. .	40	3	7.5
Mt. Tshiaberimu—Mwenga-Fizi region	40	4	10.0
Mt. Tshiaberimu—Utu region .	35	0	0
Mwenga-Fizi region—Utu region.	37	4	10.8

*Only plants which have been identified down to the species level are included in the tabulation.

TABLE 37

Presence and Utilization of Forage Species in Various Study Areas

Species	Virunga Volcanoes	Kayonza Forest	Mt. Tshiaberimu Region	Mt. Kahuzi	Mwenga-Fizi Region	Utu Region
Pennisetum purpureum . .	0	[ ]	[x]	[x]	[x]	[ ]
Arundinaria alpina	[x]	?	[x]	[x]	[x]	0
Cyathea deckenii	0	[x]	[ ]	?	[x]	0?
Marattia fraxinea	0	[ ]	0	0	[x]	[x]
Palisota sp.	0	[ ]	0	0	[x]	[x]
Aframomum sp. (stem) . .	0	[ ]	0	[x]	[x]	[x]
Piper capense	[x]	[x]	?	?	[x]	?
Urera hypselendron	[x]	[x]	[x]	[x]	[x]	?
Rubus sp.	[x]	[x]	[x]	[ ]	?	?
Pycnostachys goetzenii . .	[x]	[x]	[x]	[ ]	[ ]	?
Mimulopsis arborescens .	[x]	[x]	[x]	[x]	?	?
Xymalos monospora	[x]	[x]	?	?	[x]	?
Musa sp.	[x]	[x]	[ ]	[ ]	[x]	[x]
Galiniera coffeoides	0?	[ ]	[x]	[x]	[x]	?
Myrianthus arboreus . . .	0	[x]	0	?	[x]	[x]
Pygeum africanum	[x]	[ ]	[x]	?	?	0
Laportea sp.	[x]	?	[ ]	?	?	0

[x]: present and eaten by gorillas.
[ ]: present in area but saw no evidence of it being eaten.
0 : not seen in study area
? : not found to be eaten but status in area was not determined.

TABLE 38

Analyses of Soils Eaten by Gorillas (in Parts per Million)

	1 Mt. Mikeno Group IV	2 Mt. Mikeno Group VIII	3 Miya, Utu Region
Phosphorus....	910	not tested	not tested
Sulphur	3	not tested	not tested
Nitrogen......	1,900	not tested	not tested
Calcium	2,650	3,040	80
Magnesium	3,050	1,140	60
Potassium	8,370	2,700	40
Manganese	170	133	19
Sodium.......	3,480	10,980	160

TABLE 39

Personal Evaluation of Taste of Some Gorilla Food Plants

Species	Part Eaten	Comments on Taste
Arundinaria alpina	Shoot	At first relatively tasteless, but extremely bitter aftertaste.
Polypodium sp.	Leaf	Slightly sour.
Laportea alatipes	Leaf	General leaf or grass flavor, slightly astringent and bitter.
Helichrysum sp.	Leaf	Slightly bitter.
Rumex afromontanus . . .	Inside of stem	Like garden pea pod.
Peucedanum kerstenii . . .	Inside of stem	Like radish, but bitter resinous aftertaste.
Peucedanum linderi	Inside of stem	Some like domestic celery; some extremely bitter.
Carduus afromontanus . .	Stem	Rather tasteless, slight grass flavor.
Aframomum sp.	Fruit	Some species(?) bitter and resinous, some sour and astringent.
	Pith of stem	Bitter and resinous.
Cynoglossum geometricum and C. amplifolium	Root	Palatable—like radish.
Galium simense	Whole vine	Like garden pea pod and grass.
Droquetia iners	Whole vine	Grass-like and slightly astringent.
Rubus runssorensis	Fruit	Excellent, like domestic Rubus.
	Tender tips and leaves	Grass-like, slightly bitter and astringent.
Vernonia adolfi-frederici	Pith	First tasteless, then bitter.
	Blossoms	Slightly sweet.
Xymalos monospora	Fruit	Mealy.
Myrianthus arboreus . . .	Fruit	Sour but excellent.
Erica arborea	Bark	Slightly bitter.
Hagenia abyssinica	Bark, pith	Bitter and astringent.
Hypericum lanceolatum . .	Bark	Bitter and astringent.
Pygeum africanum	Bark	Bitter, somewhat almond-like.
Senecio erici-rosenii . . .	Pith	First tasteless, then bitter.

TABLE 40

Height of Gorilla Nests above Ground

	Kabara		Kisoro		Visoke-Sabinio Saddle		Kayonza Forest		Mt. Tshiaberimu		Mt. Kahuzi		Mwenga-Fizi		Utu		Total	
	No.	Per Cent	No.	Per Cent	No.	Per Cent	No.	Per Cent	No.	Per Cent	No.	Per Cent	No.	Per Cent	No.	Per Cent	No.	Per Cent
Total No. nests	2,488		106		14		179		85		7		23		110		3,012	
Total No. on ground	2,415		48		8		96		28		7		15		24		2,641	
Per cent on ground		97.1		45.3		57.2		53.5		32.9		100.0		65.2		21.8		87.7
Height above ground (feet)																		
2	2		2		0		5		0		0		0		1			
3	4		3		0		8		0		0		0		1			
4	14		7		1		6		1		0		1		1			
5	9		5		1		11		1		0		1		2			
6	4		3		0		12		1		0		0		0			
7	7		6		1		5		0		0		0		3			
8	7		7		1		8		6		0		1		4			
9	2		6		0		2		1		0		0		0			
10	8		14		1		6		12		0		0		3			
Total (< 10 ft.)	57	2.3	53	50.0	5	35.7	63	35.4	22	25.9	0	0.	3	13.0	15	13.6	218	7.2
11-15	6		4		1		12		24		0		1		13			
16-20	2		1		0		4		9		0		3		16			
Total (11-20 ft.)	8	0.3	5	4.7	1	7.1	16	8.9	33	38.8	0	0.	4	17.4	29	26.4	96	3.2
21-30	8		0		0		1		2		0		1		19			
31-40	0		0		0		0		0		0		0		11			
41-50	0		0		0		3		0		0		0		9			
51-60	0		0		0		0		0		0		0		3			
Total (> 21 ft.)	8	0.3	0	0.	0	0.	4	2.2	2	2.4	0	0.	1	4.4	42	38.2	57	1.9

TABLE 41

Height above Ground of Chimpanzee and Gorilla Nests in the Kayonza Forest

Height (in feet)	Chimpanzee Nests		Gorilla Nests	
	No.	Per Cent	No.	Per Cent
Ground - 1	0	0	96	53.6
2 - 10	0	0	63	35.2
11 - 20	5	12.8	16	8.9
21 - 30	11	28.2	1	0.6
31 - 40	9	23.1	0	0
41 - 50	5	12.8	3	1.7
51 - 60	5	12.8	0	0
61 - 70	3	7.7	0	0
71 - 80	1	2.6	0	0
Total	39	100.0	179	100.0

TABLE 42

Distances between the Nearest Nests in 146 Nest Sites at Kabara[*]

Age and Sex Class	Distances between Nearest Nests (in Feet)									
	0-10		11-20		21+		Total No. Obs.	Min.	Max.	Mean
	No. Obs.	Per Cent	No. Obs.	Per Cent	No. Obs.	Per Cent				
Silverbacked ♂-Silverbacked ♂.	4	16.7	5	20.8	15	62.5	24	8	110	34.0
Silverbacked ♂-Medium-sized[†].	63	61.2	26	25.2	14	13.6	103	1	85	13.4
Silverbacked ♂-Juvenile	28	34.6	26	32.1	27	33.3	81	1	90	20.7
Medium-sized - Medium-sized.	136	94.4	8	5.6	0	0	144	1	20	5.4
Medium-sized - Juvenile.	120	100.0	0	0	0	0	120	0[‡]	10	3.0
Juvenile - Juvenile.	48	63.2	20	26.2	8	10.6	76	0[§]	70	12.0

[*] Based on groups II, IV, V, VI, VII, and VIII.

[†] Medium-sized animals include females with or without infants and blackbacked males.

[‡] A juvenile slept in the nest with a medium-sized animal in 29 instances, primarily in group VII. If these 29 instances are not included, the mean distance between the nests is raised to 4.0 feet.

[§] Two juveniles slept in the same nest in 4 instances.

TABLE 43

Materials Used in the Construction of Night Nests at Kabara

Principal Nest Materials Used	Number of Nests	Percentage of Nests
Herbs*	1,068	43.8
Lobelia giberroa	432	17.7
Herbs and *Lobelia giberroa* (about half and half)	92	3.8
Vines†	235	9.6
Turf‡	173	7.1
Soil§	63	2.5
Branches¶	359	14.7
Senecio erici-rosenii	11	.5
Arundinaria alpina	6	.3
Total	2,439	100.0

*Includes primarily such species of the groundcover as *Senecio trichopterygius*, *Laportea*, *Peucedanum*, *Impatiens*, *Rumex*, etc.

†Includes mostly *Galium*.

‡Indicates the reddish, dry, and powdery earth at the base of *Hagenia* trees.

§Indicates that the animals merely slept on the bare ground, perhaps lightly covered with packed-down vegetation, without constructing a nest.

¶Indicates all nests constructed primarily of twigs and branches from shrubs and trees.

TABLE 44

Use of Night Nest Construction Materials in Three Kabara Gorilla Groups

Principal Nest Materials Used	Group IV		Group VI		Group VII	
	No. of Nests	Per Cent	No. of Nests	Per Cent	No. of Nests	Per Cent
Herbs	201	40.9	232	64.8	251	34.4
Lobelia giberroa . . .	19	3.9	22	6.2	275	37.7
Lobelia giberroa and herbs.	2	.4	2	.6	42	5.8
Vines.	73	14.9	38	10.6	31	4.3
Turf.	22	4.5	21	5.8	54	7.4
Soil	9	1.8	7	1.9	13	1.8
Branches	159	32.4	25	7.0	63	8.6
Senecio erici-rosenii	0	0	11	3.1	0	0
Arundinaria alpina . .	6	1.2	0	0	0	0
Total No. of nests checked	491	100.0	358	100.0	729	100.0
Total No. of nest sites checked . .	33		39		75	

TABLE 45

Presence and Location of Dung at the Gorilla Nest

	Kabara	Mt. Tshiaberimu	Utu	Kayonza Forest	Kisoro	Visoke-Sabinio Saddle	Mt. Kahuzi	Mwenga-Fizi
Total No. of nests checked for dung . . .	2,300	18	10	70	28	9	6	10
No. of nests with dung in nest cup lain on by the animal	1,681	2	1	6	23	5	3	0
Per cent lain on	73.1	11.1	10.0	8.6	82.2	55.6	50.0	0
No. of nests with dung on edge	523	10	6	41	3	4	3	2
Per cent on edge	22.7	55.5	60.0	58.6	10.7	44.4	50.0	20.0
No. of nests with dung outside of nest .	70	3	2	23	2	0	0	8
Per cent outside	3.1	16.7	20.0	32.8	7.1	0	0	80.0
No. of nests with no dung	26	3	1	0	0	0	0	0
Per cent with no dung	1.1	16.7	10.0	0	0	0	0	0

TABLE 46

Frequency of Self-grooming and Parts of the Body Groomed by the Various Age and Sex Classes in Gorillas

	Instances Recorded				A	B	
	Arm and Shoulder	Legs	Chest and Abdominal Region	Not Recorded	Totals	No. of Animals in Population (after Table 18B)	Grooming Index A/B
Silverbacked male . .	9	0	0	0	9	25	.36
Blackbacked male . .	3	0	0	1	4	18	.22
Female	33	3	2	2	40	65	.61
Juvenile	17	5	5	4	31	31	1.00
Infant	4	0	0	0	4	52	.07
Totals	66	8	7	7	88		

TABLE 47

Gorilla Vocalizations, Their Probable Underlying Emotions and Their Functions (for Full Description of the Vocalizations see Text)

No.	Probable Emotion	Stimulating Situation	Description of the Vocalization	Visible Response of Other Animals	Probable Function	Sex and Age Class in which Exhibited and Frequency of Occurrence				
						Silver-backed ♂	Black-backed ♂	Female	Juvenile	Infant
1	Contentment	Feeding and resting peacefully	Purr	None	Indicates that all is well	X (+)		X(X) (+)		
2	Contentment	Feeding and resting peacefully	Soft grumble	None	Indicates that all is well	X (+++)	X (++)	X(X) (+++)		
3	Contentment	Feeding and resting peacefully	Soft grunt	None	Indicates that all is well	X (+++)	X (++)	X(X) (+++)		
4	Contentment (?)	Feeding and resting peacefully	Hum (mmm)	None	(?)					
5	(?)	When moving out from rest area; when group is scattered; when animals meet in dense brush	Series of abrupt grunts of low pitch	Animals follow and join leader; or no obvious response	Aids in group cohesion; denotes "here I am"	X (+++)		X (++)	X (+)	
6	(?)	When group is scattered widely	A series of rapid, high bo-bo-bo-, or similar sound	No immediate response	Probably aids in group cohesion; denotes "here I am"			X (?)		
7	(?)	(?)	Like "neighing horse"	(?)	(?)					
8	Excitement (?)	Copulation	Rapid, loud, staccato ö-ö-ö-ö-ö	None	(?)	X (+)				
9	Excitement (?)	Copulation	Loud growling and grunting	(Heard in captivity only)	(?)	(X) (?)		(X) (?)		
10	Excitement	Given during displays in tension-producing situations: presence of man, another group, etc.	Clear but low-pitched hoots (hu-hu) ending in growl	Stimulates other males to display; animals run out of way of vocalizing male; none	Generates excitement in the male; communicates presence of animal to another group; intimidation as part of the whole display	X(X) (+++)	X (+)			
11	Excitement	Presence of man	Squeaky yip	None	Perhaps alerts others	X (+)				
12	Excitement	Given during displays in presence of man	Soft, panting ho-ho-ho	None	Perhaps intimidation			X (+)		

TABLE 47 (continued)

No.	Probable Emotion	Stimulating Situation	Description of the Vocalization	Visible Response of Other Animals	Probable Function	Sex and Age Class in which Exhibited and Frequency of Occurrence				
						Silver-backed ♂	Black-backed ♂	Female	Juvenile	Infant
13	Excitement	Presence of man	Squeaky "a"	None	Perhaps alerts others				X (+)	
14	Excitement	Play	Panting chuckle	None	(?)			(X) (?)		X(X) (++)
15	Annoyance	Given by male if females quarrel; presence of man; given by quarreling females	Harsh staccato grunts (u-u)	Females stop quarreling if given by male; look up and may join male; none	Warning	X(X) (+++)	X (++)	X(X) (++)	X (+)	(X) (?)
16	Annoyance	By quarreling females; presence of man	Short, loud barks (waw-waw)	Animals look at barking member; none	Warning		X (++)	X (+++)	X (+)	(X) (?)
17	Anger	Presence of man	Single, loud explosive roar	Animals usually cluster behind male	Intimidation, warning	X (+++)	X (++)	X (+)		
18	Anger	By quarreling females; presence of man	Harsh, fairly short screams	Male may advance and look; animals may grow nervous; none	Indicates anger; perhaps warning; (sometimes contains elements of fear)		X (+)	X (++)	X (+)	(X) (?)
19	Distress	When in danger of being abandoned or injured	Soft whine	One adult usually looks toward sound	Indicates light distress			X (+)	X (+)	X(X) (+)
20	Distress	When in danger of being left behind by mother; in situations of deprivation in captives	One or two high-pitched screeches	Female waits for infant or ignores it	Indicates distress (often contains elements of anger)					X(X) (++)
21	Fear	Presence of man	Intense screaming roar	Animals cluster around male or become alert	Warning; perhaps intimidation (contains elements of anger)	X (+++)	X (++)			
22	Fear	Presence of man	Loud, long, high-pitched screams	Animals grow uneasy; cluster by male; follow running animal	Warning of great danger		X (++)	X (++)	X (+)	X (+)

X: heard in the wild.
(X): heard in captivity.
Blank spaces indicate that I have not heard the vocalization given by that particular age class or that I was unable to determine which animal emitted it.

(+++): heard frequently.
(++): heard occasionally.
(+): heard rarely.

TABLE 48

The Number of Individual Beats per Chest-beating Display in Kabara Gorillas

No. of Individual Beats	No. of Observations
1	1
2	25
3	73
4	87
5	22
6	30
7	2
8	9
9*	1
10	1
11	1
12	1
13	1
14	0
15	1

*Numbers 9 to 15 are based on sound spectrograms only.

TABLE 49

Average Number of Beats per Chest-beating Display in Gorilla Males at Kabara

	No. of Records	Average No. of Beats per Chest Beat	Range of Variation in No. of Beats per Chest Beat
Silverbacked, group IV (No. 2 in hierarchy)	14	2.8	2 - 4
Silverbacked, group IV (No. 1 in hierarchy)	9	3.5	2 - 6
Silverbacked, group V (No. 2 in hierarchy)	11	3.4	2 - 6
Silverbacked, group VI	43	4.0	2 - 6
Silverbacked, group VII. . . .	50	4.0	2 - 9
Silverbacked, group VIII . . .	12	5.1	3 - 8
Blackbacked, group VIII . . .	8	3.5	2 - 5
Blackbacked, group VIII . . .	13	5.3	3 - 8

TABLE 50

Occurrence of the Various Acts in the Chest-beating Sequence of the Different Age and Sex Classes

		Silverbacked Male	Blackbacked Male	Female	Juvenile	Infant 1-3 Years	Infant 0-1 Years
1	Hooting	xx	x	0	0	0	0
2	"Symbolic feeding" . . .	xx	xx	x	x	x	0
3	Throwing . . .	xx	xx	xx	xx	x	0
4	Rising	xx	xx	xx	xx	xx	xx
5	Chest-beating	xx	xx	xx	xx	xx	xx
6	Kicking	xx	xx	0?	x	0	0
7	Running	xx	x	x	0	0	0
8	Slapping and tearing	xx	xx	xx	xx	x	x
9	Ground-thumping. . .	xx	xx	xx	x	0	0

xx: common.
x: infrequent.
0: not observed.

TABLE 51

Dominance Interactions between Age and Sex Classes

Dominant Animal \ Subordinate Animal	Silverbacked Male	Blackbacked Male	Female	Juvenile	Infant	Total
Silverbacked male . . .	13	1	26	9	2	51
Blackbacked male . . .	. . .	. . .	2	3	. . .	5
Female	. . .	4	12	13	11	40
Juvenile	. . .	. . .	. . .	4	8	12
Infant (one year or older) . .	. . .	. . .	. . .	. . .	2	2
Total	13	5	40	29	23	110

TABLE 52

Dominance Interactions between the Silverbacked Males of Group IV

Dominant Animal \ Subordinate Animal	Big Daddy	D.J.	The Outsider	Splitnose	Newcomer
Big Daddy	...	2	2	1	0
D.J.		...	1	1	0
The Outsider (Peripheral)			...	2(?)	0
Splitnose (Peripheral)				...	1(?)
Newcomer (Peripheral)					...

TABLE 53

Number of Observations of Mutual Grooming in the Various Age and Sex Classes

Animals Grooming \ Animals Groomed	Silverbacked Male	Blackbacked Male	Female	Juvenile	Infant	Total
Silverbacked male	0	0	0	0	3	3
Blackbacked male	0	0	0	0	0	0
Female	0	1	5	13	76	95
Juvenile	1	1	9	10	12	33
Infant	0	0	2	0	1	3
Total	1	2	16	23	92	134

TABLE 54

Mutual Grooming in *Papio ursinus* (Adapted from Bolwig, 1959b)

Animal Grooming \ Animal Groomed	Male	Female	Youngster	Total
Male	0	7	0	7
Female	33	27	5	65
Youngster	0	4	3	7
Total	33	38	8	79

TABLE 55

The Frequency of Lone and Social Play in the Various Age and Sex Classes

Animals Involved in Play	Number of Observations	
	Lone Play	Social Play
1 infant	31	
1 juvenile	7	
1 blackbacked ♂	1	
2 infants		20
1 juvenile, 1 infant		12
2 juveniles		7
1 female, 1 infant		1
1 blackbacked ♂, 1 female		1
1 blackbacked ♂, 1 juvenile		1
3 infants		3
2 infants, 1 juvenile		2
1 infant, 2 juveniles		1
3 juveniles		1
4 infants		2
3 infants, 1 juvenile		1

TABLE 56

Stages of Behavioral Development in Free-living Gorillas to the Age of 6 Months

Age in Weeks	Behavior
0-1	Arm and leg movements appear unoriented. The grasping reflex is weak—the infant is able to hold on to its mother's chest unsupported for only about 2 seconds.
4-5	The infant looks around and seems to follow the movement of other animals with its head. The grasping reflex is stronger—it holds on to its mother's chest unsupported for about 10 seconds.
6-7	The infant is able to hold on sufficiently to ride on the female's back without support. It raises its head while lying prone and looks around.
8	The infant shows a marked increase in bodily movement. It lies in the lap of the female unsupported.
10-12	The infant pulls at a vine; it east *Galium*; it bites into a herb. It crawls slowly near the sitting female; it crawls to a reclining juvenile and up on its chest; it pulls hair on another female; it crawls slowly behind its mother. It holds on to the chest of the female without support.
14	The infant crawls slowly behind the female as she walks. It grapples briefly with an older infant while held in the arm of the female.
16	The infant sits by the female on the ground, partially supported by her body. It rides frequently on its mother's back.
16-18	The infant takes its first definite steps in the typical quadrupedal gait; it crawls alone on a log; it climbs up a lobelia stalk; it rises shakily on its hindlegs and beats its chest.
20	The infant walks quadrupedally behind its mother; it sits unsupported near the female and chews on a stem; it stands bipedally with hands held above the head; it plays with other infants near its mother.
24	The infant runs quadrupedally; it climbs into low shrubs and trees.

TABLE 57

Stages of Behavioral Development in the Infant Gorilla "Jambo" at the Basel Zoo (after Lang, 1961c)

Age in Weeks	Behavior
0	The grasping reflex of the infant is very weak; its eyes have an "empty" look.
1	The female lays the infant on the floor; she attempts to carry it on her chest without support, but it hangs insecurely by its arms alone and cries.
4	The infant follows the movement of objects with its eyes.
5	The female places her infant on her head or neck, where it holds on by itself.
6	The infant raises its head and supports itself on the elbows while lying prone.
7	The infant "laughs" for the first time when tickled by its keeper.
9	The infant sits upright on the floor while holding onto the arm of its mother with its hands. It raises the upper part of its body from the floor while lying prone; it reaches for an object.
11	The infant readily reaches for objects; it shows an interest in solid foods.
12	The infant pulls itself easily into a sitting position along the bars of the cage; it crawls across the cage.
14	The infant pulls itself into a bipedal position along the bars of the cage; it crawls well.

TABLE 58

The Age of First Appearance of Various Behavior Patterns in Captive Gorillas, Chimpanzees, and Man*

Behavior Pattern	Age of Appearance (in Weeks)			
	Gorilla "Goma"	Gorilla "Colo"	Chimpanzee	Man
Raises head briefly from prone position	0-1	0	0	4
Raises head briefly from supine position	...	2.5	4	28
Arm and leg movements still unco-ordinated	3	...	...	...
Grasping reflex strong enough to hang unsupported on a person briefly	6	...	...	...
Responds to sound by turning head; begins to bite on objects, reaches for human hand, responds to environment	7-8	...	...	...
Turns from a supine to a prone position	8	10	11	24
Explores edge of basket with hand	9	...	...	...
Crawls pushing with feet	13(?)	10	14	36
Pulls itself up to look over edge of basket; great tendency to move around	12	...	...	...
Stands bipedally using support for hands	13-14	14	22	32
Crawls with legs straight but rests arm on elbow	16	...	...	...
Sits one minute erect, unsteady	...	18	25	32
Walks in typical adult quadrupedal fashion	24	20	...	...
Sits steady indefinitely	...	20-22	20	40
Climbs up 90 cm. and out of playpen	18	...	...	...
Climbs into shrubs	28	...	...	...

*Data on the gorilla "Goma" are from the series Documenta Geigy Nos. 1 to 6 (Lang 1959, Schenkel 1960a, Hauser 1960, Schenkel 1960b, Lang and Schenkel 1961a). The other material is from Knobloch and Pasamanick (1958).

TABLE 59

Estimated Dates of Birth of Infants in the Kabara Population

Month	Year				Totals
	1957	1958	1959	1960	
Jan.		1	0	1	2
Feb.		0	2	1	3
Mar.		2	2	0	4
Apr.		1	2	1	4
May		3	1		4
June		2	3		5
July	1	1	4		6
Aug.	1	2	3		6
Sept.	1	2	4		7
Oct.	1	2	1		4
Nov.	1	1	0		2
Dec.	3	2	2		7
Total					54

TABLE 60

Mean Number of Chest Beats by Silverbacked Males in Response to My Presence during the First Hour of Contact

Encounters	Male Group II		Male Group VI		Male Group VII		Male Group VIII	
	Mean No. Chest Beats	No. of Obser-vations	Mean No. Chest Beats	No. of Obser-vations	Mean No. Chest Beats	No. of Obser-vations	Mean No. Chest Beats	No. of Obser-vations
1st to 10th . .	1.0	7	2.4	7	1.3	3	7.0	1
11th to 20th . .	1.2	9	1.5	4	.9	8	3.3	6
21st to 30th . .			1.7	7	1.3	7	4.2	9
31st to 40th . .			1.4	9	.9	8		
41st to 50th . .			2.7	7	3.6	9		
51st to 60th . .					.3	7		
61st to 70th . .					.2	6		
71st to 80th . .					.2	10		

APPENDIX A

Accounts by Captain von Beringe concerning His Discovery of the Mountain Gorilla

<u>The purpose of his trip to the Virunga Volcanoes</u>

"Der Zweck der Reise war, die deutschen Posten in Tschiwitoke und Ischangi zu besuchen, durch Erledigung von Schauris die Häuptlinge in Fühlung mit der deutschen Regierung zu erhalten und die von meinen Vorgängern gepflogenen guten Beziehungen mit dem Sultan von Ruanda an Ort und Stelle weiter zu festigen. Der Marsch an den Vulkanen diente neben Sicherung der jetzt hier arbeitenden kongolesischen Grenzkommission gleichzeitig zur Stärkung der Macht und des Ansehens der deutschen Verwaltung in diesen Gebieten."

<u>The collecting of the first mountain gorilla specimen</u>

"Von unserem Lager aus erblickten wir eine Herde schwarzer, grosser Affen, welche versuchten, den höchsten Gipfel des Vulkans zu erklettern. Von diesen Affen gelang es uns, zwei grosse Tiere zur Strecke zu liefern, welche mit grossem Gepolter in eine nach Nordosten sich öffnende Kraterschlucht abstürzten. Nach fünfstündiger, anstrengender Arbeit gelang es uns, ein Tier angeseilt heraufzuholen. Es war ein männlicher, grosser, menschenähnlicher Affe von etwa 1-1/2 m Grösse und einem Gewichte von über 200 Pfund. Die Brust unbehaart die Hände und Füsse von ungeheurer Grösse. Es war mir leider nicht möglich, die Gattung des Affen zu bestimmen. Für einen Schimpansen hatte derselbe eine wohl noch nicht bekannte Grösse, und das Vorhandensein von Gorillas ist bis jetzt bis zu den Seen hin noch nicht festgestellt worden."

APPENDIX B

Weather Data

TABLE 61

Precipitation at Kisoro, Uganda (Altitude 6,200 Feet), 1940-1954 Average (after Watt, 1956)

Month	Inches of Rain
January	4.92
February	5.04
March	6.76
April	7.99
May	4.77
June	2.11
July	1.90
August	3.37
September	7.13
October	7.95
November	8.30
December	5.82
Total	66.06

TABLE 62

Weather Data, Rumangabo, Albert National Park (after Bourlière and Verschuren, 1960)

Month	Mean Monthly Temperature (°C.)	Precipitation in mm.
Jan.	19.2	87.1
Feb.	19.7	79.5
Mar.	19.9	140.9
Apr.	19.4	198.4
May	19.7	165.6
June	19.4	111.5
July	18.9	92.4
Aug.	19.0	120.9
Sept.	19.1	192.5
Oct.	19.0	210.9
Nov.	18.9	201.7
Dec.	18.9	160.6
Mean annual	19.3	
No. of years of observation	3	12
Yearly total		1,759.8
No. of days with precipitation		211

TABLE 63

Weather Data, Rumangabo, Albert National Park, during Period of Study (Altitude 5,637 Feet)

Month	Temperature °C. (shade)		Relative Humidity	Precipitation mm.
	Max. Mean	Min. Mean		
1959				
Jan.	25.5	14.1	72.7	128.2
Feb.	25.7	13.9	75.5	48.8
Mar.	26.7	14.5	71.0	141.4
Apr.	26.8	14.1	72.5	234.7
May	25.8	14.1	72.9	131.2
June	26.5	12.9	67.1	22.7
July	26.6	12.2	66.6	16.3
Aug.	25.9	12.5	69.3	140.0
Sept.	25.8	12.9	68.6	206.0
Oct.	25.1	13.3	69.4	196.2
Nov.	24.4	13.2	69.0	192.8
Dec.	25.1	13.4	69.3	156.8
1960				
Jan.	25.4	13.6	72.2	82.3
Feb.	26.1	13.7	68.7	97.5
Mar.	25.5	13.4	75.4	215.8
Apr.	25.7	13.7	70.6	238.2
May	26.0	13.2	72.7	204.5

TABLE 64

Kabara Temperatures (Altitude 10,200 Feet)*

Time of Day	Date	No. of Readings	Low °C.	High °C.	Mean °C.
0600-0630	Aug. 20-31, 1959	10	0.5	8.5	4.3
	Sept. 1-30	24	5.0	9.0	7.2
	Oct. 1-31	23	4.0	9.0	7.4
	Nov. 1-30	19	5.5	9.0	7.5
	Dec. 1-31	24	4.5	9.0	7.5
	Jan. 1-15, 1960	12	4.0	9.0	7.4
	Feb. 1-29	19	3.0	8.5	6.3
	Mar. 4-31	16	4.0	8.0	6.5
	Apr. 1-30	17	3.0	9.0	7.1
	May 1-6, May 15-31	16	5.0	9.0	7.4
	Aug. 10-17	5	3.5	5.5	4.7
1200-1230	Aug. 20-31, 1959	6	10.0	14.0	11.7
	Sept. 1-30	24	9.0	15.5	12.2
	Oct. 1-31	30	8.0	17.0	12.2
	Nov. 1-30	30	9.0	15.0	12.0
	Dec. 1-31	28	8.0	15.0	11.2
	Jan. 1-15, 1960	14	7.5	16.0	12.7
	Feb. 1-29	25	8.0	14.5	11.9
	Mar. 4-31	25	8.0	16.0	12.1
	Apr. 1-30	29	8.0	17.0	12.1
	May 1-6, May 15-31	20	9.5	17.0	14.1
	Aug. 10-17	...	...	...	...
1600-1630	Aug. 20-31, 1959	8	8.0	11.5	10.5
	Sept. 1-30	19	7.0	12.0	10.3
	Oct. 1-31	29	8.0	14.0	10.4
	Nov. 1-30	24	7.5	15.0	10.6
	Dec. 1-31	25	8.5	15.0	10.7
	Jan. 1-15, 1960	15	8.0	16.0	10.6
	Feb. 1-29	27	7.5	15.5	9.9
	Mar. 4-31	24	6.0	17.0	11.1
	Apr. 1-30	26	9.0	13.0	11.6
	May 1-6, May 15-31	19	9.5	12.5	11.2
	Aug. 10-17	5	10.0	12.0	11.2
1800-1830	Aug. 20-31, 1959	6	7.0	8.5	7.7
	Sept. 1-30	17	7.0	10.0	7.6
	Oct. 1-31	30	7.0	10.0	8.8
	Nov. 1-30	27	7.0	10.0	8.2
	Dec. 1-31	31	7.5	10.0	8.5
	Jan. 1-15, 1960	15	6.5	10.5	8.7
	Feb. 1-29	28	6.5	10.5	8.6
	Mar. 4-31	25	6.5	10.0	8.7
	Apr. 1-30	30	7.5	11.0	9.1
	May 1-6, May 15-31	21	7.5	10.5	8.9
	Aug. 10-17	3	9.0	10.0	9.0
2000-2030	Aug. 20-31, 1959	11	3.0	8.5	6.6
	Sept. 1-30	19	4.5	10.0	7.5
	Oct. 1-31	25	5.5	9.0	7.8
	Nov. 1-30	29	7.0	9.0	8.0
	Dec. 1-31	26	6.5	9.0	8.0
	Jan. 1-15, 1960	12	5.0	10.0	7.9
	Feb. 1-29	28	4.0	9.0	7.0
	Mar. 4-31	25	5.0	10.0	7.6
	Apr. 1-30	28	6.0	10.0	8.2
	May 1-6, May 15-31	18	4.5	9.5	7.9
	Aug. 10-17	...	...	...	...

*All temperatures were taken 5 - 6 feet above ground in the shade.

TABLE 65

Kabara Cloudcover

	Clear	Partially Cloudy	Cloudy
Aug. 20-31, 1959	0	7	5
Sept. 1-30	0	13	17
Oct. 1-31	0	8	23
Nov. 1-30	0	8	22
Dec. 1-31	0	6	25
Jan. 1-15, 1960	0	2	13
Feb. 1-29	0	3	26
Mar. 4-31	0	7	21
Apr. 1-30	0	6	24
May 1-6, May 15-31 . .	0	12	11
Aug. 10-17	0	1	7
Total	0	73	194

TABLE 66

Kabara Precipitation

	No. of Days	Days without Rain		Days with Rain Lasting 1/4 to 1/2 Hour		Days with Rain More Than 1/2 Hour		Total No. Hours of Rain
		No.	Per Cent	No.	Per Cent	No.	Per Cent	
Aug. 20-31, 1959	12	7	58.3	0	0	5	41.7	20
Sept. 1-30	30	3	10.0	8	26.7	19	63.3	49
Oct. 1-31	31	6	19.4	2	6.5	23	74.1	88
Nov. 1-30	30	2	6.7	10	33.3	18	60.0	78
Dec. 1-31	31	2	6.5	2	6.5	27	87.0	66
Jan. 1-15, 1960 .	15	4	26.7	1	6.7	10	66.6	22
Feb. 1-29	29	3	10.3	5	17.3	21	72.4	57
Mar. 4-31	28	4	14.2	6	21.5	18	64.3	52
Apr. 1-30	30	3	10.0	3	10.0	24	80.0	70
May 1-6, 15-31 .	23	11	47.8	3	13.1	9	39.1	21
Aug. 10-17	8	6	75.0	0	0	2	25.0	2
Total	267	51	19.1	40	15.0	176	65.9	525

TABLE 67

Weather Spells at Kabara

Mostly Sunny Spells	Rainy Spells
August 20-26, 1959	August 27-September 16, 1959
September 17-October 1	October 2-10
October 11-18	October 19-November 7
November 8-9	November 20-December 29
December 30, 1959-January 7, 1960	January 8-15, 1960
March 4-16	February 1-29
March 27-April 17	March 17-26
May 1-26	April 18-30
August 10-17	May 27-June 1
September 11-17	

APPENDIX C

Notes on Skulls of the Mountain Gorilla

The eighteen crania were arranged by age, employing the standard criteria of dental emergence, suture closure, dental attrition, and exostotic growths. No. 1, a male, has the third molars in occlusal position, as do the succeeding specimens, but the basilar suture is only partially closed. The ramification of the nasal bones into the frontal is still visible, as is the complete outline of the premaxillary, palatal, and facial bones. None of the vault sutures have closed. In the next older specimen, No. 3, and all succeeding ones, the basilar suture is closed and only traces of the other sutures remain on the external surface.

The mastoid processes of Nos. 6, 1, and 8, in descending order, are large, well delimited, and rugose. In all cases the temporal bone contacts the frontal bone in the pterionic region. The postglenoid process of No. 6 is remarkably large, and the auditory meatus is somewhat horizontally elliptical owing to the intrusion of the posterior portion of the tympanic plate. Male No. 14 displays appreciable lipping and exostotic growths on the saggital and occipital crests.

Considering all specimens from and including No. 1 as adults, there appears to be a moderate degree of dental attrition, remarkably little in the case of No. 19 from the Virunga Volcanoes. Only female No. 10 shows extensive wear into both the inner and outer cusps of the maxillary premolars. No. 18, two days old, had no teeth penetrating the gums before drying. After drying the maxillary incisors and the mandibular median incisors became visible. No. 6 displays a diminutive fourth molar, 3 mm. in maximum crown diameter and placed 10 mm. behind the third molar on the right maxilla.

All skeletal material is on file at the Department of Anthropology, University of Wisconsin, and will receive detailed study.

TABLE 68

Collecting Locations of Gorilla Skeletal Material

Location	Specimen No.*	Sex	Comments
Near Lubero, Congo; 2 km. north of Lubango	1	♂	Killed May 9, 1959, by natives
Matembe, near Lubero, Congo	2	♂	Killed by natives
	3	♀	Killed Feb., 1959, by natives
Kasinga, near Lubero, Congo	4	?	Killed Apr., 1959, by natives
	5	?	Killed May, 1959, by natives
	6	♂	Killed by natives
	7	?	Killed Nov., 1958, by natives
	8	♂	Killed by natives
Lumboku River, near Lubero, Congo	9	?	Killed by natives
	10	♀	Killed by natives
	11	♀	Killed by natives
	12	♀	Killed by natives
	13	?	Killed by natives
	14	♂	Killed by natives
Miya, near Kasese, Congo	15	?	Killed by natives
	16	♂ ?	Killed by European
	17	♀	Killed by natives
Kabara, Virunga Volcanoes, Congo	18	♂	Born April 25, 1960; died April 27, 1960; carried by female until May 1, 1960.
	19	♂	Whole skeleton found in forest

*Nos. 1-14 obtained May 23-30, 1959, by Emlen.
Nos. 15-17 obtained Aug. 30, 1960, by Schaller.
No. 18 obtained May 1, 1960, by Schaller.
No. 19 obtained Aug. 11, 1960, by Schaller.

TABLE 69

Measurements (in mm.) of Gorilla Crania*

Specimen No.	18	4	7	13	9	15	5	16	1	3	17	11	19	12	8	6	10	14
Cranial length (glabella to inion) . . .	92	129	131	152	140	144	151	157	183	150	148	168	185	161	191	192	166	189
Cranial height	64	...	...	...	98	99	110	...	114	98	107	106	117	94	122	111	102	117
Auricular height	56	...	78	79	71	76	75	74	85	72	75	75	...	69	100	...	...	101
Height of saggital crest	0	0	0	0	0	0	0	0	9	0	1	0	(15)	1	19	(35)	...	23
Orbital width	52	69	70	82	76	86	91	92	114	...	97	100	108	97	122	110	102	118
Mastoid width (on crest)	60	...	102	113	104	115	124	117	144	130	...	138	168	...	...	141	...	150
Maximum bizygomatic width	64	...	...	...	105	120	137	124	170	...	135	...	195	...	...	185	...	185
Palatal length from margin of palate to gnathion	47	...	57	...	72	84	96	98	124	105	93	108	136	101	128	130	113	134
Palatal length to line between maxillary tubercles	40	...	...	...	70	80	92	94	114	100	90	106	126	97	...	120	109	124
Nasion to basion	61	...	...	...	98	108	116	...	144	124	118	119	156	117	145	130	125	144
Bregma to lambda																		
Cord	48	64	...	71	68	69	86	66	97	82	81	100	93	92	...	...	...	...
Arc	52	69	...	75	...	74	93	69	...	84	85	104	...	96	...	...	...	...
Length of maxillary tooth row	...	...	...	...	...	...	...	...	93	83	75	87	94	83	93	85	82	97
Outside alveolar width of upper tooth row . .	...	...	...	...	...	60	60	62	70	63	66	64	66	66	73	75	64	78
Greatest nasal breadth	15	21	18	...	23	20	33	25	38	32	32	36	43	33	36	38	36	39
Height of nasal aperture	26	50	42	65	61	80	79	91	94	76	84	86	95	82	106	...	86	99

*The skulls are arranged by age.

TABLE 70

Measurements of Newborn Gorilla Infants

Measurement (in cm.)	Mountain Gorilla* (this study)	Mountain Gorilla (van den Berghe, 1959)	Lowland Gorilla Goma (Hauser, 1960)
Total length	48.0	. . .	47.5
Sitting height	32.3	30.0	28.0
Arm length	19.7	. . .	19.0
Upper arm length	10.7	10.0	10.0
Forearm length	9.0	9.0	9.0
Maximum hand length .	8.0	4.5	6.5
Leg length	19.0	. . .	21.5
Thigh length	. . .	9.0	10.0
Shank length	. . .	7.5	11.5
Maximum foot length .	9.3	. . .	9.0
Ear height	3.5	3.3	. . .

*Some measurements are perhaps slightly distorted, for the female carried the infant for 4 days after its death and several joints were disarticulated in the process. The skeleton is deposited at the University of Wisconsin Anthropology Department. The hands, feet, and hyoid were preserved separately in formalin. The heart and eyes were given to the Medical School of Makerere College, Uganda (see Hall-Craggs, 1961b).

TABLE 71

Measurements of the Deformed Arm Bones of a Mountain Gorilla

The left forearm bones of an adult male mountain gorilla (specimen no. 19), which was found dead at Kabara, are severely deformed. The distal portions of both radius and ulna are directed at an angle of almost 90° to the long axis of the shaft with the palmar surface of the hand facing the elbow. Of the various ways in which this may have occurred, the suggestion of G. W. Lasker seems most reasonable. It is likely that the radius was fractured and subsequently healed, and that the growth of the ulna conformed to the new position of the radius.

	Right Arm Total Length (in cm.)	Left Arm Total Length (in cm.)
Radius	33.9	20.2
Ulna	34.8	25.1
Humerus	39.8	39.0

TABLE 72

Sequence of Tooth Eruption in the Upper Jaw

Skull No.	Left Side								Right Side							
	M_3	M_2	M_1	Pm_2	Pm_1	C	I_2	I_1	I_1	I_2	C	Pm_1	Pm_2	M_1	M_2	M_3
4	-	-	±	+	+	S	+	S	+	+	S	+	+	±	-	-
7	0	0	0	0	0	0	0	0	S	S	S	+	+	±	-	-
13	-	S	+	+	+	0	0	0	0	0	0	+	+	S	S	-
9	-	±	S	+	+	S	S	S	S	S	S	+	+	+	±	-
15	±	S	S	S	S	S	S	S	S	S	S	S	S	S	S	±
5	±	+	+	S	+	+	+	+	+	+	+	+*	S	+	+	±
16	+*	+	+	+	+	+*	+	+	+	S	+*	+	+	+	+	+*
1	+	+	+	+	+	+	+	+	S	+	+	+	+	+	+	+
3	+	+	+	+	+	+	+	S	S	+	+	+	+	+	+	+
17	S	+	+	+	S	+	+	S	S	S	+	+	+	+	+	S
11	+	+	+	+	+	+	+	+	+	+	+	+	+	+	+	+
19	+	+	+	+	+	+	+	+	+	+	+	+	+	+	+	+
12	0	+	+	+	+	+	+	+	+	+	+	+	+	+	+	+
8	+	+	+	+	+	S	+	S	S	S	S	+	+	+	+	+
6	+	+	+	+	+	+	+	+	+	S	+	+	+	+	+	+
10	+	+	+	+	+	+	+	+	+	S	+	+	+	+	+	+
14	S	+	+	+	+	S	S	S	S	S	S	S	S	S	S	S

+ tooth present
- tooth absent
S socket intact, tooth lost
± emerging tooth below alveolar rim
+* tooth just erupted
0 socket destroyed

TABLE 73

Dental Attrition in the Teeth of the Upper and Lower Jaws*

Specimen No.	Tooth	Relative Degree of Attrition				
		Right			Left	
4	M_1	0	0	← lingual →	0	0
		0	0	side	0	0
7	M_1	0	0		...	...
		...	0		...	...
13	M_1	...	...		0	0
		...	...		0	0
9	M_1	0	0		...	...
		0	0		...	...
5	M_1	2	2		2	1
		1	1		2	1
	M_2	1	1		1	1
		1	1		1	1
16	M_1	2	3		3	2
		2	2		2	1
	M_2	1	1		1	1
		1	1		1	1
1	M_1	2	2+		2+	...
		2	2		2	...
	M_2	1	2		2	1
		1	1		1	1
	M_3	1	...		2	1
		1	1		1	1
3	M_1	...	3		3	...
		...	2		2	2
	M_2	1	2		2	...
		1	1		1	...
	M_3	1	1		1	1
		...	1		1	1
17	M_1	2	3		3	2
		2	3		3	2
	M_2	...	1		1	2-
		...	1		1	1
	M_3	...	...		...	...
		...	...		...	...
11	M_1	2	3		3	2
		2	3		3	2
	M_2	...	2		2	1
		1	1		2	2-
	M_3	1	1		1	1
		1	1		1	1
19	M_1	1	2-		2-	1
		1	2-		2-	1
	M_2	1	1		1	1
		1	1		1	1
	M_3	1	1		1	1
		1	1		1	1
12	M_1	2	3		3	...
		2	3		3	...
	M_2	2-	2		2	...
		1	1		1	...
	M_3	1	1		...	...
		1	1		...	...
8	M_1	...	3		3	...
		2	2		2	2
	M_2	...	3		...	2
		...	2		2	1
	M_3	...	2		...	...
		...	1		...	...

TABLE 73 (continued)

Specimen No.	Tooth	Relative Degree of Attrition			
		Right		Left	
6	M_1	3	4	4	3
		3	4	4	4
	M_2	2	4	4	2
		3	4	4	3
	M_3	1	3	3+	1
		...	3	3	1
10	M_1	3	4	4	3
		3	4	4	3
	M_2	2	3	3	2
		...	2	2	2
	M_3	1	2	2	1
		2	2+	2	1
- - - - -	- - - - -	- - - - -	- - - - -	- - - - -	- - - - -
5	M_1	2	1	1	2
		2	1	1	2
		2			2
	M_2	1	1	1	1
		1	1	1	1
		1			1
16	M_1	3	2-	1	3
		3	1	1	3
		3			3
	M_2	1	1	1	1
		1	1	1	1
		1			1
1	M_1	2	2-	2-	2
		2	1	2	2
		2			2
	M_2	2-	1	...	...
		2-	1	...	...
		1			...
	M_3	1	1	1	1
		1	1	1	1
		1			1
17	M_1	3	2	...	...
		3	2	...	...
		3			...
	M_2	2	1	...	...
		1	1	...	...
		1			...
	M_3	1	1	...	...
		1	1	...	...
		1			...
19	M_1	1	2-	2	1
		2-	1	1	1
		2			2
	M_2	1	1	1	2
		1	1	1	1
		2			2
	M_3	1	1	1	1
		1	1	1	1
		1			1

0: no wear
1: no dentine visible but enamel worn
2: dentine visible, forming one or more spots of color darker than the enamel
3: part of tooth worn away
4: crown worn to cement-enamel junction at cervix of tooth

*Figures above the broken line refer to the upper jaw; those below, to the lower jaw.

APPENDIX D

TABLE 74

Foot Candle Readings at Kabara

Date	Cloudcover	Time	Foot Candles
October 14, 1959	clear	0540	0
		0550	2
		0555	13
		0600	29
		0605	71
February 7, 1960	cloudy	1800	70
		1810	35
		1823	15
		1830	3
		1835	0

NOTES

Note 1 (p. 109)

I revisited Kabara from August 11 to 22, 1963, and was able to identify two of my former study groups. The composition of group VII after a period of three years was 1 silverbacked male, 1 blackbacked male, 8 females, 3 juveniles, and 4 infants. Of these 17 animals, the silverbacked male, the blackbacked male, and one female were readily identified as having been members of the group in 1960. Group VIII was considerably smaller in 1963 than in 1960, consisting of 1 silverbacked male, 5 females, 2 juveniles, and 2 infants. I was able to recognize the silverbacked male and two females as former acquaintances.

Note 2 (p. 243)

The silverbacked males of groups VII and VIII which at the beginning of the study in 1959 were the leaders and dominant animals of their respective groups still retained this position in August, 1963, a period of four years. (See note above.)

Fig. 69
(Reproduced with permission of the *St. Paul Dispatch*)

BIBLIOGRAPHY

AKELEY, C.
1923. In brightest Africa. Garden City.
AKELEY, C., and AKELEY, M. J.
1932. Lions, gorillas and their neighbors. New York.
AKELEY, M. J.
1929. Carl Akeley's Africa. New York.
ALLEN, G. M.
1939. A checklist of African mammals. Bull. Mus. Comp. Zool. **83**:1–763.
ALLEN, J. G.
1931. Gorilla hunting in southern Nigeria. Nigerian Field. **1**:4–5.
ALTMANN, S. A.
1959. Field observations on a howling monkey society. J. Mammal. **40**(3):317–30.
ANONYMOUS
1927. Annual Report of the Game Department for 1925. Uganda.
1951. Excerpts from the 1949 Report of the Uganda Game Department. Oryx. **1**(2):89–99.
1960. Laboratory examination form (Makerere College). Specimen: gorilla faeces.
ASCHEMEIER, C. R.
1922. Beds of the gorilla and chimpanzee. J. Mammal. **3**(3):176–78.
ASHTON, E. H.
1954. Age changes in some bodily dimensions of apes. Proc. Zool. Soc. London. **124**(3):587–94.
ASHTON, E. H., AND ZUCKERMAN, S.
1952. Age changes in the position of the occipital condyles in the chimpanzee and gorilla. Am. J. Phys. Anthrop. **10**:277–88.
BABAULT, G.
1928. Note sur la biologie et l'habitat du gorille de Beringe. Rev. Franç. Mammal. **2**:61–63.
BARNS, T. A.
1922. The wonderland of the eastern Congo. London.
1923. Across the great craterland to the Congo. London.
1926. An African Eldorado. London.

BARTHOLOMEW, G., AND BIRDSELL, J. B.
1953. Ecology and the protohominids. Am. Anthr. **55**(4):481–98.

BATES, G. L.
1905. Notes on the mammals of southern Cameroons and the Benito. Proc. Zool. Soc. London. **75**:65–85.

BAUMGARTEL, W.
1960. König in Gorillaland. Stuttgart.
1961. The gorilla killer. Wild Life and Sport. **2**(2):14–17.

BEATTY, E. H.
1951. A note on the behavior of the chimpanzee. J. Mammal. **32**(1): 118.

BENCHLEY, B. J.
1942. My friends, the apes. Boston.
n.d. Mountain gorillas in the zoological garden, San Diego, California, 1931 to 1940. Zool. Soc. San Diego.

BERINGE, O. VON
1903. Bericht des Hauptmanns von Beringe über seine Expedition nach Ruanda. Deutsches Kolonialblatt. pp. 234–35, 264–66, 296–98, 317–19.

BINGHAM, H. C.
1928. Sex development in apes. Comp. Psychol. Monogr. **5**(23):1–165.
1932. Gorillas in a native habitat. Carnegie Inst. Wash. Publ. **426**:1–66.

BLANCOU, L.
1950. The lowland gorilla. Animal Kingdom. **58**:162–69.
1951. Notes sur les mammifères de l'équateur africain français. Le gorille. Mammalia. **15**(4):143–56.

BLOWER, J.
1956. The mountain gorilla. Uganda Wild Life and Sport. **1**(1):41–52.

BOLWIG, N.
1959*a*. A study of the nests built by mountain gorilla and chimpanzee. S. Afr. J. of Science. **55**(11):286–91.
1959*b*. A study of the behaviour of the chacma baboon, *Papio ursinus*. Behaviour. **14**(1–2):136–63.

BOURLIÈRE, F., AND VERSCHUREN, J.
1960. L'écologie des ongulés du Parc National Albert. Fasc. 1. Inst. des Parcs Nat. du Congo Belge, Brussels.

BRADLEY, M. H.
1922. On the gorilla trail. New York.

BRANDES, G.
1930. Die Begattung des Orangs. Zool. Garten. **3**:216–17.

BREHM, A.
1916. Die Säugetiere. IV. Leipzig.

Broughton, Lady
1932. Stalking the mountain gorilla with the camera in its natural haunts. Illustrated London News. Nov. 5, pp. 701, 710–13; Nov. 12, pp. 756–57.

Burbridge, B.
1928. Gorilla. New York.

Burrows, G.
1898. Land of the Pigmies. London.

Burt, W. H.
1943. Territoriality and home range concepts as applied to mammals. J. Mammal. **24**:346–52.

Burton, R. F.
1876. Two trips to gorilla land and the cataracts of the Congo. London.

Burtt, B. D.
1934. A botanical reconnaissance in the Virunga Volcanoes of Kigezi, Ruanda, Kivu. Kew Bull. **4**:145–165.

Cameron, V. L.
1877. Across Africa. New York.

Carpenter, C. R.
1934. A field study of the behavior and social relations of the howling monkeys (*Alouatta palliata*). Comp. Psychol. Monogr. **10**(2): 1–168.
1937. An observational study of two captive mountain gorillas (*Gorilla Beringei*). Human Biol. **9**(2):175–96.
1940. A field study in Siam of the behavior and social relations of the gibbon (*Hylobates lar*). Comp. Psychol. Monogr. **16**(5):1–212.
1942. Sexual behavior of free ranging rhesus monkeys, J. Comp. Psychol. **33**:113–62.
1958. Soziologie und Verhalten freilebender nichtmenschlicher Primaten. Handbuch der Zoologie. **10**(11):1–32.

Chagula, W. K.
1961. The liver of the mountain gorilla (*Gorilla gorilla beringei*). Am. J. Phys. Anthrop. **19**(3):309–15.

Chapin, J. P.
1932. The birds of the Belgian Congo. Part I. Bull. Amer. Mus. Nat. Hist. **65**:1–756.

Chorley, C. W.
1928. Notes on Uganda gorillas seen during a visit to Mount Sabinio, Christmas, 1927. Proc. Zool. Soc. London. **98**:267–68.

Collias, N., and Southwick, C.
1952. A field study of population density and social organization in howling monkeys. Proc. Am. Phil. Soc. **96**:143–56.

COLYER, F.
1936. Variations and diseases of the teeth in animals. London.
CONANT, R.
1941. Meet the champions. Fauna. **3**(2):42–49.
COOLIDGE, H. J.
1929. A revision of the genus *Gorilla*. Mem. Harvard Mus. Comp. Zool. **50**:293–381.
1930. Notes on the gorilla. *In* The African Republic of Liberia and the Belgian Congo. Edited by R. P. Strong. II, 623–35. Cambridge, Mass.
1933. Notes on a family of breeding gibbons. Human Biol. **5**:288–94.
1936. Zoological results of the George Vanderbilt African expedition of 1934. Part IV. Notes on four gorillas from the Sanga River region. Proc. Acad. Nat. Sci., Phila. **88**:479–501.
CORNET, R.
1955. Maniema. Brussels.
COTLOW, L.
1957. Zanzabuku. London.
CURRY-LINDAHL, K.
1956. Ecological studies on mammals, birds, reptiles and amphibians in the eastern Belgian Congo. Part I. Ann. du Musée Roy. du Congo Belge, Tervuren.
1961. Contribution à l'étude des vertèbres terrestres en Afrique tropicale. Fasc. 1. Inst. des Parcs Nat. du Congo et du Ruanda-Urundi.
DART, R. A.
1961. The Kisoro pattern of mountain gorilla preservation. Current Anthrop. **2**(5):510–11.
DARWIN, C.
1873. The expression of the emotions in man and animals. London.
DEKEYSER, P. L.
1955. Les mammifères de l'Afrique noire française. Inst. Français d'Afrique Noire, Dakar.
DEROCHETTE, M.
1941. Les gorilles en territoire de Shabunda. Bull. de la Société de Botan. et de Zool. Congolaises. **4**(1):7–9.
DERSCHEID, J. M.
1927. Notes sur les gorilles des volcans du Kivu (Parc National Albert). Ann. Soc. Royale Zool. Belgique. **58**:149–59.
DEVORE, I., AND WASHBURN, S. L.
1963. Baboon ecology and human evolution. *In* African ecology and human evolution. Edited by F. C. Howell. Viking Fund Publications in Anthropology.
DEWITTE, G.
1937. Exploration du Parc National Albert, 1933–35. Fasc. 1. Inst. des Parcs. Nat. du Congo Belge, Brussels.

Didier, R.
1951. L'os penien du gorille des montagnes. Bull. Inst. Roy. Sci. Nat. Belgique. **27**(37):5.

Donisthorpe, J.
1958. A pilot study of the mountain gorilla (*Gorilla gorilla Beringei*) in South-West Uganda, February to September 1957. S. Afr. J. of Science. **54**(8):195–217.

Du Chaillu, P.
1861. Explorations and adventures in equatorial Africa. New York.

Elliot, D. G.
1913. A review of the primates. Am. Mus. Nat. Hist. III. Monogr. No. 1.

Emlen, J. T.
1960. Current field studies of the mountain gorilla. S. Afr. J. of Science. **56**(4):88–89.

Emlen, J. T., and Schaller, G. B.
1960*a*. Distribution and status of the mountain gorilla (*Gorilla gorilla beringei*)—1959. Zoologica. **45**(1):41–52.
1960*b*. In the home of the mountain gorilla. Animal Kingdom. **63**(3): 98–108.

Foley, J. P.
1935. Judgment of facial expression of emotion in the chimpanzee. J. Soc. Psychol. **65**:31–67.

Fox, H.
1929. The birth of two anthropoid apes. J. Mammal. **10**(1):37–51.
1939. Chronic arthritis in wild animals. Trans. Am. Phil. Soc. **31**:71–149.

Frechkop, S.
1938. Mammifères. Exploration du Parc National Albert, mission G. F. De Witte (1933–35). Fasc. 10. Inst. des Parcs Nat. du Congo Belge, Brussels.
1953. Animaux protégés au Congo Belge. Inst. des Parcs Nat. du Congo Belge, Brussels.

Gaffikin, P.
1949. *Gorilla gorilla Beringei* post-mortem report. East. Afr. Med. J. **26**(8):1–4.

Galloway, A., Allbrook, D., and Wilson, A. M.
1959. The study of *Gorilla gorilla Berengei* with a post-mortem report. S. Afr. J. of Science. **55**(8):205–9.

Garner, R. L.
1896. Gorillas and chimpanzees. London.

Gatti, A.
1936. Great mother forest. London.

Gavan, J.
1952. Birth order and birth weight in the chimpanzee. Am. J. Phys. Anthr. **10**(1):23–30.

GEDDES, H.

1955. Gorilla. Frankfurt.

GÖTZEN, G. A. VON

1899. Durch Afrika von Ost nach West. Berlin.

GREGORY, W. K. (ED.)

1950. The anatomy of the gorilla. New York.

GREGORY, W. K., AND RAVEN, H. C.

1937. In quest of gorillas. New Bedford.

GROGAN, E., AND SHARP, A.

1902. From the Cape to Cairo. London.

GROMIER, E.

1948. La vie des animaux sauvages de la région des Grands Lacs. Paris.

GROVE, A. T.

1959. A note on the former extent of Lake Chad. Geogr. J. **125:** 465–67.

GRZIMEK, B.

1949. Rechts und Linkshändigkeit bei Pferden, Papageien und Affen. Z. f. Tierpsych. **6**:406–32.

1957*a*. Masse und Gewichte von Flachland-Gorillas. Z. f. Säugetierk. **21**(3–4):192–94.

1957*b*. Blinddarmentzündung als Todesursache bei Gorilla-Kleinkind. Zool. Garten. **23**(1–3):249.

GYLDENSTOLPE, N.

1928. Zoological results of the Swedish expedition to Central Africa 1921: Vertebrata, 5. Mammals from the Birunga Volcanoes, north of Lake Kivu. Ark. Zool., Uppsala. **20A**:1–76.

HAAS, G.

1958. Händigkeitsbeobachtungen bei Gorillas. Säugetierk. Mitteil. **6** (2):59–62.

HADDOW, A. J.

1952. Field and laboratory studies on an African monkey, *Cercopithecus ascanius Schmidti* Matschie. Proc. Zool. Soc. London. **122**(2):297–394.

1958. Uganda's chimpanzees. Uganda Wild Life and Sport. **1**(3): 3 pp.

HADDOW, A. J., AND ROSS, R. W.

1951. A critical review of Coolidge's measurements of gorilla skulls. Proc. Zool. Soc. London. **121**(1):43–54.

HALL-CRAGGS, E.

1961*a*. The skeleton of an adolescent gorilla (*Gorilla gorilla Beringei*). S. Afr. J. Science. **57**(11):299–302.

1961*b*. The blood vessels of the heart of *Gorilla gorilla beringei*. Am. J. Phys. Anthrop. **19**(4):373–77.

HARRISON, R. J.
1958. Man the peculiar animal. London.
HAUSER, F.
1960. Goma das Basler Gorillakind; ärztliche Berichte über "Goma." Documenta Geigy. Bull. No. 4.
HAYES, K., AND HAYES, C.
1954. The cultural capacity of the chimpanzee. Hum. Biol. **3**:288–303.
HEDBERG, O.
1951. Vegetation belts of the East African mountains. Svensk Bot. Tidskrift. **45**(1):140–202.
HEDIGER, H.
1950. Wild animals in captivity. London.
HEIM, A.
1957. Auf den Spuren des Berg-gorillas. Mitt. Naturf. Gesell. Bern. **14**:87–96.
HEINZELIN DE BRAUCOURT, J. DE
1953. Les stades de recession du glacier Stanley occidental. Fasc. 3. Inst. des Parcs Nat. du Congo Belge, Brussels.
HINDE, R. A., AND TINBERGEN, N.
1958. The comparative study of species-specific behavior. pp. 251–68. *In* Behavior and evolution. Edited by A. Roe and G. G. Simpson. New Haven.
HOIER, R.
1955*a*. A travers plaines et volcans au Parc National Albert. 2d ed. Inst. des Parcs Nat. du Congo Belge, Brussels.
1955*b*. Gorille de volcan. Zooleo. **30**:8–12.
HORNADAY, W. T.
1885. Two years in the jungle. London.
IMANISHI, K.
1957. Social behavior in Japanese monkeys, *Macaca fuscata*. Psychologia. **1**(1):47–54.
1958. Gorillas: a preliminary survey in 1958. Primates. **1**(2):73–78.
1960. Social organization of subhuman primates in their natural habitat. Current Anth. **1**(5–6):393–407.
JAY, P.
1962. The social behavior of the langur monkey. Ph.D. Thesis. Department of Anthropology, University of Chicago.
JENKS, A. E.
1911. Bulu knowledge of the gorilla and chimpanzee. Amer. Anth. **13**:56–64.
JOHNSON, M.
1931. Congorilla. New York.

KAWAI, M., AND MIZUHARA, H.
1959*a*. J.M.C. (Japan Monkey Center), the second gorilla expedition, 1959. Mimeographed report. 4 pp.
1959*b*. Printed in March, 1962. An ecological study on the wild mountain gorilla (*Gorilla gorilla beringei*)—report of the Japan Monkey Center second gorilla expedition 1959. Primates. **2**(1):1–42.

KEAY, R. W.
1959. Vegetation map of Africa south of the Tropic of Cancer. Oxford. 24 pp.

KNOBLOCH, H., AND PASAMANICK, B.
1958. Gross motor behavior in an infant gorilla. Mimeographed report of talk presented at the A.P.A. meeting, Sept. 3.

KÖHLER, W.
1957. The mentality of apes. London.

KOPPENFELS, H. VON
1877. Meine Jagden auf Gorillas. Gartenlaube. pp. 416–20.

KRAFT, L.
1952. J'ai vu au Congo le gorille géant des montagnes. Zooleo. **13**: 187–93.

KRAMPE, F.
1960. Gorillas auf der Spur. Der Kreis. **4**:104–12.

LANG, E. M.
1959. Goma das Basler Gorillakind; die Geburt. Documenta Geigy. Bull. No. 1.
1960*a*. Goma das Basler Gorillakind; Gomas Fortschritte. Documenta Geigy. Bull. No. 3.
1960*b*. Goma, das Basler Gorillakind, ein Jahr alt. Zolli (Bull. Zool. Gart., Basel). **5**:8–13.
1961*a*. Goma—das Gorillakind. Zurich.
1961*b*. Goma das Basler Gorillakind; Rückblick und Ausblick. Documenta Geigy. Bull. No. 8.
1961*c*. Jambo—unser zweites Gorillakind. Zolli (Bull. Zool. Gart., Basel). **7**:1–9.

LANG, E. M., AND SCHENKEL, R.
1961*a*. Goma das Basler Gorillakind; die Reifung der Kontaktweisen im Umgang mit den Dingen. Documenta Geigy. Bull. No. 6.
1961*b*. Goma das Basler Gorillakind; die Entwicklung der sozialen Kontaktweisen. Documenta Geigy. Bull. No. 7.

LEBRUN, J.
1935. Les essences forestières des régions montagneuses du Congo Oriental. Inst. Nat. pour l'Étude Agronomique du Congo Belge. Ser. Scient. No. 1.
1942. La vegetation du Nyiragongo. Fasc. 3–5. Inst. des Parcs Nat. du Congo Belge, Brussels.

LEBRUN, J., AND GILBERT, G.
1954. Une classification écologique des forêts du Congo. Inst. Nat. pour l'Étude Agronomique du Congo Belge. Ser. Scient. No. 63.
LEGGAT, G. J., AND OSMASTON, H. A.
1961. Working plan for the Impenetrable Central Forest Reserve, Kigezi District. Uganda Forest Dept.
LE GROS CLARK, W.
1960. The antecedents of man. Chicago.
LEQUIME, M.
1959. Sur la piste du gorille. La Vie des Bêtes. **14**:7–8.
LIBURNAU, L. L. VON.
1917. Beitrag zur Kenntnis der Affen und Halbaffen von Zentralafrika. Ann k. k. Naturhistorischen Hofmuseums, Vienna. **31**: 169–241.
LIND, E. M.
1956. The natural vegetation of Buganda. Uganda J. **20**(1):13–16.
LIVINGSTONE, D.
1874. The last journals of David Livingstone in Central Africa. London.
LIZ FERREIRA, A. J. DE, ATHAYDE, A., AND MAGALHAES, H.
1945. Gorilas do Maiombe Portugues. Mem. Jta. Miss. Geogr. Colon., Ser. Zool., Lisbon.
LÖNNBERG, E.
1917. Mammals collected in Central Africa by Captain Arrhenius. K. Svenska Vetensk. Akad. Handl., Uppsala. **58**:3–110.
LORENZ, K.
1957. King Solomon's Ring. London.
MALBRANDT, R., AND MACLATHY, A.
1949. Faune de l'équateur africain français. Paris.
MARCH, E. W.
1957. Gorillas of eastern Nigeria. Oryx. **4**(1):30–34.
MASON, W. A.
1960. The effect of social restriction on the behavior of rhesus monkeys: I. free social behavior. J. Comp. and Phys. Psych. **53**(6): 582–89.
MATHIS, M.
1954. Vie et moeurs des anthropoïdes. Paris.
MATSCHIE, P.
1903. Über einen Gorilla aus Deutsch-Ostafrika. Sitzber. Ges. naturf. Fr. Berlin. pp. 253–59.
1914. Neue Affen aus Mittelafrika. Sitzber. Ges. naturf. Fr. Berlin. pp. 323–42.
MAXWELL, M.
1928. The home of the eastern gorilla. J. Bombay Nat. Hist. Soc. **32**:436–49.

MECKLENBURG, A. F. HERZOG ZU
1910. In the heart of Africa. London.
MERFIELD, F. G., WITH MILLER, H.
1956. Gorilla hunter. New York.
MEYER, A.
1955. Aperçu historique de l'exploration et de l'étude des régions volcaniques du Kivu. Fasc. 1. Inst. des Parcs Nat. du Congo Belge, Brussels.
MEYER, H.
1923. Hochturen im tropischen Afrika. Leipzig.
MILDBRAED, J.
1909. Die Vegetationsverhältnisse der zentral-afrikanischen Seenzone vom Viktoria-See bis zu den Kiwu-Vulkanen. Bericht über die botanischen Ergebnisse der Expedition des Herzogs Adolf Friedrich zu Mecklenburg, 1907–1908. Sitzb. K. Preuss. Akad. Wiss., Berlin. **34**:989–1017.
MILLS, J.
1960. Juvenile male gorilla rescued and adult male gorilla found dead on Mountain Mgahinga in the gorilla sanctuary on 23rd February, 1960. Mimeographed report by game ranger, southern range, Uganda.
MILTON, O.
1957. The last stronghold of the mountain gorilla in East Africa. Animal Kingdom. **60**(2):58–61.
MIYADI, D.
1959. On some new habits and their propagation in Japanese monkey bands. Proc. Intern. Zool. Congress, London. pp. 857–60.
MONTAGU, M. F. A.
1951. An introduction to physical anthropology. Springfield.
MOOREHEAD, A.
1959. No room in the ark. London.
MOREAU, R. E.
1954. The distribution of African evergreen-forest birds. Proc. Linn. Soc. London. **165**:35–46.
MORRIS, D., AND JARVIS, C. (EDS.)
1960. The international zoo yearbook. I. London.
MOYNIHAN, M.
1955. Remarks on the original sources of displays. Auk. **72**:240–46.
NEWMAN, K.
1959. Saza chief. African Wildlife. **13**(2):137–42.
NISSEN, H. W.
1931. A field study of the chimpanzee. Comp. Psychol. Monogr. **8**(1):1–122.

NOBACK, C. R.
1939. The changes in the vaginal smears and associated cyclic phenomena in the lowland gorilla (*G. gorilla*). Anat. Record. **73**: 209–25.
OERTZEN, J. VON
1913. In Wildnis und Gefangenschaft. Kameruner Tierstudien. Berlin.
O'REILLY, J.
1960. The amiable gorilla. Sports Illustrated. **12**(25):68–76.
OSBORN, R. M.
1957. Observations on the mountain gorilla, Mt. Muhavura, S. W. Uganda. Unpublished manuscript.
OWEN, R.
1859. On the classification and geographical distribution of the mammalia. London.
1865. Memoir on the gorilla (*Troglodytes gorilla* Savage). London.
PEDEN, C.
1960. Report on post-mortem performed on adult mountain gorilla. Mimeographed report by the Department of Veterinary Services and Animal Husbandry, Uganda.
PETIT, L.
1920. Notes sur le gorille. Bull. Soc. Zool. Fr. **45**:308–13.
PHILIPPS, T.
1923. Mfumbiro: the Birunga Volcanoes of Kigezi-Ruanda-Kivu. Geogr. J. **61**(4):233–58.
1950. Letter concerning: Man's relation to the apes. Man. **272**:168.
PITMAN, C. R. S.
1931. A game warden among his charges. London.
1935. The gorillas of the Kayonsa Region, Western Kigezi, S. W. Uganda. Proc. Zool. Soc. London. **105**(2):477–99.
1942. A game warden takes stock. London.
PRETORIUS, P. J.
1947. Jungle man. London.
PURSEGLOVE, J. W.
1950. Kigezi resettlement. Uganda J. **14**(2):139–52.
RANDALL, F. E.
1943–44. The skeletal and dental development and variability of the gorilla. Human Biol. **15**:236–54, 307–37; **16**:23–76.
RAVEN, H. C.
1931. Gorilla: the greatest of all apes. Nat. Hist. **31**(3):231–42.
1936. Genital swelling in a female gorilla. J. Mammal. **17**(4):416.
REED, T. H., AND GALLAGHER, B.
In press. Gorilla birth at National Zoological Park. Zool. Garten.

REICHENOW, E.
1920. Biologische Beobachtungen an Gorilla und Schimpanse. Sitzber. Ges. naturf. Fr. Berlin. pp. 1–40.

RIESS, B. F., ROSS, S., LYERLY, S. B., AND BIRCH, H. G.
1949. The behavior of two captive specimens of lowland gorilla, *Gorilla gorilla gorilla*. Zoologica. **34**(3):111–18.

ROBYNS, W.
1947–55. Flore des spermatophytes du Parc National Albert. I-III. Inst. des Parcs Nat. du Congo Belge, Brussels.
1948*a*. Les territoires biogéographiques du Parc National Albert. Inst. des Parcs Nat. du Congo Belge, Brussels.
1948*b*. Les territoires phytogéographiques du Congo Belge et du Ruanda-Urundi. Fasc. 1. Inst. Roy. Col. Belge, atlas général du Congo Belge et du Ruanda-Urundi.

ROSS, R.
1954. Ecological studies on the rain forest of Southern Nigeria. III. Secondary succession in the Shasha Forest Reserve. J. Ecol., **42**:259–82.

RUCH, T. C.
1941. Bibliographia primatologica. Part I. Baltimore.
1959. Diseases of laboratory primates. Philadelphia.

RYHINER, P., WITH MANNIX, D.
1958. The wildest game. London.

SABATER PI, G.
1960. Beitrag zur Biologie des Flachland Gorillas. Z. f. Säugetierk. **25**(3):133–41.

SABATER PI, G., AND LASSALETTA, L. DE
1958. Beitrag zur Kenntnis des Flachland Gorillas (*Gorilla gorilla* Savage und Wyman). Z. f. Säugetierk. **23**:108–14.

SABEN'S COMMERCIAL DIRECTORY AND HANDBOOK OF UGANDA 1960–61
1959. Kampala.

SANDGROUND, J. H.
1930. Notes and descriptions of some parasitic helminths collected by the expedition. pp. 462–86. *In* The African Republic of Liberia and the Belgian Congo. Edited by R. P. Strong. Cambridge, Mass.

SAVAGE, T. S., AND WYMAN, J.
1847. Notice of the external characters and habits of *Troglodytes gorilla*, a new species of orang from the Gaboon River; osteology of the same. Boston J. Nat. Hist. **5**:417–43.

SCHÄFER, E.
1960. Über den Berggorilla (*Gorilla gorilla beringei*). Z. f. Tierpsych. **17**(3):376–81.

SCHALLER, G. B.
1960. The conservation of gorillas in the Virunga Volcanoes. Current Anth. **1**(4):331.
1961. The orang-utan in Sarawak. Zoologica. **46**(2):73–82.
1962. The mountain gorilla. New Scientist. **13**(268):16–18.
SCHALLER, G. B., AND EMLEN, J. T.
1963. Observations on the behavior of the mountain gorilla. *In* African Ecology and Human Evolution. Edited by F. C. Howell. Viking Fund Publications in Anthropology.

SCHENKEL, R.
1960*a*. Goma das Basler Gorillakind; Nesthocker oder Nestflüchter. Documenta Geigy. Bull. No. 2.
1960*b*. Goma das Basler Gorillakind; die Reifung artgemässen Fortbewegung und Körperhaltung. Documenta Geigy. Bull. No. 5.
SCHOUTEDEN, H.
1927. Gorille de Walikale. Rev. Zool. Afr. **15**:47.
1947. De zoogdieren van Belgisch-Congo en van Ruanda-Urundi. Fasc. 1–3. Ann. Mus. van Belgisch Congo.
SCHULTZ, A. H.
1927. Studies on the growth of gorilla and of other higher primates with special reference to a fetus of gorilla, preserved in the Carnegie Museum. Mem. Carneg. Mus. **11**(1):1–87.
1930. Notes on the growth of anthropoid apes, with especial reference to deciduous dentition. Rep. Lab. Mus. Zool. Soc. Phila. **58**:34–45.
1934. Some distinguishing characters of the mountain gorilla. J. Mammal. **15**(1):51–61.
1937. Proportions, variability and asymmetries of the long bones of the limbs and the clavicles in man and apes. Human Biol. **9**:281–328.
1938. Genital swelling in the female orang-utan. J. Mammal. **19**(3): 363–66.
1939. Notes on diseases and healed fractures of wild apes and their bearing on the antiquity of pathological conditions in man. Bull. Hist. of Med. **7**:571–82.
1942. Morphological observations on a gorilla and an orang of closely known ages. Am. J. Phys. Anth. **29**:1–21.
1950. Morphological observations on gorillas. pp. 227–51. *In* The anatomy of the gorilla. Edited by W. K. Gregory. New York.
SCHWARZ, E.
1927. *Gorilla gorilla rex-pygmaeorum*. Rev. Zool. Afr. **14**:373.
SHARP, N. A.
1927. Notes on the gorilla. Proc. Zool. Soc. London. **97**:1006–9.
1929. The Cameroon gorilla. Nature. **123**:525.

SNOWDEN, J. D.
1933. A study in altitudinal zonation in South Kigezi and on Mounts Muhavura and Mahinga, Uganda. J. Ecol. **21**:7–27.

SOUTHWICK, C. H., BEG, M. A., AND SIDDIQI, M. R.
1961. A population survey of rhesus monkeys in northern India: II. Transportation routes and forest areas. Ecology. **42**(4):698–710.

SPEKE, J. H.
1864. Journal of the discovery of the source of the Nile. London.

STANLEY, H. M.
1872. How I found Livingstone. London.
1878. Through the dark continent. II. London.
1890. In darkest Africa. 2 vols. London
1893. My dark companions and their strange stories. London.

STECKER, R. M.
1958. Osteoarthritis in the gorilla. Description of a skeleton with involvement of knee and spine. Lab. Invest. **7**(4):445–57.

STEINER, P.
1954. Anatomical observations in a *Gorilla gorilla*. Am. J. Phys. Anth. **12**:145–79.

STRAUS, W. L.
1930. The foot musculature of the highland gorilla (*Gorilla beringei*). Quart. Rev. Biol. **5**:261–317.
1950. The microscopic anatomy of the skin of the gorilla. pp. 213–21. *In* The anatomy of the gorilla. Edited by W. K. Gregory. New York.

STUHLMANN, F.
1894. Mit Emin Pascha ins Herz von Afrika. Berlin.

THOMAS, W. D.
1958. Observations on the breeding in captivity of a pair of lowland gorillas. Zoologica. **43**(3):95–104.

THORPE, W. H.
1956. Learning and instinct in animals. Cambridge, Mass.

TINBERGEN, N.
1951. The study of instinct. Oxford.

TOBIAS, P. V.
1961. The work of the gorilla research unit in Uganda. S. Afr. J. Science. **57**(11):297–98.

TROUESSART, E.
1920. La pluralité des espèces des gorilles. Bull. Mus. Nation. d'Hist. Nat. Paris. **26**:102–8, 191–96.

ULLRICH, W.
1961. Zur Biologie und Soziologie der Colobusaffen (*Colobus guereza caudatus* Thomas 1885) Zool. Garten. **25**:305–68.

URBAIN, A.
1940. L'habitat et moeurs des gorilles. Sciences. **35**:35.
VAN DEN BERGHE, L.
1959. Naissance d'un gorille de montagne à la station de zoologie experimentale de Tshibati. Folia Scientifica Africae Centralis. **4**:81–83.
VAN DEN BERGHE, L., AND CHARDOME, M.
1949. Une microfilaire du gorille, *Microfilaria gorillae*. Ann. Soc. Belge Med. Trop. **29**:495–99.
VAN STRAELEN, V.
1960. Sanctity of gorilla fastnesses threatened. Wild Life. **2**(2):10–11.
VERHAEGHE, M.
1958. Le Volcan Mugogo. Fasc. 3. Inst. des Parcs Nat. du Congo Belge, Brussels.
VOGEL, C.
1961. Zur systematischen Untergliederung der Gattung *Gorilla* anhand von Untersuchungen der Mandibel. Z. f. Säugetierk. **26** (2):65–76.
WAINWRIGHT, G. A.
1952. The coming of the banana to Uganda. Uganda J. **16**(2):145–47.
WALKER, E. P.
1954. The monkey book. New York.
WALLACE, A.
1869. The Malay Archipelago. 2 vols. London.
WALLIS, W. D.
1934. A gorilla skull with abnormal denture. Amer. Nat. **68**:179–83.
WASHBURN, S. L., AND AVIS, V.
1958. Evolution of human behavior. pp. 421–36. *In* Behavior and evolution. Edited by A. Roe and G. G. Simpson. New Haven.
WASHBURN, S. L., AND DEVORE, I.
1961. The social life of baboons. Scient. Amer. **204**(6):62–71.
WATT, A.
1956. Working plan for Mgahinga Central Forest Reserve Kigezi District, Uganda. Mimeographed report from the Forest Department, Uganda.
WESTLEY, B.
1951. To Africa for chimpanzees. London.
WILHELM, PRINCE OF SWEDEN
1923. Among Pygmies and gorillas. London.
WILSON, A. M.
1958. Notes on a gorilla eviscerated at Kabale, August 1958. Labora-

tory report on faeces and stomach contents. Unpublished report from the Animal Health Research Centre, Entebbe.

WILSON, G.
1951. The Nyakyusa of South-western Tanganyika. *In* Seven tribes of British Central Africa. Edited by E. Colson and M. Gluckman. Oxford.

WISSMANN, H. VON
1891. My second journey through Equatorial Africa. London.

WORDSWORTH (DONISTHORPE), J.
1961. Gorilla mountain. London.

WRIGHT, A. C.
1949. Maize names as indicators of economic contacts. Uganda J. **13** (1):61–81.

WYKES, A.
1960. Snake man. London.

YERKES, R. M.
1927. The mind of a gorilla. Parts I and II. Genetic Psych. Monogr. **2**(1 and 2):1–191; **2**(6):377–551.
1928. The mind of a gorilla. Part III. Comp. Psych. Monogr. **5**(2): 1–92.
1943. Chimpanzees. New Haven.
1951. Gorilla census and study. J. Mammal. **32**(4):429–36.

YERKES, R. M., AND YERKES, A. W.
1929. The great apes. New Haven.

ZAHL, P.
1960. Face to face with gorillas in Central Africa. Natl. Geogr. Mag. **117**(1):114–37.

ZEDTWITZ, Z.
1930. Beobachtungen in Zoologischen Garten Berlin. Zool. Garten. **2**:278–85.

ZUCKERMAN, S.
1932. The social life of monkeys and apes. New York.
1933. Functional affinities of man, monkeys, and apes. New York.

INDEX

Activity cycle, 138–48, 330
  bedding time, 146–48
  daylight schedule, 139–46
  rising time, 138–39
Age classes
  description, 66–68, 328
  differences in chest-beating, 226
  frequency of self-grooming, 207
  interactions in dominance, 242–44
  nesting positions, 183–85
  response to man, 317–18
  response to other groups, 120
  vocalizations, 212
Aggressive behavior, 23, 114, 119, 120, 129, 228, 242, 257, 289–92, 334
  charges, 116, 119, 290–91, 306, 334
  toward man, 228–29, 290, 292, 293, 306–8, 315, 316, 317, 334
  quarrels, 116, 117, 119–20, 129, 211, 216, 217, 218, 219, 252, 259, 260, 291–92, 331, 334
  redirected aggression, 291
Air sacs, 225
Albert National Park, 2, 4, 5, 9, 16, 18, 22, 48, 307, 321–23, 325
Albertine Rift, 8, 9, 17, 33, 36, 37, 48
Angumu, 35, 321
Appeasement gesture, 292–93

Baboon, 121, 128, 136, 207, 239, 240, 245, 248, 279, 291
Baiting gorillas, 22, 168
Bamboo, 18, 37, 41–42, 45–46, 47, 51, 52–53, 54, 55–56, 57, 60–61, 88, 130, 134, 136, 150, 151, 152, 153, 157, 168, 180, 186, 188, 191, 301, 305, 314, 322, 323, 324, 328, 330
Banana, 8, 38, 39, 151, 152, 154, 165, 168
Bantu tribes, 5, 8, 9, 18–19, 20, 33, 38, 167, 307
Barriers to dispersal, 29, 30, 32, 327
Basel zoo, 5, 20, 74, 76, 80, 96, 231, 274, 275, 277, 282, 283, 287–88
Batwa, 5, 9, 18–19, 20, 36, 167, 307, 322, 324
Bees, 167, 322
Birth rates, 97, 329
Births, records of, 100, 101, 105, 106, 283, 286, 287, 288, 289
Bishitsi, 57, 94, 128, 132, 322
Blackbacked male, interactions of
  with blackbacked male, 259
  with female, 114, 120, 241, 243, 245, 254, 259
  with infant, 243, 260
  with juvenile, 114, 116, 117, 241, 243, 259–60
  see also Mortality; Silverbacked male
Breeding season, 289, 334
Budongo Forest, 305
Buffalo, 9, 299, 300, 301–2, 321, 334
Bukavu, 5, 10, 18, 36, 44, 151
Burping, 155, 205, 331
Bushbuck, 300

Captive gorillas, 19–20, 67, 80, 81, 91, 96–97, 157, 160, 163, 165, 167, 168, 169, 199, 200, 208, 209, 215, 216, 220, 226, 231, 235–36, 241, 274, 275, 276–77, 278, 279–82, 286–90, 291, 292, 293, 313, 333–34
Capturing gorillas, 18, 102, 273, 320, 321
Cattle, 9, 34, 302, 321, 322, 323
Cave, response to, 221
Cercopithecus monkeys, 66, 146, 157, 207, 245, 289, 304–5, 334
Chameleon, 299
Chest-beating sequence, 112, 114, 115, 116, 117, 118, 123, 124, 138, 215–16, 222–35, 250, 252, 255, 293, 309, 310, 311, 313, 314, 319, 331, 332
  comparisons with other primates, 234–35
  duration, 222

Chest-beating sequence—*Continued*
ethological interpretations, 233–34
frequency, 313, 318
functions and causation, 233
individual acts, 222–29; combinations of, 229–30
individual variation, 230
at night, 148
ontogeny, 231–32
response to throwing, 230–31
situations eliciting, 232
Chicago zoo, 99
Chimpanzee, 2, 14, 16, 74, 77, 80, 81, 83, 84, 90, 121, 181, 195–96, 198, 199, 207, 208, 214, 216, 234–35, 248, 275, 276, 279, 281, 282, 289, 304, 305–6, 334
Cincinnati zoo, 99
Climbing, 81, 83–84, 112, 116, 158, 194, 195, 249, 250, 253, 254, 264, 266, 309, 333
Colobus monkey, 128, 207
Columbus zoo, 5, 20, 74, 96, 155, 167, 169, 215, 220, 226, 231, 236, 252, 274, 275, 276, 278, 279, 280, 286, 291, 292, 293
Congo Basin, 4, 10, 29, 37
Congo independence, 302, 321–23
Congo (Lualaba) River, 8, 9, 13, 14, 29, 30
Conservation, 320–26, 335
Co-ordination of group, 139, 147, 183, 211, 272
through dominance hierarchy, 243–44
through internal organization, 239–40
through leadership, 237–39
through vocalizations, 213–21
Copulation, 236, 242, 252, 275, 278, 279–86, 292
positions during, 280–86, 333
vocalizations during, 214–15, 283, 284
Coughing, 204, 331
Cultivated plants, 8, 9, 38, 151, 152, 168, 199
Cultural differences between populations, 153–54, 181
Curiosity, 310–11, 318, 334

Defecation, 197, 201–3, 310, 331
Diarrhea, 81, 160, 273
Digging, 157, 159, 166, 191
Discovery of mountain gorilla, 12–15
Diseases, 98–99, 329
Displacement activity, 204, 207, 234, 248, 332
Displays, 80, 119, 221–36, 239, 293, 306, 316, 317, 332, 339
branch-shaking, 116, 229, 230
compared to man; *see* Man
ground-thumping, 114, 115, 117, 222, 229, 230, 232, 309, 314, 332
hooting, 115, 211, 215, 222–23, 227, 229, 230, 231, 232, 233, 234, 279, 314, 331, 332; *see also* Vocalizations
leg-kicking, 222, 227–28, 229, 230, 231, 233
running, slapping, and tearing, 222, 228–29, 230, 231, 232, 233, 234, 319, 332
"symbolic feeding," 222, 223–24, 229, 230, 231, 232, 233, 234, 332
throwing, 116, 222, 224–25, 228, 229, 230–31, 232, 233, 234, 332
Distribution, 25–36, 327
distributional dynamics, 30–34
factors limiting, 29
local, 34–35
lowland gorilla, 25
mountain gorilla, 27, 320
past distribution, 27–30
stragglers, 36, 327
Dog, 217, 220, 222, 298, 307
Dominance, 112, 114, 115, 116, 122, 138, 139, 146, 147, 154, 182, 237, 238, 239, 240–44, 249, 254, 255, 257, 259, 260, 261, 269, 270, 271, 272, 283, 309, 332, 333
gestures used in, 241–42
hierarchy, 121, 243, 244, 291; positions in, 242–44
situations eliciting, 241–42
Drinking, 169, 331
Duiker, 167, 191, 301, 303, 304, 321
Dung, 88–90, 92, 99, 112, 164, 170, 171, 182, 191, 195, 196–97, 201–3, 261, 329

Ears, description of, 71
Edema, 286, 288
Eggs, of birds, 167
Elephant, 9, 299, 301, 334
Elephant grass, 130, 154, 157, 168
Elila River, 6, 44
Erotic behavior, 277–79, 333
Estrous cycle, 275–77, 287, 288, 333
Eye
description, 71
expression, 208–10

Facial expressions, 208–10, 217, 219, 221, 222, 252, 272, 279, 283, 284, 290, 292

Feeding behavior, 45, 48, 54, 64, 79, 82, 112, 114, 116, 117, 118, 123, 138, 139–41, 142, 146, 147, 154–56, 213–14, 238, 242, 248, 274, 295, 310, 311, 312, 314, 319, 330, 333
descriptions, 155–65
diameter of feeding group, 155
feeding on bark and wood, 150, 164
feeding on fungus, 164
of infants, 163, 165, 262
periods, 154
sharing of food, 155
Feet, description of, 68–69
Female, interactions of, 184, 218, 220
with female 217, 242, 244, 245, 247, 260, 269, 291, 292
with infant: feeding, 261–63; transport, 220, 263–65, 269; protection, 218, 265–67, 295; social, 245–46, 252, 267–68; response to wounded infant, 268–69, 333; response to dead infant, 269, 333; with infants not her own, 246, 252, 262, 265, 269–70
with juvenile 184, 219, 230, 241, 243, 246, 260–61, 271, 310, 333
see also Blackbacked male; Mortality; Silverbacked male
Fetus, aborting of, 287
Fizi, 27, 30, 41, 44, 151
Flight distance, 316
Fog, response to, 298
Foods, 149–69, 306, 330
acquisition of new habits, 168–69
animal matter, 167, 331
comparisons between areas, 153–54
honey, 167
nutritional value, 153
preferences, 149–52
soil, 166, 330
species of plants, 149–52; total, 152–53
Forests
dry colonizing woodland, 54, 55, 57, 59
Hagenia and *Hypericum* woodlands, 53, 56–57, 61–62, 65, 136, 137, 150, 186, 305
lowland rain forest, 6, 16, 17, 21, 34–35, 37–40, 43–44, 151, 182, 325–26, 328
mountain rain forest, 29, 37, 40–41, 42, 44, 45, 46, 48, 51–52, 151, 182, 328, 325
mountain woodland, 51–52, 62, 134, 150
see also under individual names
Francolin, 300

Gastroenteritis, 98, 273
Genet, 300
Genital swelling, 275, 276
Genitals, 277, 278
Gestation period, 288–89, 333
Gibbon, 2, 66, 77, 84, 128, 136, 211, 234, 283, 298
Grooming, 79, 82, 141, 146, 254, 259, 260, 261, 270, 271, 272, 310
mutual, 245–48, 332
self-grooming, 205–7, 246, 331
Groups of gorillas
attachments of individuals, 120
change between members, 108, 109, 120
composition, 104, 330; changes in, 105–9, 330
contacts between, 111–21, 330
factors affecting size, 103
formation, 124
internal organization, 239–40
joining, 105, 106, 109, 120–21, 330
lone males joining, 105, 106, 122–24, 243
size, 102–4, 330

Habituation of gorillas to man, 22, 140, 225, 311–13, 316, 331
Hail, 59, 294, 298
Handedness, 77
Hands, description of, 68, 328
Hawk, 299, 334
Head, features of, 71
Hearing, sense of, 79, 329
Height of gorillas, 76
Hiccuping, 205, 331
Home range, 124–29
boundaries, 128–29
centers of activity, 128
familiarity with, 129
of lone males, 128
overlap, 128
size, 126–27
Homosexual behavior, 278, 286
Horse, 205, 214
Howler monkey, 66, 211, 240
Hunting gorillas, 1, 12, 15, 16, 17, 72, 98, 103, 272, 291, 294, 304, 306–8, 320–21, 322, 325, 326, 334, 335
Hyena, 300
Hyrax, 167, 301, 304

Impenetrable Central Forest Reserve, 324
Individual variation
in chest-beating, 230, 318
in morphology, 68–77
in response to man, 309, 310, 311, 312, 317-18
in temperament, 80, 240
Infant
development, 273–75
interactions with infant, 247, 252, 253, 254, 333
newborn, 110, 111, 255, 260, 263, 269, 286, 287, 288, 333
see also Blackbacked male; Feeding behavior; Female; Juvenile; Mortality; Nests; Silberbacked male
Injuries, 77–78, 97–98, 119-20, 269, 328, 333
Insects, 167, 299
I.R.S.A.C. station, 5, 17, 18, 74, 98, 149, 151, 157, 165, 275, 282, 286–87
Ischial callosities, 76–77, 328

Juvenile, interactions of
with infant, 114, 242, 247, 252, 254, 267, 270–71
with juvenile, 114, 164, 184, 214, 247, 253, 254, 270, 295
see also Blackbacked male; Female; Mortality; Silverbacked male

Kabale, 33, 46
Kabara, 20, 21, 23, 50, 57–64, 70, 85, 93, 94, 96, 99, 100, 101, 102, 103, 104, 105, 109, 111, 121, 122, 124, 126, 127, 128, 130, 132, 134, 136, 149, 150, 152, 153, 158, 159, 160, 164, 165, 167, 169, 172, 175, 180, 181, 182, 185, 186, 187, 191, 196, 197, 201, 205, 273, 289, 294, 295, 298, 300, 301, 302, 304, 305, 308, 313, 316, 318, 323, 328, 331, 334
Kasese, 32, 34, 35, 37, 43
Kayonza Forest, 10, 11, 12, 13, 16, 20, 21, 32–33, 41, 46–48, 90, 93, 94, 95, 103, 108, 111, 134, 136, 149, 150, 154, 155, 157, 161, 162, 165, 167, 169, 171, 178, 179, 180, 181, 186, 196, 197, 203, 214, 295, 299, 301, 305, 307, 320, 324-25, 326, 329, 331, 335
Kigezi district, 13, 33
Kisoro, 4, 17, 20, 21, 50, 51–55, 56, 63, 98, 99, 102-3, 104, 105, 108, 119, 120, 127, 128, 129, 130, 134, 136, 150, 153, 159, 164, 167, 168, 169, 170, 171, 172, 175, 178, 180, 181, 196, 197, 239, 273, 289, 295, 302–4, 306, 314, 323–24
Kob antelope, 9

Lakes
Albert, 8, 9, 10
Chad, 30
Edward, 8, 9, 10, 13, 45, 46
Kivu, 8, 9, 14, 18, 30, 34, 36, 46, 48, 151, 275
Tanganyika, 8, 9, 10, 13, 14, 15, 16, 41, 151, 325
Victoria, 20
Langur monkey, 128
Lava, 54, 55, 56, 57, 64–65
Leadership, 123, 237–39, 296, 332
Learning
to build nests, 195
to care for infant, 288
to climb on female, 261
to eat certain food plants, 165, 262–63
to integrate into group, 267
to respond to displays, 223, 224
to return to female, 265
Leopard, 98, 109, 273, 294, 302–4, 334
Lion, 300
Lobelia, giant, 53, 55, 61, 62, 150, 160, 172, 186, 187, 193, 250, 267
Locomotion, 81–85, 199, 329
bipedal, 82, 281, 315
brachiation, 84
during displays, 224–29
on ground, 81–83
jumping, 84–85, 228, 230, 253, 254
quadrupedal, 81–82, 83–84, 274, 333
in trees, 83–85
when carrying infant, 263–65, 333
London Convention of 1933, 320
London zoo, 20, 236, 279
Lone males, 36, 100, 105, 106, 121–24, 138, 148, 201, 232, 290, 315, 327, 329, 330; *see also* Groups; Home range
Longevity, 99–100, 329
Lowa River, 6, 14, 27, 30, 32, 35, 43
Lowland gorilla, 1, 11, 12, 19, 25, 74, 75, 76, 77, 78, 81, 96, 98, 102, 121, 128, 129, 138, 139, 146, 152, 167, 171, 177, 196, 203, 273–81, 286–89, 299, 308, 321
Lubero, 10, 27, 30, 32, 307
Lubutu, 14, 27
Lugulu River, 27, 30, 32, 35, 43
Luka River, 30

Macaca monkey, 128, 153, 168, 207, 211, 239, 240, 278, 279, 289
Maiko River, 27, 32
Maize, 9, 151, 168
Makerere College, 3, 4
Man
  comparisons with gorilla, 68, 78, 80, 82, 157, 204, 205, 207, 208, 209, 219, 221, 230, 234–35, 252, 268, 274, 275, 276, 289, 290, 293, 329, 331, 333
  injured by gorillas, 228–29, 307, 325, 334
  reaction to gorilla foods, 165–66
  response of gorillas to, 79, 309–19, 334
  see also Aggressive behavior; Capturing gorillas; Displays; Habituation; Hunting gorillas
Manioc, 8, 38, 151, 152, 168
Manipulation of strange objects, 200, 331
Maramagambo Forest, 305
Masturbation, 277, 279, 333
Mating behavior, 275–89, 333–34
Meadows, 53, 56
  Rukumi meadow, 57, 63, 64, 129, 301, 322
Menstruation, 276
Milk, 90, 287, 288
Milwaukee zoo, 20, 236, 277, 279
Mock biting, 252, 291, 292
Mongoose, 300
"Moods" of gorillas, 318–19
Mortality
  through disease, 98–99, 108, 120, 273, 329
  of infants, 100, 269, 304, 321, 329, 333
  through injuries, 97, 269, 303, 119
  through predation, 98, 109, 302–3, 306–8, 321
  rate of, 100, 101
  response to dying and dead gorillas, 269, 272–73, 333
Mt. Kahuzi, 10, 98, 103, 151, 307, 326
Mt. Tshiaberimu, 10, 13, 20, 21, 36, 41, 45–46, 88, 93, 95, 98, 103, 130, 151, 154, 157, 168, 180, 197, 203, 301, 302, 320, 325, 326, 335
Mouse, 300
Movement, 129–37
  daily distance, 134–37
  daily pattern, 132–34, 146, 330
  general pattern, 130–32
  seasonal, 130, 330
  speed, 137
Mulenge, 151, 187
Mwenga, 34, 41, 44, 151
Mwenga-Fizi region, 44–45, 93, 103, 151, 154, 187, 325–26

Nests and nesting behavior, 23, 92, 104, 111, 112, 116, 118, 124, 126, 134, 138, 139, 146–48, 154, 155, 169–98, 201, 203, 253, 261, 269, 270, 301, 302, 305, 317, 329, 331, 334
  aging of nests, 171
  construction behavior, 187–94, 146–47
  construction by infants, 194–95
  construction time, 193, 194, 195, 196
  day nests, 170–71
  distance between nests, 185–86
  frequency of tree nesting, 181
  functions of nests, 197–98
  ground and tree nest positions, 177–81
  height of nests above ground, 180–81
  location of dung in nest, 196–97
  location of individual nests in group, 182–86
  location of sites, 171
  materials, 186–87
  night nests, 171–98
  persistence of nests, 85
  positions of animals in nest, 196
  in response to rain, 297–98
  size and shape of nests, 174–75, 182
  time of building, 170
  unused nests, 182, 195
New York (Bronx) zoo, 5, 20, 75, 77, 96, 99, 200, 276, 277, 278, 279, 291
Nile River, 9, 13
Nose
  description, 71–72
  picking at, 79, 207
Numerical estimates of gorillas, 93–95, 229
Nzombe, 151

Odor, 90–91
Olive pigeon, 167, 300
Orang-utan, 2, 3, 74, 77, 84, 196, 234, 282, 289
Orgasm, 282, 284
Oso River, 14, 32

Parasites, 98–99
Parturition, 286–88, 334
Pelage, 69–71, 328
Penis, 245, 279, 282
Peripheral males, 121, 122, 123, 138, 142, 146, 183, 232, 239, 243, 255, 283, 284
Philadelphia zoo, 99
Photographing gorillas, 24, 306, 314

Pig, 300, 324
Placenta, 287, 288
Play, 114, 141, 216, 232, 233, 240, 242, 248–54, 257, 259, 268, 270, 271, 274, 321, 332
  lone play, 249–50
  sex play, 278, 286
  social play, 250–54
Population
  density, 92–93, 95, 329
  structure, 100–101, 329
Postures, 78
Pregnancy, 286–88
Protection
  of group by male, 231, 309, 314, 316–17
  of juvenile by male, 257
  of young by female, 242, 265–67, 333; *see also* Female

Queen Elizabeth National Park, 5, 9

Rain, response to, 82, 139, 140, 142, 148, 241, 270, 295–98, 309, 317, 334
Raven, 218, 300, 334
Rebuffs of young by female, 261, 262, 268
Recognition of individual gorillas, 68, 102
Resting, 79, 80, 111, 112, 114, 117, 118, 123, 139, 140–46, 154, 158, 213–14, 217, 237, 238, 248, 265, 283, 295, 311, 312, 313, 319, 330
Rumangabo, 10
Rutshuru River, 9
Ruwenzori Mountains, 9, 10, 13, 182, 306
Ruzizi River, 9
Rweru, 57, 85, 130, 132, 305, 323

Sagittal crest, 66, 71
Salt, 166, 330
San Diego zoo, 76, 169, 236, 278, 291
Scratching, 207, 331
Secondary forest succession, 39–40, 41–42, 44, 95, 151, 294
Semliki Forest, 305
Semliki River, 9, 14, 30
Senecio, giant, 53, 55, 62, 88, 150, 163, 172, 294, 328
Senses, 78–80
Sex classes
  behavior in response to man, 317–18
  description, 66–68
  differences in chest-beating, 226
  frequency of self-grooming, 207
  interactions in dominance, 242–44
  nesting positions, 183–85
  response to other groups, 120
  vocalizations, 212
Sexual behavior, 254, 275–86
Sexual cycles, 96, 275, 276–77, 286
Sexual maturity, 96–97, 329
Shaking
  of branches; *see* Displays
  when wet, 298
Siamang, 2, 84
Sight, sense of, 78–79, 329
Silverbacked male, interactions of
  with blackbacked male, 116, 123, 223, 255
  with female, 114, 120, 184, 215, 217, 218, 228, 241, 242, 255–57, 278, 283, 384, 333
  with infant, 220, 245, 257, 259, 266, 273, 333
  with juvenile, 108, 228, 239, 241, 242, 257, 333
  with silverbacked male, 112, 114, 115, 116, 120, 123, 124, 223, 241, 243, 254–55, 284, 290, 291
  see also Lone male; Mortality; Peripheral male
Skin, 69, 328
Smell, sense of, 79, 329
Snake, 272, 299
Sneezing, 204, 331
Snoring, 148
Snow, 58, 59
Soils, 51, 159
  feeding on, 166, 330
  nesting on, 171, 174, 187, 188, 191, 197
Spiders, 167
Spoor of gorillas, 85–91
Squirrel, 301
Stanleyville, 43, 44
Strutting walk, 221, 235–36, 280
Study methods, 18–24, 327
  equipment used, 24
  finding gorillas, 20–21
  observing gorillas, 21–23
Subgroups, 108, 109–11, 239, 330
Submissive behavior, 235, 241, 280, 290, 292–93, 334
Suckling, 261–62, 287, 288, 333
Sugar cane, 152, 168
Sun, response to, 140, 142, 265, 294–95, 317, 334
Sunbird, 299
Sweating, 295

Swimming, 29, 298–99
"Symbolic feeding"; *see* Displays

Taxonomy of gorillas, 10–12
Teeth
 caries, 71, 99
 description, 71, 79
 eruption, 274
 picking at, 208
 tartar, 71
Temperament, 80–81, 317–18, 329
Territory, 129, 330
Threat behavior, 23, 112, 116, 124, 208, 216, 217, 257, 259, 290, 292, 293, 317, 334
Throwing; *see* Display
Thumb-sucking, 262
Timber-felling, 324–25
Tolerance distance, 312, 316
Tool-using, 199–200, 331
Topi antelope, 9
Tracks, 86, 87–88, 200
Trails, 21, 85, 87, 93, 110, 112, 116, 122, 124, 149, 151, 154, 160, 200, 305, 329
Transport
 of food objects, 198, 331
 of infants by females, 263–65, 333
 of infants by juveniles, 271, 333
 of infants by males, 257
 of nest materials, 186, 187, 331
 of play objects, 198–99, 253
Tree fern, 41, 46, 48, 150, 152, 154, 157, 299
Tree heath, 46, 53, 57, 62, 55
Turaco, 300

Uele River, 25, 29, 30, 32
Ulindi River, 6, 13
Umbilical cord, 287, 288
Urination, 203–4, 331
Utu region, 5, 16, 17, 21, 35, 37, 43, 44, 69, 93, 95, 98, 102, 103, 151, 154, 155, 157, 160, 164, 165, 166, 168, 171, 172, 180, 186, 191, 197, 203, 215, 295, 307, 325–26, 329, 331

Vegetation types, listing of, 37, 50, 59
Virunga Volcanoes, 2, 4, 9, 12, 13, 14, 15, 16, 17, 18, 20, 32–33, 41, 42, 49–65, 69, 76, 77, 93, 94, 95, 108, 149, 153, 154, 197, 203, 300, 301, 306, 320, 321–24, 326, 327, 328, 329, 335
Vocalizations, 21, 210–21, 238, 267, 272, 281, 282, 290, 312, 319, 331–32
 in alarm-producing situations, 220–21
 by copulating animals, 214–15, 283, 284
 by displaying animals, 215–16
 in disturbing situations, 216–20
 by feeding and resting animals, 212–13
 frequency, 211
 hooting; *see* Displays
 by moving animals, 213–14
 by playing young, 216
 roaring, 123, 209, 218–19, 299, 300, 309, 310, 311, 312, 314, 315, 316, 317, 334
 screaming, 118, 209, 219, 265, 283, 291–92, 299, 300, 312, 315, 316, 319
 total number, 211
 in various age and sex classes, 212
 whining, 209

Walikale, 30, 35, 36
Washington zoo, 74, 96, 200, 275, 276, 277, 279, 280, 288
Water, response to, 298–99
Watutsi, 9, 34, 321, 322, 323
Weather, 8, 10, 37, 43, 45, 46–47, 51, 58–59, 294–98, 328
Weights of gorillas, 72, 74–76, 328
Wind, response to, 298, 317
Wind-breaking, 138, 141, 205
Wolfram, 324
Woodpecker, 300
Wrestling, 249, 252, 253, 254, 259, 280–81, 290, 291, 293, 333

Yawning, 138, 139, 204, 310, 311, 331